Green Energy and Technology

Climate change, environmental impact and the limited natural resources urge scientific research and novel technical solutions. The monograph series Green Energy and Technology serves as a publishing platform for scientific and technological approaches to "green"—i.e. environmentally friendly and sustainable—technologies. While a focus lies on energy and power supply, it also covers "green" solutions in industrial engineering and engineering design. Green Energy and Technology addresses researchers, advanced students, technical consultants as well as decision makers in industries and politics. Hence, the level of presentation spans from instructional to highly technical.

Indexed in Scopus.

Indexed in Ei Compendex.

Eriola Betiku · Mofoluwake M. Ishola
Editors

Bioethanol: A Green Energy Substitute for Fossil Fuels

Editors
Eriola Betiku ⓘ
Department of Chemical Engineering
Obafemi Awolowo University
Ile-Ife, Nigeria

Department of Biological Sciences
Florida Agricultural and Mechanical University
Tallahassee, FL, USA

Mofoluwake M. Ishola
Department of Research and Development
Scanacon AB
Stockholm, Sweden

ISSN 1865-3529   ISSN 1865-3537 (electronic)
Green Energy and Technology
ISBN 978-3-031-36541-6   ISBN 978-3-031-36542-3 (eBook)
https://doi.org/10.1007/978-3-031-36542-3

© The Editor(s) (if applicable) and The Author(s), under exclusive license to Springer Nature Switzerland AG 2023

This work is subject to copyright. All rights are solely and exclusively licensed by the Publisher, whether the whole or part of the material is concerned, specifically the rights of translation, reprinting, reuse of illustrations, recitation, broadcasting, reproduction on microfilms or in any other physical way, and transmission or information storage and retrieval, electronic adaptation, computer software, or by similar or dissimilar methodology now known or hereafter developed.
The use of general descriptive names, registered names, trademarks, service marks, etc. in this publication does not imply, even in the absence of a specific statement, that such names are exempt from the relevant protective laws and regulations and therefore free for general use.
The publisher, the authors, and the editors are safe to assume that the advice and information in this book are believed to be true and accurate at the date of publication. Neither the publisher nor the authors or the editors give a warranty, expressed or implied, with respect to the material contained herein or for any errors or omissions that may have been made. The publisher remains neutral with regard to jurisdictional claims in published maps and institutional affiliations.

This Springer imprint is published by the registered company Springer Nature Switzerland AG
The registered company address is: Gewerbestrasse 11, 6330 Cham, Switzerland

In Memoriam: Prof. Bamidele Ogbe Solomon
† *May 10, 2023*

Preface

Energy demand will continue to increase in order to meet consumption due to the increasing world population, urbanization, and industrialization. Although fossil sources continue to dominate the energy supply around the globe, particularly in the transportation sector, its challenges remain environmental pollution, availability in limited quantity, and its use as a political weapon. The listed concerns have led scientists to seek alternative energy sources to abate these problems. Biofuels, including bioethanol, biodiesel, biogas, and biohydrogen, have all been touted as substitutes for fossil fuels because they are renewable, sustainable, and environmentally benign.

Among all the biofuels, bioethanol is by far the most advanced in terms of production, adoption, and utilization across the globe. Yet, it is at a far distance behind fossil fuels in production. Presently, it is combined in various ratios with gasoline and sold commercially as E10–E25 across the World. Despite its many advantages over gasoline, it is not typically used 100% as a transportation fuel, mainly because of its limited quantity. This book, Bioethanol: A green energy substitute for fossil fuels, contains thirteen chapters that address the various aspects of these challenges. Chapter "Introduction: Benefits, Prospects, and Challenges of Bioethanol Production" reviews ethanol production, its prospects as a biofuel, and the challenges of full adoption as a transportation fuel. One of the greatest problems facing bioethanol production is feedstock.

Chapter "Novel and Cost-Effective Feedstock for Sustainable Bioethanol Production" addresses old and new materials regarding the first, second, third, and fourth-generation substrates. Waste streams and mixed feedstocks as substrates for bioethanol production were also discussed. Chapter "Feedstock Conditioning and Pretreatment of Lignocellulose Biomass" focuses on one of the most consequential issues militating against high bioethanol production and its full adoption across the globe because lignocellulose biomass is everywhere. The physical, biological, chemical, and intensification assistance of pretreating lignocellulose biomass used in bioethanol production were comprehensively reviewed. Chapter "Current Status of Substrate Hydrolysis to Fermentable Sugars" review feedstock classification, pretreatment methods, and hydrolysis types. Also, the chapter review factors affecting the processing of lignocellulosic biomass, biowastes, and starch materials to fermentable sugars. Chapter "Bioethanol Production Using Novel Starch Sources" is an overview of novel starches used for bioethanol production. Strategies for improving bioethanol production from novel starches were also discussed. Chapter "Bioethanol Production from Lignocellulosic Wastes: Potentials and Challenges" provides information on the potential and challenges of utilizing lignocellulosic wastes for bioethanol production. The process, including pretreatment, hydrolysis, and fermentation, was discussed. Chapter "Bioethanol Production from Microalgae: Potentials and Challenges" focuses explicitly on bioethanol production from microalgae. The various aspects of cultivation, harvesting, extraction, pretreatment, and fermentation of microalgae were covered in this chapter. Chapter "Bioethanol Production via Fermentation: Microbes, Modeling and Optimization" reviews processes involved in bioethanol production. The different microorganisms used for the fermentation step were evaluated. Also, a significant part of the chapter concentrates on modes of fermentation used for bioethanol production. The modeling and optimization of the critical factors for the processes were reviewed. Applications of machine learning in modeling the process were also discussed.

The ethanol produced via fermentation must be recovered through various processes to be usable as a transportation fuel. Thus, Chapter "Bioethanol Recovery and Dehydration Techniques" centers on recent advances in bioethanol recovery and dehydration technologies, and their economic implications were also addressed. Chapter "Ethanol Utilization in Spark-Ignition Engines and Emission Characteristics" emphasizes the application of ethanol in spark-ignition engines. It provides an overview of the commercial mixing ratios in different countries, such as E10 and E20. The benefits of these ethanol-gasoline products in terms of emission characteristics and engine performance were also reported. Commercialization of ethanol production is crucial to its full adoption. Presently, only a limited quantity is produced globally. Chapter "Overview of Commercial Bioethanol Production Plants" addresses the present status and challenges confronting the commercialization of the ethanol industry. Chapter "Techno-Economic and Life Cycle Analysis of Bioethanol Production" deliberates on different scenarios of bioethanol production using different feedstock. The life cycle aspect of the chapter assesses the environmental impacts of bioethanol plants. The overall conclusion and future directions were presented in Chapter "Concluding Remarks and Future Directions."

Although the editors recognized that there are books on ethanol production, various aspects of the problems confronting its full adoption as a fuel were discussed extensively in this book and will undoubtedly form part of the puzzle toward the realization of this objective worldwide. Both editors believe this book is a significant contribution to the knowledge of bioethanol production with a view to meeting the world's energy demand and may serve as a useful resource for researchers and graduate students pursuing careers/research in the areas of bioethanol production.

Ile-Ife, Nigeria   Eriola Betiku
Stockholm, Sweden   Mofoluwake M. Ishola

Acknowledgements

The editors thank all scientists who contribute to this book by sharing their wealth of experience with the world on bioethanol and for their patience and cooperation during the long and rigorous review process. The chapters provided in this book were subjected to a double-blind peer-review process before acceptance for publication. So, we would like to thank all our reviewers for their time, critical, and constructive feedback, which in no small measure, improved the quality of the book. The editors acknowledge the opportunity and support provided by Springer Nature Switzerland AG throughout the preparation of this book. Dr. Eriola Betiku is grateful to his colleagues at the Department of Chemical Engineering and the management of the Obafemi Awolowo University for granting him a leave of absence to work on this book. He is equally thankful to the Florida Agricultural and Mechanical University, Tallahassee, USA, for the Visiting Faculty position given to him and the enabling environment provided while working on this book. The editors acknowledge the love, care, and support received from their families while working on the publication of this book.

Contents

Introduction: Benefits, Prospects, and Challenges of Bioethanol Production 1
Olayomi Abiodun Falowo and Eriola Betiku

Novel and Cost-Effective Feedstock for Sustainable Bioethanol Production 21
Atilade A. Oladunni and Mofoluwake M. Ishola

Feedstock Conditioning and Pretreatment of Lignocellulose Biomass 47
Adeolu A. Awoyale, David Lokhat, Andrew C. Eloka-Eboka, and Adewale G. Adeniyi

Current Status of Substrate Hydrolysis to Fermentable Sugars 69
Olayomi Abiodun Falowo, Abiola E. Taiwo, Lekan M. Latinwo, and Eriola Betiku

Bioethanol Production Using Novel Starch Sources 103
Gabriel S. Aruwajoye, Daneal C. S. Rorke, Isaac A. Sanusi, Yeshona Sewsynker-Sukai, and Evariste B. Gueguim Kana

Bioethanol Production from Lignocellulosic Wastes: Potentials and Challenges 123
Esra Meşe Erdoğan, Pınar Karagöz, and Melek Özkan

Bioethanol Production from Microalgae: Potentials and Challenges 161
Mallika Boonmee Kongkeitkajorn

Bioethanol Production via Fermentation: Microbes, Modeling and Optimization 193
Adebisi Aminat Agboola, Niyi Babatunde Ishola, and Eriola Betiku

Bioethanol Recovery and Dehydration Techniques 229
Babatunde Oladipo, Abiola E. Taiwo, and Tunde V. Ojumu

Ethanol Utilization in Spark-Ignition Engines and Emission Characteristics 255
Roland Allmägi, Marcis Jansons, Kaie Ritslaid, and Risto Ilves

Overview of Commercial Bioethanol Production Plants 279
Bárbara P. Moreira, William G. Sganzerla, Paulo C. Torres-Mayanga, Héctor A. Ruiz, and Daniel Lachos-Perez

Techno-Economic and Life Cycle Analysis of Bioethanol Production ... 305
Ana Belén Guerrero and Edmundo Muñoz

Concluding Remarks and Future Directions 339
Eriola Betiku and Mofoluwake M. Ishola

Editors and Contributors

About the Editors

Prof. Dr. Eriola Betiku holds B.Sc. and M.Sc. degrees in Chemical Engineering from the Obafemi Awolowo University (Nigeria) and a Doctor of Natural Sciences (Dr. Rer. Nat.) degree in Biotechnology from the Technical University Carolo-Wilhelmina, Braunschweig, Germany. He is a Full Professor of Chemical Engineering at the Obafemi Awolowo University and is currently a Visiting Faculty at the Department of Biological Sciences, Florida A&M University, USA. He was a DAAD scholar and a Guest Scientist at the National Research Centre for Biotechnology (now Helmholtz Centre for Infection Research (HZI)), Germany. He has served as Interim Chair of the Department of Chemical Engineering at Obafemi Awolowo University, Ile-Ife, Nigeria. He is a recipient of several scholarships, awards, and grants, including World University Service (Germany), Tertiary Education Trust Fund (TETFUND), DAAD, and Carnegie of New York. His research focuses on biofuel development, catalyst synthesis and applications, and bioconversion of crops and wastes to valuable products as well as the application of microbiomes to biological systems such as the natural environment. He has collaborated with other scientists around the world and supervised over 50 graduate students at masters and doctoral levels. He has participated as a presenter, session chair, and organizer in several conferences and professional meetings. He is an Associate Editor of *Frontiers in Energy Research* (Bioenergy and Biofuels) and an editorial board member of *Heliyon*, *Materials Today Communications*, *Processes*, and *Future Energy* journals. He is a Guest Editor on special topics in *Sustainability* and *Frontiers in Energy Research* journals. He has authored over 160 publications in journal articles, book chapters, and conference proceedings. He is an Ad-hoc reviewer for many scientific journals and grant agencies. He is very active professionally; his membership includes bodies such as the Nigerian Society of Chemical Engineers (NSChE), Nigerian Society of Engineers (NSE), American Institute of Chemical Engineers (AIChE), American Chemical Society (ACS), American Society for Microbiology (ASM), Society for Biological

Engineering (SBE), and Sigma Xi. He has been registered as a Professional Engineer with the Council for the Regulation of Engineering in Nigeria (COREN) since 2001.

Dr. Mofoluwake M. Ishola graduated from Ladoke Akintola University of Technology, Ogbomoso, in 2002 with B.Tech. Chemical Engineering, Second class honors upper division. She later did her M.Sc. studies in Chemical Engineering at Obafemi Awolowo University, Ile-Ife, Nigeria, and graduated with Distinction in 2008. She proceeded to the Swedish Centre for Resource Recovery, University of Borås, Sweden, for her Ph.D. studies. Her Ph.D. research was focused on the development of a novel process to produce biofuel ethanol from lignocellulosic feedstocks. She defended her Ph.D. thesis in 2014 and became the first Ph.D. candidate to finish in a record time of three years and seven months in the university. She later did her Postdoctoral research at the Department of Chemistry and Chemical Engineering of Chalmers University in 2017–2018. She has previously worked as a University Lecturer and Researcher, as an Environmental Engineer in the energy industry, and has served as a reviewer for several high-ranking academic journals. She supervised several postgraduate students and has authored about thirty academic publications including articles in high-ranking international reputable journals, chapters contributions in published books and a published book. She has been a recipient of different scholarships, awards, and grants including, Carnegie Scholarship of the UK and Fredrika Bremer Scholarship of Sweden as well as ÅForsk grants of Sweden. She is currently a Senior Research and Development Engineer at Scanacon AB, Sweden, a world-leading international environmental cleantech company. She is also a Visiting Professor at the Department of Chemical Sciences, Ajayi Crowther University Oyo, Nigeria.

Contributors

Adewale G. Adeniyi Chemical Engineering Department, University of Ilorin, Ilorin, Nigeria

Adebisi Aminat Agboola Department of Chemical Engineering, Obafemi Awolowo University, Ile-Ife, Osun State, Nigeria

Roland Allmägi Institute of Forestry and Engineering, Estonian University of Life Sciences, Tartu, Estonia

Gabriel S. Aruwajoye School of Life Sciences, University of KwaZulu-Natal, Pietermaritzburg, South Africa

Adeolu A. Awoyale Petroleum and Natural Gas Processing Department, Petroleum Training Institute, Effurun, Nigeria

Eriola Betiku Department of Chemical Engineering, Obafemi Awolowo University, Ile-Ife, Osun State, Nigeria;
Department of Biological Sciences, Florida Agricultural and Mechanical University, Tallahassee, FL, USA

Andrew C. Eloka-Eboka Centre of Excellence in Carbon based Fuels, School of Chemical and Mineral Engineering, North-West University, Potchefstroom Campus, South Africa

Esra Meşe Erdoğan Department of Environmental Engineering, Gebze Technical University, Gebze-Kocaeli, Turkey

Olayomi Abiodun Falowo Department of Chemical Engineering, Landmark University, Omu-Aran, Kwara, Nigeria

Ana Belén Guerrero Faculty of Environmental and Agricultural Sciences, Universidad Rafael Landivar, Guatemala, Guatemala;
Trisquel Consulting Group, Quito, Ecuador

Risto Ilves Institute of Forestry and Engineering, Estonian University of Life Sciences, Tartu, Estonia

Mofoluwake M. Ishola Department of Research and Development, Scanacon, Stockholm, Sweden

Niyi Babatunde Ishola Department of Chemical Engineering, Obafemi Awolowo University, Ile-Ife, Osun State, Nigeria

Marcis Jansons Mechanical Engineering Department, College of Engineering, Wayne State University, Detroit, MI, USA

Evariste B. Gueguim Kana School of Life Sciences, University of KwaZulu-Natal, Pietermaritzburg, South Africa

Pınar Karagöz Department of Biochemical Engineering, University College of London, London, UK

Mallika Boonmee Kongkeitkajorn Department of Biotechnology, Faculty of Technology, Khon Kaen University, Khon Kaen, Thailand

Daniel Lachos-Perez Department of Chemical Engineering, Federal University of Santa Maria, Santa Maria, RS, Brazil

Lekan M. Latinwo Department of Biological Sciences, Florida Agricultural and Mechanical University, Tallahassee, FL, USA

David Lokhat Reactor Technology Research Group, School of Engineering, University of KwaZulu-Natal, Durban, South Africa

Bárbara P. Moreira Department of Chemical Engineering, Federal University of Santa Maria, Santa Maria, RS, Brazil

Edmundo Muñoz Center for Sustainability Research, Universidad Andres Bello, Santiago, Chile

Tunde V. Ojumu Department of Chemical Engineering, Cape Peninsula University of Technology, Bellville, Cape Town, South Africa

Babatunde Oladipo Department of Chemical Engineering, Cape Peninsula University of Technology, Bellville, Cape Town, South Africa

Atilade A. Oladunni Department of Chemical Sciences, Ajayi Crowther University, Oyo, Nigeria

Melek Özkan Department of Environmental Engineering, Gebze Technical University, Gebze-Kocaeli, Turkey

Kaie Ritslaid Institute of Forestry and Engineering, Estonian University of Life Sciences, Tartu, Estonia

Daneal C. S. Rorke School of Life Sciences, University of KwaZulu-Natal, Pietermaritzburg, South Africa

Héctor A. Ruiz Biorefinery Group, Food Research Department, Faculty of Chemistry Sciences, Autonomous University of Coahuila, Saltillo, Coahuila, Mexico

Isaac A. Sanusi Fort Hare Institute of Technology, University of Fort Hare, Alice, South Africa

Yeshona Sewsynker-Sukai Fort Hare Institute of Technology, University of Fort Hare, Alice, South Africa

William G. Sganzerla Food Engineering School, University of Campinas (UNICAMP), São Paulo, Brazil

Abiola E. Taiwo Faculty of Engineering, Mangosuthu University of Technology, Durban, South Africa

Paulo C. Torres-Mayanga Universidad Tecnológica del Perú, Lima, Peru; Innovative Technology, Food and Health Research Group, Departamento de Ingeniería de Alimentos Y Productos Agropecuarios, Facultad de Industrias Alimentarias, Universidad Nacional Agraria La Molina, Lima, Peru

Introduction: Benefits, Prospects, and Challenges of Bioethanol Production

Olayomi Abiodun Falowo and Eriola Betiku

Abstract This chapter discussed the importance of bioethanol as a renewable energy fuel. The underlying factors driving the promotion of renewable fuels were highlighted to acknowledge several initiatives, studies, and policies impacting the energy sector. An overview of global bioethanol output was conducted to document the dynamism of the industry and its acceptance by different stakeholders. The benefits of bioethanol as a transportation fuel were delineated under various categories to highlight its advantages to the environment and humanity. The strategic implementations of policies, technological innovation, improved production process, and discovery of sustainable feedstocks capable of transforming the bioethanol industry were discussed. The challenges plaguing large-scale bioethanol production were reviewed. Under this category, feedstock-associated challenges affecting the output of the bioethanol manufacturing industries were underscored.

Keywords Biofuel · Bioethanol · Fermentation · Hydrolysis · Feedstock · Cellulosic biomass · Pretreatment

1 Introduction

Humanity faces numerous challenges today, including energy crises, population explosion, food shortages, climate change, and environmental pollution. These challenges are interconnected as population growth means an increase in food demand and

O. A. Falowo
Department of Chemical Engineering, Landmark University, P.M.B. 1001, Omu-Aran, Kwara, Nigeria

E. Betiku (✉)
Department of Chemical Engineering, Obafemi Awolowo University, Ile-Ife 220005, Osun, Nigeria
e-mail: ebetiku@oauife.edu.ng; eriola.betiku@famu.edu

Department of Biological Sciences, Florida Agricultural and Mechanical University, Tallahassee, FL 32307, USA

© The Author(s), under exclusive license to Springer Nature Switzerland AG 2023
E. Betiku and M. M. Ishola (eds.), *Bioethanol: A Green Energy Substitute for Fossil Fuels*, Green Energy and Technology,
https://doi.org/10.1007/978-3-031-36542-3_1

energy demand by the people, resulting in increased environmental pollution due to different activities to meet vital needs and ultimately affecting the climate. One such activity is industrialization; the expansion of industries due to population explosion means that fossil fuel utilization is increasing daily and subsequently increases greenhouse gas (GHG) emissions. The interaction between human needs and activities is proportional to the purity of our environment. To date, fossil fuels play a prominent role in meeting the energy demand of most developed and developing nations. Fossil fuels are the major contributor to transportation fuels worldwide. Fuels from fossil sources have led to an increased emission of GHG into the atmosphere. A direct link has been established between the level of environmental pollution and fuel utilization from fossil sources [30]. This development has raised serious local and global concerns as it endangers the present and future environment. The high CO_2 concentration in the atmosphere is correlated with the increased emission due to the overuse of fuels from fossil sources, which has led to global warming and climate change [44]. The implications of this phenomenon are felt by the natural resources (air, soil, and water) in our environment. The global temperature in recent years has increased by 0.8 °C, leading to a 15–20 cm rise in sea level due to the melting of polar ice [43].

Besides growing concerns for climate change and global warming due to fossil fuel utilization, the over-dependence of the economies of several nations on fossil oil is another worrisome issue. Many such economies have suffered due to the price fluctuation of fossil oil in the international market. This instability has led to several global economic crises, with the most affected nations being the developing nations. Economic recession and domestic inflation are notable phenomena associated with crude oil fluctuation in the world market. To avoid negative consequences on the economy due to the high price of foreign crude oil, the United States raised the level of investment in plant-derived sustainable biofuel sources [31]. Each time crude oil prices become a public subject due to the increase in domestic commodities, an increased interest in energy from renewable sources takes center stage.

The fundamental issues such as global warming, energy crises, and climate change associated with fossil fuel energy necessitate establishing more promising alternatives. An alternative renewable and sustainable fuel is a win for humanity and the environment. Currently, there is international cooperation to limit emissions and avert further global warming [43]. Different meetings and conferences have initiated international frameworks to mitigate emission effects. Biofuel is considered a suitable replacement since it is generated from renewable sources and solves the depletion of fossil fuels. Similarly, the automobile industry has embraced this new direction due to forces including low or zero emission, health concerns, consumers' choices, and increasing mineral oil prices [41].

Different governments are working towards implementing renewable energies under different policies. Advanced countries such as the United States, China, Australia, European Union, and Canada, as well as developing countries such as Brazil and India, are making tremendous progress in ensuring that renewable energy sources partially or fully replace fossil oil. For instance, the United States formulated renewable energy standards and regulations to ensure the application of renewable feedstocks for energy production purposes. Also, a different set of strategies and regulations have been upheld as outlined by Policy Energy [37] to ensure that fuel from biomass sources is price-competitive with mineral fuels in internal combustion engines. Following this agreement, the Energy Independence and Security Acts [17] proposed 36.0 billion gallons in the applicable volume of renewable fuel to be introduced into transportation fuel by 2022. In 2006, the first law on the utilization of renewable energy from biomass was issued in China. Through different government restrictions on using first-generation feedstocks, biorefinery companies are now exploring second-generation feedstocks [22]. European Union is ranked third in bioethanol production after US and Brazil, with the leading countries being France and Germany [11]. Other member states have continued integrating bioethanol into their energy mix while gradually reducing the net import of fossil fuels [4]. Developing countries are included in the effort to shift towards green renewable energies. For example, Pakistan is working to shift 5% of energy from fossil sources to renewable sources under Vision 2030 [26]. Brazil has been utilizing bioethanol as a transportation fuel based on the creation of the Brazilian Alcohol Program in 1975 [41]. Several institutions and government organizations are working on resources, methodologies, and technologies associated with renewable energies to make them cost-competitive with the current fossil energy.

Biofuels are biomass fuels, including bio-oil, biohydrogen, bioethanol, biodiesel, and biogas. There has been a growing worldwide interest in using bioethanol as a transport fuel since the mid-1990s. Bioethanol, a renewable biofuel, is a clean and safe fuel that can be produced from fermentable sugars. Bioethanol is an oxygenated fuel with 35% oxygen, reducing particulate and NO_x emissions from its combustion [8]. Bioethanol properties include a high octane number, high heat of vaporization, lower heating value, and low cetane number [8]. The persisting trend in the high price of oil coupled with environmental concerns has stimulated several developments to emancipate bioethanol as a current and future fuel. Bioethanol requires 68% less energy to generate than high-octane gasoline [9]; thereby, it remains a promising alternative to substitute fossil fuels partially. To this effect, global bioethanol production has continued to increase yearly (Fig. 1). According to data from the renewable fuel association, the global output of bioethanol reached 29,330 million gallons in 2019.

The coronavirus outbreak in 2019 affected the global output the following year, and only 26,470 million gallons of bioethanol were produced due to the halting of several production activities. Global bioethanol production increased again in 2021, and a subsequent increase in global bioethanol output is expected in 2022 and beyond.

Currently, many countries have medium- and long-term planning for biofuel utilization. However, only some countries produce bioethanol. The United States and Brazil contributed over 80% of the world's bioethanol production (Fig. 2). To date, bioethanol is classified into four generations depending on the source of feedstocks. The raw materials for first-generation bioethanol include grains and starchy foods such as cassava, corn, sugarcane, sweet potatoes, and beet. Most bioethanol-producing countries obtained fuel from these feedstocks until recently when the food versus fuel debate arose. The United States, the leading bioethanol producer in the world, uses corn, while Brazil, the second-largest ethanol producer, frequently uses sugarcane and molasses. Likewise, China produces bioethanol from

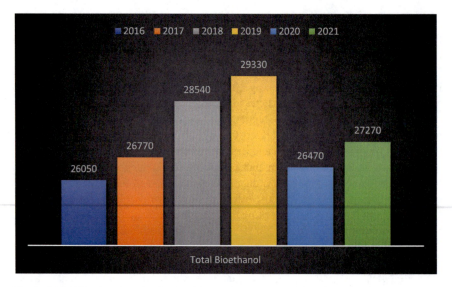

Fig. 1 Global bioethanol production in million gallons from 2016 to 2021. *Data source* Renewable fuels Association. https://ethanolrfa.org/markets-and-statistics/annual-ethanol-production

sweet sorghum and cassava. The contribution of major ethanol-producing nations has increased every year until the COVID-19 outbreak (Fig. 3). New countries like Thailand and Indonesia have initiated several programs to sustain bioethanol production development.

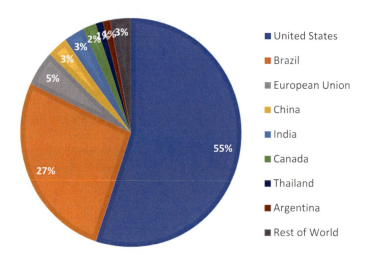

Fig. 2 Percent contribution of ethanol-producing countries to date. *Data source* Renewable fuels Association. https://ethanolrfa.org/markets-and-statistics/annual-ethanol-production

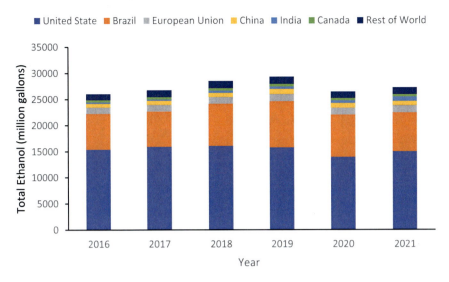

Fig. 3 Contributions of ethanol-producing nations to global production. *Data source* Renewable fuels Association. https://ethanolrfa.org/markets-and-statistics/annual-ethanol-production

Due to the competition between food and fuel from common feedstocks, which invariably led to high food prices, other sources were explored to produce bioethanol. The second-generation bioethanol came from agricultural wastes and residues, known as lignocellulose biomass. Lignocellulose biomass includes corn stover, wheat straw, sugarcane bagasse, coconut husks, and peanuts husks. These feedstocks differ from the first-generation feedstocks because they cannot be consumed as they contain mainly cellulose, hemicellulose, and lignin polymer. Lignin contents must be removed to maximize bioethanol production from these feedstocks. Undoubtedly, bioethanol can be obtained from lignocellulosic materials. However, the technology to produce bioethanol from this feedstock category has yet to be perfected, and it is currently the focus of intensive research. Hence, pretreatment of cellulosic biomass is necessary (Fig. 4). Feedstocks such as Cyanobacteria, macroalgae, and microalgae are used to produce third-generation bioethanol, while fourth-generation bioethanol is produced from genetically modified crops and algae [16].

In converting starchy foods, sugar-rich crops, and grains to fuel bioethanol, feedstock hydrolysis is required to generate fermentable sugars. The fermentable sugars are converted to fuel bioethanol using yeasts, particularly *Saccharomyces cerevisiae* [10]. The enzymatic fermentation of first-generation feedstocks has advanced, resulting in bioethanol commercialization. This feedstock category constitutes the major raw material for bioethanol production in many countries (Table 1). However, technology to produce bioethanol from lignocellulosic material is in the infant phase, so the question remains whether the current technology can produce large-scale ethanol from these feedstocks. A recalcitrant nature of lignocellulosic material means that pretreatment of these feedstocks is needed before microbial fermentation

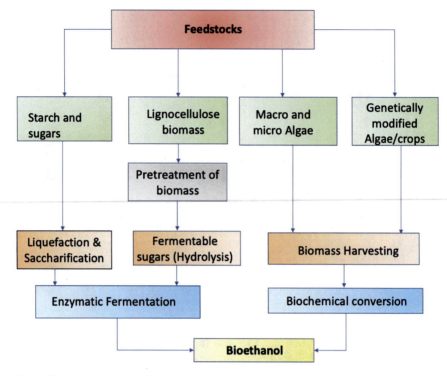

Fig. 4 Bioethanol production from various generational feedstocks

occurs. The pretreatment step decoupled lignin from the biomass, leaving behind the fermentable sugars, which can be converted by biological fermentation.

Three approaches commonly used in the fermentative conversion of lignocellulosic feedstock to bioethanol include separate hydrolysis and fermentation (SHF), simultaneous saccharification and fermentation (SSF), and consolidated bioprocessing (CBP) [35]. SHF is a sequential process involving enzymatic hydrolysis of cellulosic and hemicellulose biomass to reducing sugars, then converting the sugars to bioethanol through fermentative microorganisms. Each conversion stage can be operated at its optimal condition, which is the main advantage of this configuration. The activity and environment in one process are not influencing or interfering with another process. The major drawbacks have been the high economic investment in this approach and the inhibition of enzyme activity due to the increased sugar concentration in the hydrolysis.

In SSF, both enzymatic hydrolysis and sugar fermentation are carried out in a single stage to produce bioethanol. Compared to SHF, the SSF approach is less capital intensive, requires less reactor volume and processing time, overcomes

Table 1 Major feedstocks in ethanol-producing nations

Country	Feedstock	References
United States	Corn, corn stover	[2]
Brazil	Sugarcane juice and molasses, sugarcane bagasse	[13]
China	Cassava (18.50%), corn, sweet sorghum, wheat (grains 72.03%), corn and wheat straw (9.47%)	[45]
Germany	Maize (65%), rye, barley, wheat, and triticale (28%)	[40]
Poland	Maize (55.61%) and waste starch (30%)	[38]
India	Sugarcane juice and molasses (90%), corn, rice, and wheat	[39]
France	Wheat and sugar beets	[11]
Canada	Corn (87%), cereals (12.7%)	[42]
Sweden	Wheat, maize, barley, triticale, sugarcane, and sugar beet	[4]
Italy	Wheat and rice straw	[7]
UK	Wheat and sugar beets	[11]
Australia	Grain sorghum, sugarcane molasses, wheat, and waste wheat starch	[3]
Argentina	Sugarcane (47%) and Corn starch	[20]
Thailand	Molasses, cassava, and sugarcane juice	[27]

enzymatic inhibition, and requires lower enzyme amount and sterile conditions [15]. Meanwhile, it is difficult to maintain the reaction temperature for both enzymatic hydrolysis and fermentation processes. These two processes have different optimal temperatures for their efficiency. One way to overcome this problem is to perform the pre-saccharification step [19]. This strategy produced a high bioethanol yield compared to when only SSF was applied. CBP is an integrated approach involving enzyme production, saccharification, and fermentation. The advantage of this approach is reduced capital cost, reduced energy requirement, simplified reaction steps, improved hydrolysis efficiency, and low contamination risk [12].

2 Benefits of Bioethanol Production

The advantages of utilizing fuel ethanol from biomass can be categorized into three. These benefits can be discussed in economic, environmental, and social terms.

2.1 Substantial Contribution to Gross Domestic Product (GDP)

Mitigating trade deficits is a huge benefit for bioethanol-producing nations. A country that largely depends on the agrarian economy could produce a large volume of bioethanol that can be exported to other countries. Ethanol export to other countries reduces the gap created in the economy by importing other commodities. The economic prospect of bioethanol production in Pakistan revealed that the GDP of the country could be increased, and bioethanol could also be exported to other countries [26]. Likewise, the Brazilian government initiated ProAlcool Program to reduce the country's dependence on imported oil [5]. Besides, bioethanol production can increase the GDP of a country and household income. In agricultural-dominated countries, most people reside in rural areas far more than in urban areas. This rural population can, directly and indirectly, affect the volume of cellulosic materials generated, and with the proper policy incentives, such practice can boost the economy of the country.

2.2 Reduction in GHG Emissions

Introducing bioethanol as an alternative transport fuel can mitigate CO_2 emissions into the atmosphere. Besides, other pollutants such as NO_x and particulate emissions have been reported to reduce significantly [30, 32]. The potential benefit of bioethanol to the environment has been highlighted by several studies as concern for global warming grows. A significant reduction in GHG emissions has been speculated when replacing mineral fuels with biofuel. In Brazil, ethanol production does not affect the food supply or impact deforestation [22]; instead, it reduces the net contribution of CO_2 emitted. The carbon footprint of corn-based and sugarcane bioethanol has been estimated in two leading ethanol-producing nations [34]. The GHG emissions assessed were from crop transport, fertilizer, farming, residual biomass, ethanol plant, and land use change. The study reported a carbon footprint of 38.5 gCO_2e/MJ and 44.9 gCO_2e/MJ for Brazilian sugarcane and U.S. corn bioethanol, respectively. A comparison study on the life cycle assessment of carbon emission from ethanol from coal and cassava shows that carbon emission from coal is much higher than the cassava-conversion route [30]. Significant CO_2 reduction was observed in crop-based ethanol compared to coal-derived ethanol.

2.3 Bioethanol as a Substitute for Fossil Fuels

Bioethanol has emerged as one of the promising alternative fuels to blend fossil fuels as a transportation fuel. The major environmental pollution is from fossil fuels, particularly from transportation fuels. Approximately 70% and 19% of CO and CO_2 are emitted from motor vehicles globally [14]. The introduction of the volume of bioethanol into gasoline has been reported to improve fuel combustion and reduce GHG emissions [23]. Due to their environmental benefits and sustainability, bioethanol can reduce the dependence on fossil oil. It can be used alone in spark ignition engines or as an additive to gasoline. Using bioethanol as a transportation fuel can improve engine performance, reduce exhaust emissions and improve combustion properties [46]. Since bioethanol production is not at the level to compete with mineral fuels in terms of cost, several countries have adopted a partial blending of bioethanol with gasoline. The Pakistan government permitted the using E10 (10% substitution of gasoline with bioethanol) [18]. E25 of gasoline is sold and used in Brazil as a transportation fuel, as over 95% of cars use a blend of bioethanol and gasoline [5]. The E10 is used regularly in the US, while E5 and E85 are the adopted blends in Europe.

2.4 Enhanced Energy Security

Bioethanol is locally produced because local feedstocks used for its production are readily available in ethanol-producing nations. Using bioethanol reduces the country's dependence on importing foreign crude oil, which is unreliable. Many factors are responsible for the energy instability obtained from fossil fuels; crises between nations, wars, oil price fluctuation, and depleted crude oil reserves have had serious implications on the economy in the past. There is a linear relationship between a country's economic growth and energy consumption; particularly in many developing countries, the economy is built on crude oil proceeds. To this effect, crude oil has become an economic weapon for many developed countries since bans or restrictions can be placed on crude oil importation from countries with strained relationships. This scenario has played out repeatedly on the international stage, and the economic growth of such a country heavily built on crude oil is therefore hindered. Bioethanol production offers a route for countries to secure energy security and economic prosperity.

3 Prospects of Bioethanol

Various factors that can make sufficient bioethanol production possible are addressed in this section.

3.1 Financial Support and Subsidies

For the sustainability of bioethanol fuel, different states have introduced long-term planning instruments in their national legislation, which is in line with the Paris Agreement. This united effort sends the right signal to stakeholders and can stimulate the expansion of the bioethanol industry. Provision of subsidies, financial support, and incentive to all players involved in the bioethanol production chain are some ways governments are adopting for its expansion. The Chinese government offers substantial financial support to farmers cultivating sugarcane to mitigate the decline in sugarcane production [22]. In Europe, companies producing biofuel and farms were awarded subsidies to mitigate the production cost under the EU infrastructure and environmental program [38]. These gestures encourage players in the industry and could further increase bioethanol fuel.

3.2 Introduction of Preferential Policies

Reducing taxes and increasing the subsidies for bioethanol-producing companies would invariably expand the bioethanol industry. The development of government and legislative measures has continued to influence bioethanol production and expand its market positively. Government direction on biofuel utilization would receive public support once the policy is supported with good strategies. Such policy helps China to gradually shift from grain-based (corn) bioethanol to 1.5th and second-generation feedstocks (sweet sorghum, cassava, and fibers) [22]. Both administrative and fiscal regulations strongly impact the bioethanol sector as their contribution influences the pace and scale of the sector. Government policy influences the acceptance of bioethanol at each level of the supply chain. A successful policy will positively impact land use, water use, agricultural practices, environmental conservation, the biofuel market, and other stakeholders, while such policies must mitigate the risk of involvement in developing a bioeconomy [29].

3.3 Job Creation Opportunities

Establishing and implementing a bioethanol program is a great opportunity to create employment. Direct and indirect jobs are created at the downstream and upstream sectors of ethanol processing plants. The expansion of bioethanol benefits farmers and uplifts their economic status. The proposed bioethanol project in Cradock, South Africa, shows that its implementation will reduce poverty and strengthen food security [36]. The stakeholders in the biofuel supply chain are rural communities, biorefineries, energy crop producers, government, biofuel consumers, blending facilities workers, and storage facilities workers. The activities of each stakeholder have to be synchronized to ensure the maximum operability of the biofuel plant. Multiple jobs can be created at each level to sustain biofuel development [29].

3.4 Availability of Bioethanol Market

Many countries are committed to renewable energy, and there is an obligation to include a specific ratio of bioethanol in transportation fuel. The ratio of bioethanol included in transportation fuel is expected to increase yearly and ultimately achieve a 50% reduction in GHG emitted. A strategic goal of EU energy policy is establishing a biofuel market [38]. Member states are committed and obligated to include a 10% share of renewable fuel in the final transportation fuel consumed to qualify for assistance. The United States and Brazil are the highest bioethanol consumers, and legislative support has been implemented to grow the bioethanol market. The market offers flex-fuel (a mixture of fuel containing 51% petrol and 83% ethanol) for car users. Currently, most cars on American and Brazilian roads are equipped with flex-fuel technology [28].

3.5 Discovery of Novel Feedstocks

Bioethanol development is expected to be sustained economically by the discovery of fourth-generation feedstocks since it has no impact on food security. Apart from genetically modified crops and lignocellulosic biomass, fruit waste is another biomass generated in large quantities. This poses challenges for waste management in different regions. The volume of fruit waste generated suggests that it can serve as raw materials for bioethanol fermentation. The prospect of bioethanol production from lignocellulosic biomass and fourth-generation feedstocks presents a huge opportunity for bioethanol production growth [38]. Feedstocks that can be cheaply processed using current technology are researched daily to ensure bioethanol sustainability. These feedstocks can be used alone or combined with others to increase bioethanol yield from its blend. In the US, several farmers are motivated by off-farm incomes to grow perennial grasses as feedstock options. Miscanthus and switchgrass could be grown on marginal land as supplemental crops [29].

3.6 Improvement in Processing Technology

Technology that can take advantage of purification and concentration of pre-hydrolysate, enzymatic saccharification process, cell cycle, fermentation process, and bioethanol recovery process to enhance the economics of bioethanol production is desirable. The production of first-generation bioethanol was discouraged, mainly due to food security. However, second-, third-, and fourth-generation bioethanol has promising potential in this regard but with little commercial prospect. The emergence of new technologies to process lignocellulosic biomass into bioethanol provides opportunities to scale up global production. Advanced membrane technologies could facilitate lignocellulosic bioethanol production at the commercial scale. The membrane-based technologies have been designed to make the overall process continuous and successful [15]. Likewise, genetically modified crops and algae have continued to receive great attention in fourth-generation bioethanol. Considering the inherent advantages of microalgae, future bioethanol production at the commercial level may be from this feedstock.

3.7 Integration of Bioethanol Processing Technology

An integrated bioethanol facility with the installation of combined heat and power plant (CHP) can extract maximum benefit from the lignocellulosic biomass. Bioethanol from lignocellulosic materials is yet to be commercialized due to the imperfection in the used technology. Using lignin, a waste product generated from the lignocellulosic feedstock, to produce heat and power could enable economic bioethanol fuel production. A lignin-fired biomass boiler can be installed with the main hydrolysis installations to supply heat and energy. This development will consequently improve the net energy balance of the process.

4 Challenges Associated with Bioethanol

Significant progress has been made in the last few years in advancing bioethanol production in a manner that would make it a commercial fuel in several countries. The discovery of new feedstocks, improvement in bioethanol technology, and favorable legislation have aided some of the notable progress recorded in bioethanol industries. However, different challenges across the bioethanol supply chain have limited its expansion as an alternative to mineral fuels. These challenges can be social, economic, technical, and environmental. Limitations, including high-cost involvement, possible equipment corrosion, poor solvent recovery, and formation of inhibitory products, are technically associated with bioethanol processing. Moreover, bioethanol production is considered a high-energy consuming process; among other challenges are the ineffectiveness of conventional pretreatment techniques, inefficient saccharification processes, and ineffective separation and purification steps [15]. Meanwhile, feedstocks-related challenges facing bioethanol production have impeded the development and expansion of renewable biofuel, especially in sub-Saharan countries [1]. This includes a lack of bioenergy policy, insufficient water supply, food insecurity, lack of incentive to local industries, land availability, and lack of financial support. The same pattern of challenges is common in Middle

East countries. For example, Iran has not produced or consumed bioethanol since feedstocks cannot be produced locally. Different scenarios were modeled in Iran to assess the economic impacts of bioethanol production. Three feedstocks, including Salicornia, sweet sorghum, and low-quality corn, have been investigated to establish their feasibility in producing bioethanol [6].

4.1 High Capital and Energy Investment

The cost of feedstock represents approximately 80% of the total production cost. Though second-generation feedstocks are abundant in nature and available through human activities, their processing into fuel bioethanol is complex using current technologies. The pretreatment of lignocellulosic feedstock is energy and cost-intensive. Lignocellulosic materials are complex polymers with a recalcitrant nature. During pretreatment, harsh conditions such as high temperature, high pressure, and high molar solvent concentration are needed to deconstruct the feedstock. Moreover, the chemicals used during pretreatment are sometimes difficult to recycle; some constitute health and environmental hazards [21]. The production cost of bioethanol from different energy crops (corn, wheat, sweet sorghum, sugarcane, and cassava) has been compared to ascertain the bioeconomy potential of these feedstocks. The bioethanol production cost per meter cube is low for sugarcane and corn and highest for cassava and wheat. Bioethanol conversion from a feedstock of 5 or 6 carbon atoms is easier than starchy and lignocellulosic feedstocks [8]. Meanwhile, the application of expensive commercial cellulase enzymes in lignocellulosic fermentation is common in the large-scale production of bioethanol. Identifying and properly isolating cellulase-producing microorganisms for bioethanol production may improve bioethanol economics.

4.2 Unavailability of Feedstocks

Raw materials for bioethanol production are obtained through farming practices in different geographical regions. The availability of this feedstock for bioethanol production depends on prevailing climatic conditions and arable land. Realistically, these feedstocks are only available during specific periods of the year as their yield varies from season to season. Besides feedstock availability, feedstock price is highly volatile and can subsequently affect the volume of bioethanol produced in a specific location.

4.3 Formation of Inhibitory Compounds

Various inhibitory products are formed during the pretreatment of lignocellulosic feedstocks. Other generational feedstocks do not have this constraint, but future bioethanol at a large scale requires the utilization of cellulosic biomass since it is the most abundant reproducible resource. Inhibitory compounds at various concentrations can affect the hydrolysis step and influence fermentation efficiency in an enzymatic process, consequently affecting the bioethanol yield obtained. Examples of inhibitory compounds formed during pretreatment are furans, phenols, and carboxylic acids. Inhibitory products produced from the second-generation feedstock lower the efficiency of the bioethanol production process and add extra cost to the whole process. The mechanism of inhibitory compounds formed during pretreatment has been explored to reduce its effect on fermentation [24].

4.4 Poor Performance of Fermentation Microorganisms

Inefficient fermentation of sugars due to selected microorganisms is a major drawback in the commercialization of bioethanol. The conversion of fermentable sugars to bioethanol is usually through biological pathways with either bacteria or yeasts. One of the technical challenges of the fermentation process involved a selection of suitable microorganisms. For large-scale bioethanol production, ideal microorganisms must have the ability to convert various sugars substrate, generate high bioethanol yield, withstand high processing temperatures, and tolerate high concentrations of bioethanol and yield [15]. The sugars generated from the pretreatment process involved a significant amount of pentose, which cannot be easily fermented by *Saccharomyces cerevisiae*. Moreover, microbial contaminants are present in the cellulosic materials and can produce toxic bye products or compete with the fermenting yeast for substrates [31]. To optimally improve the efficiency of this process, microorganisms capable of fermenting various sugars are usually desirable. On the other hand, genetically modified microorganisms can be applied during sugars fermentation to obtain maximum bioethanol. A mixed microbial population may be used to produce bioethanol by utilizing all sugars (pentose and hexose) obtained from lignocellulose hydrolysate. Implementing a coculture system can better utilize total sugars and maximize bioethanol production.

4.5 High Bioethanol-Water Separation Cost

The separation of the bioethanol-water mixture is technically challenging and cost-intensive. To blend bioethanol with regular gasoline as a vehicle fuel, it must be in an anhydrous form. Distillation is a common technique to recover bioethanol from the first-generation feedstock. However, bioethanol recovery from other generational feedstock using distillation is not economical due to the low ethanol concentration in the mixture. A sufficient amount of sugar is required in the hydrolysate to reduce the cost associated with bioethanol distillation [15]. Lignocellulosic feedstocks with low sugar concentration would increase processing costs since several separation steps would be required to obtain pure bioethanol. An effective technique for separating bioethanol is through pervaporation, and it does not depend on the relative volatility of the mixture [25]. Even with the advancement in separation technologies, distillation combined with dehydration is the dominant separation method in biorefineries due to the uncommercialized level of pervaporation technologies.

4.6 Political Instability and Changes in Government Policies

Changes in the political environment of a particular country may impact the bioethanol sector dynamics. Government plays a crucial role between bioethanol producers, farmers, and consumers as they formulate strategies to sustain their commitment to renewable fuel utilization. The government, one of the major stakeholders in renewable fuel, may remain relatively inactive due to low oil prices. Besides, changing government may cause a big deviation from the energy policies initiated by the previous government. The interruption of renewable energy policies affects other stakeholders and leads to a lack of interest in ethanol-dedicated vehicles. The combination of these constraints can affect the fortune of bioethanol fuel.

4.7 Limited Land Use for Energy Crops

As a globally adopted fuel, not all countries have the capacity to expand bioethanol production due to the limitations in their agriculture sector. Some semi-arid countries do not have feedstocks to support the large-scale production of bioethanol. For example, the agriculture sector in Iran suffers from soil salinity and water scarcity problems, which affect crop productivity [6]. Though the country has the potential to produce 3600 million liters of bioethanol from sweet sorghum, no bioethanol fuel has been consumed either as fuel or fuel enhancer. Arable and fallow land to grow energy crops is a serious problem for Middle East countries. In assessing bioethanol potential in sub-Saharan countries, issues such as small farm size, the land tenure system, land availability, and rights were listed to prevent large-scale exploitation of energy crops [1]. In sustaining biofuel, the land is vital to feedstock availability through agricultural practices. However, access to fallow and arable land is a major bottleneck in developing countries. In Nigeria, land use is under individual ownership or community, which generates much dispute, thereby preventing mechanization or large-scale use. The implication is that massive arable farmland to grow feedstocks for bioethanol production is unavailable. Apart from the fragmentation of farmland, the continuous invasion of grassing herders has threatened the sustainability of agricultural practices in Nigeria. Farmers in rural areas suffered a major loss of feedstocks, resulting in the incapacitation of renewable energy sectors.

4.8 Lack of Technological Advancement

Even with vast resources for biofuel generation, some countries fail due to low scientific development. Due to maintenance costs, access to improved new technology for bioethanol production is a challenge in most West African nations [1]. There are over 210 biorefineries in the United States for processing bioethanol, and less than 6% of total biorefineries can process lignocellulosic feedstock. The mandate to construct new biorefineries capable of converting cellulosic biomass using current technology is given priority [29]. It is worth noting that biorefinery establishment requires huge capital investment. Therefore, challenges, including technology limitations, resources, infrastructure limitations, and lack of private investment in biorefinery construction, are barriers to its rise.

4.9 Demand for Bioethanol

Though bioethanol is a promising alternative, not all vehicles can use more than 5% bicomponent fuel. Due to the age of the cars, the demand for bioethanol can be affected since existing cars in some countries can only utilize bioethanol on a small scale. The average age of cars in Poland was 14 years in 2018. Due to technical failure, these cars cannot use E10, the adopted bicomponent ratio by the EU [38]. Bioethanol-fuel-advanced countries have vehicles specifically designed to utilize flex-fuel. However, developing countries still use gasoline-fuel vehicles, though this scenario can change quickly.

4.10 Water Unavailability

Much water is used in the biorefinery processing feedstock to bioethanol. The required amount of water at the processing stage may not be a threat, but emerging agricultural practices can affect the water supply [31]. Water may not pose a problem in some countries, but only some nations in the Middle East would encounter water scarcity issues when the biofuel plant was fully implemented. Therefore, water availability may vary from region to region due to the nature of the feedstock used. Applying water (irrigated water) to energy crops for increased productivity would pressure the water supply, leading to water shortage and, eventually, a water crisis.

Water networks from consumption, recycling, reuse, and pretreatment are considered to optimize water consumption in bioethanol production. It is worth noting that the water consumption was in the range or below the amount required for equal-volume gasoline production [33]. Moreover, agricultural practices such as herbicides, pesticides, and fertilizer application can result in a dead zone in the water body and consequently affect water quality. The inefficient utilization of water from farming practices through bioethanol processing may constitute a major barrier to bioethanol industry expansion.

References

1. Adewuyi A (2020) Challenges and prospects of renewable energy in Nigeria: a case of bioethanol and biodiesel production. Energy Rep 6:77–88
2. Aghaei S, Alavijeh MK, Shafiei M, Karimi K (2022) A comprehensive review on bioethanol production from corn stover: worldwide potential, environmental importance, and perspectives. Biomass Bioenerg 161:106447
3. Akbar D, Subedi R, Rolfe J, Ashwath N, Rahman A (2019) Reviewing commercial prospects of bioethanol as a renewable source of future energy—an Australian perspective. In: Advances in eco-fuels for a sustainable environment. Elsevier, pp 441–458
4. Amiandamhen SO, Kumar A, Adamopoulos S, Jones D, Nilsson B (2020) Bioenergy production and utilization in different sectors in Sweden: a state of the art review. BioResources 15(4):9834
5. Amorim HV, Lopes ML, de Castro Oliveira JV, Buckeridge MS, Goldman GH (2011) Scientific challenges of bioethanol production in Brazil. Appl Microbiol Biotechnol 91(5):1267–1275
6. Araghi MK, Barkhordari S, Hassannia R (2023) Economic impacts of producing bioethanol in Iran: a CGE approach. Energy 263:125765
7. Avanthi A, Mohan SV (2022) Emerging innovations for sustainable production of bioethanol and other mercantile products from circular economy perspective. Bioresour Technol:128013
8. Balat M, Balat H, Öz C (2008) Progress in bioethanol processing. Prog Energy Combust Sci 34(5):551–573
9. Bender LE, Lopes ST, Gomes KS, Devos RJB, Colla LM (2022) Challenges in bioethanol production from food residues. Bioresour Tech Reports:101171
10. Betiku E, Alade O (2014) Media evaluation of bioethanol production from cassava starch hydrolysate using *Saccharomyces cerevisiae*. Energy Sour, Part A: Recov Util Environ Effects 36(18):1990–1998
11. Broda M, Yelle DJ, Serwańska K (2022) Bioethanol production from lignocellulosic biomass—challenges and solutions. Molecules 27(24):8717
12. Choudhary J, Singh S, Nain L (2016) Thermotolerant fermenting yeasts for simultaneous saccharification fermentation of lignocellulosic biomass. Electron J Biotechnol 21:82–92
13. de Souza Abud AK, de Farias Silva CE (2019) Bioethanol in Brazil: status, challenges and perspectives to improve the production. Bioethanol Prod Food Crops:417–443
14. Deenanath ED, Iyuke S, Rumbold K (2012) The bioethanol industry in Sub-Saharan Africa: history, challenges, and prospects. J Biomed Biotechnol
15. Dey P, Pal P, Kevin JD, Das DB (2020) Lignocellulosic bioethanol production: prospects of emerging membrane technologies to improve the process–a critical review. Rev Chem Eng 36(3):333–367
16. Dutta K, Daverey A, Lin J-G (2014) Evolution retrospective for alternative fuels: first to fourth generation. Renew Energy 69:114–122
17. EISA (2007) Energy independence and security act of 2007, Federal and state incentives and laws

18. Ghani HU, Gheewala SH (2021) Environmental sustainability assessment of molasses-based bioethanol fuel in Pakistan. Sustain Prod Consum 27:402–410
19. Goncalves FA, Ruiz HA, dos Santos ES, Teixeira JA, de Macedo GR (2015) Bioethanol production from coconuts and cactus pretreated by autohydrolysis. Ind Crops Prod 77:1–12
20. Grellet MAC, Dantur KI, Perera MF, Ahmed PM, Castagnaro A, Arroyo-Lopez FN, Gallego JB, Welin B, Ruiz RM (2022) Genotypic and phenotypic characterization of industrial autochthonous Saccharomyces cerevisiae for the selection of well-adapted bioethanol-producing strains. Fungal Biol 126(10):658–673
21. Halder P, Azad K, Shah S, Sarker E (2019) Prospects and technological advancement of cellulosic bioethanol ecofuel production. In: Advances in eco-fuels for a sustainable environment. Elsevier, pp 211–236
22. Huang J, Khan MT, Perecin D, Coelho ST, Zhang M (2020) Sugarcane for bioethanol production: potential of bagasse in Chinese perspective. Renew Sustain Energy Rev 133:110296
23. Jhang S-R, Lin Y-C, Chen K-S, Lin S-L, Batterman S (2020) Evaluation of fuel consumption, pollutant emissions and well-to-wheel GHGs assessment from a vehicle operation fueled with bioethanol, gasoline and hydrogen. Energy 209:118436
24. Kausar F, Irfan M, Shakir HA, Khan M, Ali S, Franco M (2021) Challenges in bioethanol production: effect of inhibitory compounds. In: Bioenergy research: basic and advanced concepts. Springer, pp 119–154
25. Khalid A, Aslam M, Qyyum MA, Faisal A, Khan AL, Ahmed F, Lee M, Kim J, Jang N, Chang IS (2019) Membrane separation processes for dehydration of bioethanol from fermentation broths: recent developments, challenges, and prospects. Renew Sustain Energy Rev 105:427–443
26. Khan MU, ur Rehman MM, Sultan M, ur Rehman T, Sajjad U, Yousaf M, Ali HM, Bashir MA, Akram MW, Ahmad M (2022) Key prospects and major development of hydrogen and bioethanol production. Int J Hydr Energy 47(62):26265–26283
27. Kumakiri I, Yokota M, Tanaka R, Shimada Y, Kiatkittipong W, Lim JW, Murata M, Yamada M (2021) Process intensification in bio-ethanol production-recent developments in membrane separation. Processes 9(6):1028
28. Kupczyk A, Mączyńska J, Redlarski G, Tucki K, Bączyk A, Rutkowski D (2019) Selected aspects of biofuels market and the electromobility development in Poland: current trends and forecasting changes. Appl Sci 9(2):254
29. Leibensperger C, Yang P, Zhao Q, Wei S, Cai X (2021) The synergy between stakeholders for cellulosic biofuel development: perspectives, opportunities, and barriers. Renew Sustain Energy Rev 137:110613
30. Li J, Cheng W (2020) Comparison of life-cycle energy consumption, carbon emissions and economic costs of coal to ethanol and bioethanol. Appl Energy 277:115574
31. Limayem A, Ricke SC (2012) Lignocellulosic biomass for bioethanol production: current perspectives, potential issues and future prospects. Prog Energy Combust Sci 38(4):449–467
32. Lisboa CC, Butterbach-Bahl K, Mauder M, Kiese R (2011) Bioethanol production from sugarcane and emissions of greenhouse gases–known and unknowns. Gcb Bioenergy 3(4):277–292
33. Martín M, Ahmetovic E, Grossmann IE (2011) Optimization of water consumption in second generation bioethanol plants. Ind Eng Chem Res 50(7):3705–3721
34. Mekonnen MM, Romanelli TL, Ray C, Hoekstra AY, Liska AJ, Neale CM (2018) Water, energy, and carbon footprints of bioethanol from the US and Brazil. Environ Sci Technol 52(24):14508–14518
35. Melendez JR, Mátyás B, Hena S, Lowy DA, El Salous A (2022) Perspectives in the production of bioethanol: a review of sustainable methods, technologies, and bioprocesses. Renew Sustain Energy Rev 160:112260
36. Nasterlack T, von Blottnitz H, Wynberg R (2014) Are biofuel concerns globally relevant? Prospects for a proposed pioneer bioethanol project in South Africa. Energy Sustain Dev 23:1–14
37. PEA (2005) Energy policy energy act of 2005
38. Piwowar A, Dzikuć M (2022) Bioethanol production in Poland in the context of sustainable development-current status and future prospects. Energies 15(7):2582

39. Ramesh P, Selvan VAM, Babu D (2022) Selection of sustainable lignocellulose biomass for second-generation bioethanol production for automobile vehicles using lifecycle indicators through fuzzy hybrid PyMCDM approach. Fuel 322:124240
40. Röder LS, Gröngröft A, Grünewald M, Riese J (2023) Assessing the demand side management potential in biofuel production; a theoretical study for biodiesel, bioethanol, and biomethane in Germany. Biofuels, Bioprod Biorefin 17(1):56–70
41. Rosillo-Calle F, Walter A (2006) Global market for bioethanol: historical trends and future prospects. Energy Sustain Dev 10(1):20–32
42. Saini R, Osorio-Gonzalez CS, Brar SK, Kwong R (2021) A critical insight into the development, regulation and future prospects of biofuels in Canada. Bioengineered 12(2):9847–9859
43. Schleussner C-F, Lissner TK, Fischer EM, Wohland J, Perrette M, Golly A, Rogelj J, Childers K, Schewe J, Frieler K (2016) Differential climate impacts for policy-relevant limits to global warming: the case of 1.5 C and 2 C. Earth Syst Dyn 7(2):327–351
44. Singh S, Kumar A, Sivakumar N, Verma JP (2022) Deconstruction of lignocellulosic biomass for bioethanol production: recent advances and future prospects. Fuel 327:125109
45. Wu B, Wang Y-W, Dai Y-H, Song C, Zhu Q-L, Qin H, Tan F-R, Chen H-C, Dai L-C, Hu G-Q (2021) Current status and future prospective of bio-ethanol industry in China. Renew Sustain Energy Rev 145:111079
46. Yusoff M, Zulkifli N, Masum B, Masjuki H (2015) Feasibility of bioethanol and biobutanol as transportation fuel in spark-ignition engine: a review. RSC Adv 5(121):100184–100211

Novel and Cost-Effective Feedstock for Sustainable Bioethanol Production

Atilade A. Oladunni and **Mofoluwake M. Ishola**

Abstract Bioethanol is a biofuel from renewable energy sources that can be used in different blends with gasoline in spark ignition engines. It has been commercially produced from starch and sugar-based energy crops through yeast fermentation. However, due to the competition with food and feed sources, several efforts have been made to utilize other feedstocks from various sources that are not being used for food or feed production. Consequently, novel and cost-effective feedstocks such as lignocellulosic, algae, genetically modified biomass, and captured carbon dioxide (CO_2) are currently being investigated for commercial-scale applications. According to the feedstocks, the sources are categorized into first, second, third, and fourth generations. This chapter provides an overview of all the different generations of feedstocks, those currently in use, and those under investigation for bioethanol production. Research progress and their status regarding large-scale applications are also evaluated and discussed.

Keywords Bioethanol · Biofuels · Bioenergy · Cost-effective feedstocks · Novel feedstocks

1 Introduction

About a decade ago, bioethanol has been embraced as an alternative fuel to gasoline since it can be produced from renewable resources [15, 84]. It is used as automobile fuel in spark ignition engines or blended with gasoline in different mixture percentages of 10% ethanol (E10), 15% ethanol (E15), 25% (E25), 30% (E30), and 51–83% ethanol (E85) [26]. The use of biofuel ethanol has been reported to improve air quality through a complete combustion process [59]. Apart from being used as fuel,

---

A. A. Oladunni
Department of Chemical Sciences, Ajayi Crowther University, Oyo, Nigeria

M. M. Ishola (✉)
Department of Research and Development, Scanacon, Stockholm, Sweden
e-mail: mofoluwake.ishola@scanacon.se

© The Author(s), under exclusive license to Springer Nature Switzerland AG 2023
E. Betiku and M. M. Ishola (eds.), *Bioethanol: A Green Energy Substitute for Fossil Fuels*, Green Energy and Technology,
https://doi.org/10.1007/978-3-031-36542-3_2

ethanol is also be used as disinfectants in the medical industries, as a solvent in the pharmaceutical, cosmetic, and biomedical industries [36]. Furthermore, it is being used in gas turbines to lower the inlet temperature when injected into the inlet air [44].

Production of bioethanol involves fermentation of monomeric sugars with ethanol-producing microorganisms such as yeast as represented in Eqs. 1 and 2.

$$C_6H_{12}O_6(glucose) \xrightarrow{microorganisms} 2C_2H_5OH(ethanol) + 2CO_2 \qquad (1)$$

$$3C_5H_{12}O_5(xylose) \xrightarrow{microorganisms} 5C_2H_5OH(ethanol) + 5CO_2 \qquad (2)$$

The choice of feedstock for bioethanol is an important factor because this controls the whole conversion process and the ethanol yield. Different feedstocks have been evaluated for ethanol production. The feedstocks have been classified into first, second, third and fourth generations. The first-generation feedstocks include sugar and starch-based crops such as sugarcane, corn, barley, wheat, potato, rice, etc. [46]. This category of feedstock is currently used in countries like USA, Brazil, India, Belgium, Germany, France, UK, Hungary, and Netherlands for commercial-scale bioethanol production [35, 53, 78]. However, they have been continuously criticized owing to their competitive effects on the food supply and availability of arable lands [7]. Besides, the complete life cycle analysis has shown that the whole production process of ethanol from first-generation feedstocks is often associated with significantly high emissions of greenhouse gases (GHG) [38].

Second-generation bioethanol is produced from a wide collection of feedstocks, mainly lignocellulosic biomass [38]. Examples include energy crops, woody biomass, and agricultural by-products, which are affordable and available. The production process requires lower energy input compared to the energy required for first-generation ethanol [27]. Although the raw materials for production are abundant, the cost of processing is high, and the economic sustainability of the conversion process on a large scale is still being investigated [20].

Third-generation bioethanol has algae biomass as their sources. Over time, they have been regarded as the best option for demeriting the two earlier generations of bioethanol [12]. It also does not compete with food and feed production [21]. Microalgae are photosynthetic organisms which mainly use few nutrients, carbon dioxide and light for growth. They produce carbohydrates and lipids for bioethanol production [14, 60]. In third-generation bioethanol, hydrolysis and fermentation steps are more manageable than in second-generation ethanol because of low hemicellulose and zero lignin composition [72]. Although third-generation bioethanol production is well established on a small-scale level, expansion to a commercial scale is not yet economically feasible due to the slow growth rate of the strains and low lipid content [75].

Fourth-generation bioethanol is produced from captured CO_2 and genetically modified organisms [71]. Bioengineering procedures are used to transform the properties of algal in capturing CO_2 and its metabolism to produce bioethanol [6].

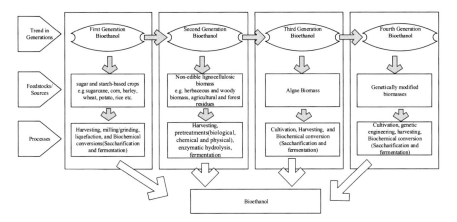

Fig. 1 Trends in bioethanol production from novel feedstocks

This genetic alteration helps reduce the stages involved in converting energy from the Sun into biofuel, making the process more economical than third-generation bioethanol production [6]. However, this currently exists only in small-scale research and needs further understanding. Figure 1 shows the summary of trends in bioethanol production from these novel feedstocks.

2 Overview of Bioethanol Production Process

Bioethanol production can be simply categorized into four different units, pretreatment, extraction of fermentable sugars into solutions, bioconversion of the fermentable sugars to ethanol as well as separation and purification of the product. The production process depends on the physiological activity and species of yeast, cell density, and the configuration of the hydrolysate [89].

2.1 Modes of Fermentation

The fermentation process proceeds either by continuous, batch, or fed-batch methods. The batch process involves the introduction of a substrate, yeast culture and nutrients at the commencement of fermentation only to collect the product at the end. The operation is relatively easy with low operating costs. However, the yield is always low due to the lag phase present at the commencement [39]. The ethanol product is distilled out and dehydrated to > 99.5% concentration. Another benefit derived from this method is low production cost resulting from lesser control mechanism required and ease of feedstock administration compared to other methods [35].

Fed-batch or semi-continuous methods involve the addition of yeast to a portion of the substrate at the commencement of the fermentation process. The other parts are introduced intermittently at a pre-determined time without removing any product till the fermentation process is completed [82]. This method avoids catabolic repression and exposure of the cells to high concentration of inhibitors at the start of the process. It therefore yield production of high biomass concentration and also circumvents the challenges of wash-out and contamination commonly associated with continuous processes [38].

Continuous fermentation involves the simultaneous addition of substrate and product withdrawal at the same rate. The rate of supply of fresh media to the culture is made to balance with the rate of withdrawal of ethanol, toxic metabolites, and byproducts. If these rates are maintained, the fermentation process will eventually reach a steady state. The dilution rate will equal the cell growth rate, and cell wash-out will be prevented and made fermentable over a long time [94]. This method increases productivity, reduces vessel cleaning downtime, makes room for a high degree of control, and allows for more cost-effective operations [87].

2.2 Fermentation Methods

Three main types of fermentation processes employed in the commercial-scale production of bioethanol include solid-state fermentation, submerged/liquid-state fermentation and very high-gravity fermentation [87].

2.2.1 Solid State Fermentation

This fermentation method involves the development of microorganisms on solid substrate in a limited supply of free water. Example of such organisms include bacteria and yeasts which are single-celled organisms, as well as filamentous fungi [69]. The seed culture can be pure (containing a specific microorganism strain) or mixed (containing different microorganisms). Reducing sugars are constantly released through enzymatic hydrolysis of cellulose and subsequently undergo simultaneous saccharification and fermentation to bioethanol [69, 87]. Several factors, including aeration, temperature, substrate type, microorganism, bioreactor, and water activity influence the process [56]. This method is commonly applied in processing agricultural residues to bioethanol and the advantages include but are not limited to reduced effluent generation and increased product yield. There is, however, possible release of volatile organic compounds (VOCs) such as NH_3, N_2O and CH_4 from the process [56].

2.2.2 Liquid State (Submerged) Fermentation

This involves the growth of microbes in a liquefied substrate. It is mostly employed in first-generation bioethanol production. Water is added to pulverized sugar- or starch-rich substrate. The mixture is cooked and hydrolysed with enzymes to obtain a medium containing various nutrients and sugars in dissolved and suspended forms [69]. The disadvantages of this method include the need for water and energy as inputs, production of a large volume of wastes as wet distillers' grains and thin stillage, as well as the need for a large volume of distillation columns and bioreactors [69].

2.2.3 Very High Gravity Fermentation

This fermentation method is still evolving and makes use of feedstocks with high sugar content (≥ 250 g/L) to produce bioethanol at a reduced cost [16, 30]. The advantages include reduced water and energy consumption as well as reduced waste generation, resulting in cheap production costs and better environmental sustainability. Nevertheless, the process conditions, including osmotic stress, temperatures, acidic conditions and presence of metal ions are often increased and inadvertently subject yeast to a series of stress [87].

3 Novel and Cost-Effective Feedstock Sources

Various novel and cost-effective feedstocks used in bioethanol production cut across first, second, third and fourth generations. These feedstocks and different processes commonly applied for transformation to bioethanol will be discussed in detail in the following sections. When sugar-containing materials like fruits, sugar beet, sugarcane juice and molasses are used, pretreatment stage is bypassed, and direct fermentation takes course. However, when fermentable sugar is to be extracted from starchy sources; liquefaction, granulating and saccharification processes are applied. Meanwhile, extraction of fermentable sugar from lignocellulosic biomass requires an additional pretreatment to break down lignin structure.

3.1 First-Generation Feedstocks

First-generation feedstocks have been the mainstay for bioethanol production because of the large commercial-scale applications. Sugarcane, fruits, sweet sorghum, sugar beet, and starchy feedstocks, such as feed grains, food grains, and tubers, are feedstocks applicable for first-generation bioethanol production [79]. Globally, sugarcane, sugar beet and corn rank highest in productivity per unit area compared to other first-generation feedstocks [55]. On a commercial scale, sugarcane is mostly used as feedstock for bioethanol production in Brazil, India, The Philippines, and Thailand. Corn, on the other hand, is mostly used by US, UK and China. France, Spain, Germany, and the UK largely use wheat. While cassava is used in China and Thailand and sweet sorghum and sugar beet are used in China and France, respectively [53, 78].

3.1.1 Sugarcane

Sugarcane (*Saccharum officinarum*) is an important industrial crop, rich in sucrose as the fermentable sugar which can be converted to ethanol. It is a semi-perennial plant of the Poaceae (grass) family. The sucrose is concentrated in the stalks. Hence, readily available for ethanol production. The harvest is mechanically carried out by slicing the stalks into billets while leaving the leafy parts as residue and the roots to re-grow. The billets are cleaned in screens and cyclones to remove dirt and afterward positioned in mills for processing to obtain juice and bagasse. Phosphoric acid is added to the juice and heated to 70 °C. Lime juice is subsequently added and further heated to 105 °C. The juice is degassed, filtered, and concentrated in evaporators. The final juice has about 22 wt% sucrose concentration. This is further sterilized at 130 °C before fermentation [18].

3.1.2 Sugar Beets

Sugar beet (*Beta vulgaris*) is a temperate root crop that thrives in deep soils with good aeration, water retention and a neutral pH [55]. It is mainly planted for sugar production. Beet molasses or syrup is often used for bioethanol production because of domestic sugar demand and the high cost of sugar beet juice. The sucrose content is concentrated in the root, unlike sugarcane. The roots are mechanically harvested and loaded into silos through channels with circulating water to wash away leaves, small roots, and stones. The beet is then thinly sliced into cossettes in the choppers system and afterward propelled to rotary drums containing hot water (between 70 and 80 °C) to extract the sugars by rupturing the proteins in the cell walls. Ethanol production follows a similar step to sugarcane [18, 55, 91].

3.1.3 Corn

Corn or Maize (*Zea mays*) is a grain plant cultivated on hectares worldwide. It is widely consumed in the diet by humans as food and by animals as feed due to its nutritional contents. It also finds application in processes like bioethanol production [31]. Corn grows within a limited period and could be affected by solar radiation, temperature, and water. It thrives between 10 and 30 °C in the presence of light and humidity. The harvesting is done mechanically to separate the corn cob from the stem. The grains are also reaped mechanically and transported for storage in silos, which are taken for various processes, including ethanol production.

3.1.4 Bioconversion Process of Feedstocks to Bioethanol

The various bioconversion processes undergone by different feedstocks to bioethanol depend on the nature of the material. First-generation feedstocks are usually in the form of fermentable sugars found in juices and starch found in grains.

Fermentable Sugars

All sugars are fermented to ethanol by yeast strain in a similar pattern. The most important sugar in fermentable juices is sucrose. It is readily converted through hydrolysis by invertase enzyme released by yeast to give C_6 sugars which are subsequently fermented to ethanol (Eqs. 3 and 4) [55].

$$C_{12}H_{22}O_{11} + H_2O \rightarrow 2C_6H_{12}O_6 \; (Sucrose \; inversion) \tag{3}$$

$$C_6H_{12}O_6 \rightarrow 4C_2H_5OH + 4CO_2 + biomass + byproducts \; (fermentation) \tag{4}$$

Grains

Grains are converted to bioethanol through wet or dry milling [35]. Although the two processes have different initial treatments, they both involve the initial hydrolysis of starch to obtain glucose (Eq. 5), which is then fermented to ethanol by yeast. The wet-milling produces animal feed and oil alongside bioethanol, while the dry-milling only produces dried distillers grain with soluble [22].

$$(C_6H_{10}O_5)_n + nH_2O \rightarrow nC_6H_{12}O_6 \tag{5}$$

Dry-milling process of grains to ethanol

Grains are crushed to a meal, mixed with water to form a mash, and cooked with α-amylase at 60–70 °C (if warm-cooked) or 80–90 °C (if hot-cooked) to extract starch. The corn mash is forced through jet cooker at 140–150 °C to hydrolyze the starch to short-chain monosaccharides. The resulting 6-carbon sugars at 32 °C and pH 4.5 are then fermented to ethanol [55]. Glucoamylase, yeast culture and Urea (as a source of Nitrogen) are added to the fermented mash for simultaneous saccharification and fermentation (SSF). The mixture is left for 50–60 h to obtain approximately 15% v/v ethanol. Ethanol is then separated from the stillage through distillation [22, 55].

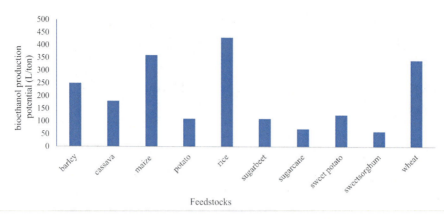

Fig. 2 Bioethanol production potential from some first-generation feedstocks. *Source* Patni et al. [64]

Wet-milling process of grains to ethanol

The grains are submerged in water containing 0.1 to 0.2% SO_2 at 52 °C for 24 to 40 h to give a mixture of softened grains and light steep water (LSW). The LSW is further concentrated to heavy-steep water (HSW) using an evaporator [81]. On the other hand, the softened grain is mildly ground to remove kernels and release germs that are less dense and contain 40–50% oil. The slurry without germ is grounded to slack off the gluten and starch from the fibre left. The fibre is removed, washed and hard-pressed to 60% moisture content. It is afterward dried with HSW and sold as corn gluten feed (containing about 21% protein) to the livestock industry. The gluten-starch mixture is centrifuged to separate starch (a denser fraction) from gluten (a lighter fraction). The starch residue is washed, and impurities are removed by processing in hydro cyclones to obtain about 99.6% starch. The starch is hydrolysed to yield glucose units (Eq. 5), which are fermented to ethanol. The separated gluten at 35 °C can be filtered using a rotary vacuum filter to yield gluten cake containing 60% moisture. This can be further dried to gluten meal (10% moisture content) and applied in poultry feed [35]. Bioethanol yield from some common first-generation feedstocks is shown in Fig. 2. Rice has the highest bioethanol production potential of about 430L/ton. This is followed by maize and wheat having bioethanol production potentials of 360 and 340L/ton respectively. The production potentials of barley, cassava, potato, sugar beet, sugarcane, sweet potato, and sweet sorghum are 250, 180, 110, 110, 70, 125, and 60 L/ton respectively.

3.2 Second-Generation Feedstocks

These sources are essentially non-edible lignocellulosic biomass which does not compete with food sources. The term lignocellulosic biomass cuts across herbaceous and woody biomass and agricultural and forestry residues. Examples include cane and sweet sorghum bagasse, corn stover, different straw types, rice hulls, olive stones, pulp, aspen and poplar (hardwoods), pine and spruce (softwoods), alfalfa hay, switchgrass, and other types of grasses (herbaceous biomass), as well as food wastes. They do not compete with food sources, and the cost of procurement is low and economical compared to first-generation feedstocks, which involve full cultivation [15]. They are also estimated to have global production potential of 442 billions litres ethanol per year [4].

3.2.1 Herbaceous Biomass

Herbaceous biomasses are otherwise called invasive weeds. They are economic feedstock used for bioethanol production because they are easily cultivated on water bodies and degraded lands [13]. Similarly, they are sustainable because they possess the ability to survive harsh conditions, including low nutrients, drought, and extreme atmospheric temperatures. In addition, they do not require pesticides and fertilizers for survival. Hence, they are available in all seasons [10]. They grow aggressively, and when not harnessed, they often result in environmental challenges [13]. The bioconversion process to ethanol involves pretreatment, hydrolysis, fermentation, and distillation, commonly used in second-generation ethanol processes. Some weedy plants with their typical compositions and corresponding estimated theoretical bioethanol yields are shown in Table 1.

3.2.2 Woody Biomass

This is essentially divided into hardwood and softwood. They are both forest biomass that could be harvested anytime because their availability does not hinge on seasonal variations. Hardwoods are obtained from angiosperms and grow very slowly. Hence, they have high density. Softwoods, on the other hand, are from gymnosperms and grow rapidly. Hence, the low densities [25]. Hardwood stems typically contain 40–50% cellulose, 24–40% hemicellulose and 18–25% lignin, while the stem of a typical softwood contains 45–50% cellulose, 25–35% hemicellulose and 25–35% lignin. The relatively high cellulose and hemicellulose contents make them suitable as feedstock for bioethanol production [25]. Examples of hardwoods are aspen and poplar, while pine and spruce are examples of softwoods.

Table 1 Weedy plants typical compositions and corresponding estimated theoretical bioethanol yields [66]

	Imperata cylindrica	Sida acuta	Amaranthus viridis	Rottboellia cochinchinensis	Pennisetum polystachyon	Sorghum halepense
Ash (%)	6.9 ± 0.00	5.1 ± 0.1	16.9 ± 0.2	10.5 ± 0.3	7.5 ± 0.3	8.8 ± 0.3
Cellulose (%)	44.4 ± 0.1	56.0 ± 0.3	37.4 ± 0.1	41.6 ± 0.7	40.0 ± 0.0	44.4 ± 0.1
Lignin (%)	6.7 ± 0.00	6.8 ± 0.1	5.1 ± 0.1	7.5 ± 0.1	6.2 ± 0.2	6.6 ± 0.5
Hemicellulose (%)	31.1 ± 0.0	16.0 ± 0.4	34.2 ± 0.0	28.6 ± 0.4	23.3 ± 0.1	25.8 ± 0.2
EtOH theoretical yield (L/Ton)	548.4 ± 1.4	520.3 ± 5.4	521.0 ± 0.9	509.7 ± 8.1	459.2 ± 0.6	508.8 ± 2.6

3.2.3 Agro-industrial Wastes

Agro-industrial wastes are usually produced as large industrial residues that are often disposed in the environment without proper treatment. Hence, they often constitute an environmental nuisance to human and animal health [70]. These usually contain high nutritional content consisting of proteins, sugars, and minerals. As a result, they are often categorized as raw materials to produce other value-added products like ethanol. Waste conversion from crops to ethanol enhances the economic benefits that farmers derive from their products and helps manage waste in Agro-industries [29]. They can be divided into agricultural or crop residues and industrial or process residues [70].

Agricultural or crop residues refer to the remains of plants after the removal of the economic part, such as stubbles and stocks. They do not include the root because of its insignificant organic matter [73]. Industrial or process residues include the leftovers after the processing of crops into other valuable products. Examples are dregs, husks, or hull from cereals, as well as molasses and bagasse from sugarcane [70]. Most crop residues are rich in hemicellulose and cellulose, which can be broken down to xylose and glucose, then converted to bioethanol. Those used as feedstocks are, therefore, crop residue with high cellulose and hemicellulose [68]. Common examples include, cereal waste comprising 20–30% hemicellulose, 40–50% cellulose and 10–20% lignin [65]. The typical percentage for cellulose, hemicellulose and lignin constituents of common Agro-industrial wastes are shown in Table 2 and ethanol yield from some wastes are shown in Table 3.

Table 2 Typical percentage constituents of common cereal wastes [48, 69]

Agro-industrial waste	Cellulose (% dry wt.)	Hemicellulose (% dry wt.)	Lignin (% dry wt.)
Cornstalk	39–47	26–31	3–5
Corn cobs	45	35	15
Corn stover	38–40	28	7–21
Rice straw	28–36	23–28	12–14
Wheat straw	33–41	26–32	13–19
Wheat husk	36	18	16
Barley straw	31–45	27–38	14–19
Sorghum stalks	27	25	11
Sorghum straw	32	24	13
Rye husk	26	16	13
Sweet sorghum bagasse	34–45	18–28	14–22
Potato peel waste	2.2%	–	–
Orange peel	9.21	10.5	0.84
Coffee skin	23.77	16.68	28.58
Pineapple peel	18.11	–	1.37

Table 3 Ethanol yield from some selected lignocellulosic feedstocks

Lignocellulosic feedstock	Bioprocess mode	Microorganism	Pretreatment/ hydrolysis	Ethanol yield (g/L)	References
Corn cob	Batch SSF	*Saccharomyces cerevisiae*	Diluted acid (H_2SO_4) follows by sterilization and alkaline delignification (NaOH)	3820	[4]
Sugar cane bagasse	Batch SSF	*Saccharomyces cerevisiae*	Diluted acid (H_2SO_4) follows by sterilization and alkaline delignification (NaOH)	7680	[4]
	Batch SSF	Recombinant *Saccharomyces cerevisiae* containing the β-glucosidase gene from *Humicola grisea*	Diluted acid (H_2SO_4) followed by alkaline delignification (NaOH), cellulase complex obtained from Trichoderma reesei (MULTIFECT®)	51.7	[15]
Bagasse	SHF SSF	*Z. mobilis* ATCC 29191 immobilized in Ca-alginate (CA) and polyvinyl alcohol (PVA) gel beads	Acid (H_3PO_4), Accellerase® 1500 enzyme	SHF (5.52–6.24) SSF (5.44–5.53)	[93]
Eucalyptus globulus Wood	SSF	*S. cerevisiae* IR2T9-a	Organosolv (50% EtOH, 200 °C, 45 min), cellulase *(Celluclast)* and *β-glucosidase (Novozym 188)*	~42	[95]
Rice straw	Batch SSCF	*S. cerevisiae*, *Candida tropicalis*, *S. stipitis*	Alkali (NaOH), Accellerase® 1500 enzyme	28.6	[80]
Rice husk	Batch SSF		Diluted acid (H_2SO_4) follows by sterilization and alkaline delignification (NaOH)	4170	[4]

(continued)

Table 3 (continued)

Lignocellulosic feedstock	Bioprocess mode	Microorganism	Pretreatment/ hydrolysis	Ethanol yield (g/L)	References
Corn stover	SHF, SSCF		AFEX commercial enzymes mixture (Ctec 2, Htec 2 and Multifect pectinase)	SHF (45.5), SSCF (51.3)	[43]
Corn stover	CBP		AFEX	7.0	[15]
Cellulosic material, β-glucan	CBP		–	4.24	[42]

3.2.4 Municipal Solid Wastes

Generally, municipal solid waste (MSW) consists of compostable and non-compostable organics, as well as inorganics from various sources such as households, restaurants, canteens, offices, markets and retail premises. Various organic wastes from these sources include kitchen, paper, and green organic wastes [62]. Although these wastes could also be subjected to non-isothermal solid-state fermentation (SSF) to produce ethanol, the process could be challenging due to the heterogeneity of the sources [57]. The yield obtainable depends on the sorting system, socio-economic conditions, season, population, and dietary routines of the people [49].

3.2.5 Conversion Process

Lignocellulosic bioethanol production passes through four main stages; pretreatment, enzymatic hydrolysis, fermentation and bioethanol purification [90]. Pretreatment processes are meant to remove lignin to access biomass cellulose and hemicellulose fractions for conversion to bioethanol [77]. The different pretreatment methods, such as biological, chemical and physical methods, avoid the breaking down of pentoses, reduce the formation of inhibitors and energy requirement for a cost-effective operation, as well as recovers lignin for transformation to valuable products [48, 77]. Enzymatic hydrolysis releases the hexoses and pentoses of the cellulose and hemicellulose by breaking down their H_2 bonds.

Owing to the recalcitrant nature of the biomass sources, fermentation and enzymatic hydrolysis are either carried out by separate hydrolysis and fermentation (SHF) or simultaneous saccharification and fermentation (SSF) [23].

SHF makes it possible to independently carry out hydrolysis and fermentation at optimum pH and temperature. However, an increase in sugar concentration in the bioreactor during hydrolysis inhibits the hydrolysis products and reduces the effectiveness of the cellulose enzyme being. This diminishes the effectiveness of hydrolysis [83].

In SSF, hydrolysis of pretreated biomass and fermentation simultaneously occur at compromised optimum conditions between fermenting organisms and enzymatic reaction [41]. The fermenting organism immediately consumes the glucose released from hydrolysis, thereby preventing the inhibition of products by the enzymes during hydrolysis [92]. SSF does not allow for recirculation of the fermenting organism since it is already mixed with the biomass [86].

An improvement in SHF and SSF is the simultaneous saccharification, filtration and fermentation (SSFF). It averts end-product inhibition, allows for the simultaneous use of fermenting organisms and enzyme cocktails at independent optimal conditions, and enables several recirculation of fermenting organisms [41]. Enzyme is introduced to lignocellulosic slurry which has been pretreated. Crossflow filtration conveys filtrate rich in sugar to the fermentation medium. The flocculating yeast strain is retained by settling, while the fermented fluid is propelled back to the hydrolysis basin [38]. The current trend in the fermentation and hydrolysis methods of lignocellulosic biomass is further summarized in Table 4.

3.2.6 Limitations of Second-Generation Feedstocks

Although bioethanol has been produced from lignocellulosic biomass on research and large pilot scale, it still faces technical and economic difficulties. The high cost of enzymes for hydrolysis of the feedstocks to fermentable sugars makes it expensive compared to first-generation ethanol. There is also the possibility of having a sustainability problem in the future if there is a reduction in the available land [15, 9].

3.3 Third-Generation Feedstocks

The third-generation sources are algae biomass and adequately cater for the inadequacy of the first- and second-generation sources. They also do not compete with food and feed supplies [12].

Algae are otherwise called thallophytes i.e., plants having no leaves, stem and roots. They are categorized as prokaryotic and eukaryotic photosynthetic organisms. They could be unicellular (such as coccoid, chlorella, palmelloid, and diatoms); filamentous and colonial or multicellular (such as free-living or giant kelp); motile (having flagella); immotile (with no flagella); macroscopic or microscopic; terrestrial or aquatic; and aerial or subaerial [88]. Algae could also be autotrophs (photosynthetic and require only inorganic compounds for their growth), heterotrophs (non-photosynthetic and depend on an external source for nutrients and energy) or mixotrophs [75].

Algae are classified by size and morphology. The microscopic ones existing as unicellular or multicellular filaments and colonies are called microscopic algae or microalgae. Macro algae, on the other hand, are known to have a defined multicellular

Table 4 Current trend in the fermentation and hydrolysis methods of lignocellulosic biomass

Fermentation method	Application	Limitations	Recommendations	References
SHF	• Independent optimized conditions for fermentation and enzymatic hydrolysis	• Inhibition of hydrolysis products and reduction in the effectiveness of cellulose enzyme • Increase cost of investment from the use of separate reactors • Sugar loss resulting from physical separation of biomass from hydrolysate	SSFF should be embraced to improve fermentation efficiency	[23, 40, 54]
SSF	• Increase hydrolytic activity • Higher solid loadings • It circumvents the necessity for physical separation of biomass from hydrolysate thereby preventing sugar loss • Faster period of operation because the simultaneous combination of two steps in a single reactor • Reduction in investment cost as a result of a reduction in a number of reactors needed	• Difficulty in getting an optimal combination of microorganisms and enzymes • Fitness of temperature and pH range for the combination of hydrolytic enzymes and microorganisms • No recirculation of fermenting organism because it gets mixed with the biomass • The two bioconversion steps compete for chemical resources		[24, 40]

(continued)

Table 4 (continued)

Fermentation method	Application	Limitations	Recommendations	References
SSFF	• prevents inhibition by enzymes • allows simultaneous combination of fermenting organism and enzyme cocktail at independent optimal conditions • enables several recirculation of fermenting organism	Not much investigation on industrial scale for production of bioethanol	Membrane pervoration and distillation systems should be integrated to attain high purification and concentration of bioethanol	[24, 40]

structure comparable to plant leaves. They are further sub-classified as brown, red and green algae [5].

Microscopic algae can complete its growing cycle using photosynthesis within a few days. The doubling time for its biomass within the exponential growth phase can be as low as 3.5 h [17]. It requires nitrogen, potassium, phosphorus, carbon dioxide and light for growth to manufacture carbohydrates and lipids, which can be converted to bioethanol and other valuable by-products [14, 60]. The fermentation and hydrolysis efficiency of algal biomass is reportedly high owing to its reduced hemicelluloses and lignin content [85]. Although some microalgae species grow heterotrophically on renewable organic carbon sources, they are not as efficient as photosynthetic growth because the organic carbon sources are also produced from photosynthetic plants [63].

Third-generation sources attracted huge attention due to the promising high yield of biofuel, absorption of CO_2 (thereby reducing greenhouse effects), non-competition with food sources, and comparatively simple production processing [3].

3.3.1 Limitations of Third-Generation Feedstocks

Crucial issues in the production of third-generation bioethanol include enhancement of microalgae growth rate and synthesis of product, algae culture dewatering for production of biomass, pre-treatment of biomass, and optimization of the algal fermentation process to bioethanol. In addition, though several microalgae species show high potential for large-scale bioethanol production, there is no sufficient information to run commercial trials [45, 50].

3.4 Fourth-Generation Feedstocks

The quest to improve the yield of third-generation biofuel sources necessitates the advent of fourth-generation feedstocks. The feedstocks for fourth-generation bioethanol are obtained by applying genetically modified algae biomasses, such as microalgae, macroalgae and cyanobacteria [1, 6].

Genetically modified microalgae (GMM) can be cultivated in two types of systems, namely contained and uncontained systems. Contain systems have tight, controlled conditions, causing reduced exposure to the environment and less contamination. Uncontained systems, on the other hand, are the opposite. The contained system is costly and less economical because of the complexities involved [1]. Carbon captured in fourth-generation bioethanol production is more than the carbon produced during the process. The process is, therefore, carbon negative and environment friendly. The major demerit of fourth-generation bioethanol is the potential health and environmental risks that toxic genetically modified algae strains pose when released into the environment [1].

In recent studies plant cell wall engineering approach have been used to increase biomass and bioethanol production. The gene of plants are modified to improve the synthesis of polymer of interest in the cell wall (e.g., cellulose), transform the lignin structure through alteration of phenolic metabolism as well as decrease the crystalline ability of cellulose [25]. Modifying metabolic pathways of lignin may help in getting feedstock with reduced lignin recalcitrant and improved yield of bioethanol [2].

3.4.1 Limitations of Fourth-Generation Feedstocks

The technologies involved in fourth-generation bioethanol feedstocks are often complex and uneconomical. In addition, the applications are still in the early stages and need a lot of research and improvements [32].

3.5 Advances in Bioethanol Production from Waste Streams

In addition to the common routes for bioethanol production, waste streams of CO_2 emanating from industrial processes could also be utilized through various routes. For example, waste streams of CO_2 from corn dry-milling could be captured and converted into bioethanol using a bio-electrochemical process. In this process, CO_2 and water are electrolyzed differently to produce carbon monoxide (CO) and hydrogen (H_2). The two new gas species are then fermented to ethanol (Eq. 6). Similarly, CO can also be fermented in the absence of H_2 (Eq. 7), and ethanol can be produced from solventogenesis (Eq. 8) [33, 37].

$$4H_{2(g)} + 2CO_{(g)} \rightarrow C_2H_5OH_{(l)} + H_2O_{(l)} \qquad (6)$$

$$2H_2O_{(g)} + 4CO_{(g)} \rightarrow C_2H_5OH_{(l)} + 2CO_{2(g)} \tag{7}$$

$$2CO_{2(g)} + 6H_{2(g)} \rightarrow C_2H_5OH_{(l)} + 3H_2O_{(l)} \tag{8}$$

4 Mixed Feedstocks

This involves the extraction of bioethanol from simultaneous processing and conversion of mixtures of feedstocks [61]. This approach continuously gains acceptance because it increases yield and makes the process more economical. The components of the mixture may be of the same or different origin, similar or different features, similar or different methods of processing, or from different generations. As a result of synergistic interactions of the combined biomass, mixing hydrolysates from different biomass often reduce the concentration of the inhibitor and the corresponding cost of detoxification [11]. The choice of feedstocks to be mixed for bioethanol production is influenced by the drive to produce high quantity and quality ethanol. Hence, most of the feedstocks employed contain an appreciable fermentable sugar content. Other factors include the nearness of the potential mixture component, proximity to processing facilities and rich in Nitrogen [11, 34].

Examples of mixed feedstock approaches in forest residues include; mixtures of Bark-rich sawmill residues from dissimilar plants and hardwoods from different plants [47, 51]. Among common crop residues, examples of mixed feedstock include rice straw mixed with wheat bran; sugar cane bagasse mixed with wheat bran; and oil palm empty fruit bunch mixed with oil palm frond and rice husk [61, 74]. Furthermore, for biomass from the same sources, examples include mixtures of Sugarcane bagasse and sugarcane straw, wheat bran and wheat straw [8, 58].

5 Economic Feasibility of Bioethanol Production from Novel Feedstocks

The major cost incurred in bioethanol production are from energy consumed during the processes as shown in Table 5 [19]. This can be largely influenced by the type of pretreatment method employed [76]. In cellulosic bioethanol production, pretreatment process cost analysis accounts for 18% of the total project cost [76]. Steam explosion treatment, requires about 0.2–0.6 MJ/kg of energy, while milling procedure requires about 4.0–12.5 MJ/kg of substrate [52]. At 2.8 MPa, the energy required to treat with steam explosion is almost 22 times larger than the energy required to generate milled biomass with 0.5 mm particle size [52].

Table 5 Principal operating cost of some common pretreatment methods

Pretreatment method	Steam explosion with SO_2	Liquid hot water	Soaking in aqueous solution	Ammonia fiber explosion (AFE)	Dilute acid	Lime
Principal operating cost	Energy consumption	Energy consumption	Cost of aqueous solution and energy consumption	Cost of ammonia and energy consumption	Cost of acid and energy consumption	Cost of lime and energy consumption
Average fixed capital cost ($/m^3.day)	17,500	6125	22,500	28,350	17,500	19,825

Source Cuong and Tabil [19]

In order to have an economical and sustainable production of bioethanol from biomass, innovative integration, use of cheap raw materials, adoption of methods utilizing reduced amount of chemicals and strain improvement via bioengineering and mutation in biological treatments must be engaged [28, 67].

6 Status Summary of Bioethanol Production from Different Feedstocks

The status of the four generation bioethanol feedstocks discussed in this chapter is presented in Table 6.

Table 6 Status summary of different feedstocks

Feedstocks	Status summary	Advantages	Challenges	Recommendation	References
1st generation	• Advanced in technologies and applications • Commercial scale applications	• High ethanol yields from the crops • Ease of process of feedstock to bioethanol	• High cost of feedstocks • Competitiveness with food and feed sources • Unavailability of arable lands • Higher emission of Greenhouse gases • Non-sustainability	Discontinue application to embrace the newer bioethanol technologies	[55]
2nd generation	• Technologies are semi-advanced • Used on commercial scale	• Low-cost feedstock, • It does not contend with food and feed supplies	• Still faces technical and economic difficulties • There is possibility of sustainability problem in the future if there's reduction in the available lands • Heterogeneity of municipal solid wastes	Optimization of processes to reduce technical and economic challenges	[15, 57]
3rd generation	• Technologies are still developing • High potential for commercial scale applications	• Promising high yield of bioethanol • Absorption of CO_2, thereby reducing greenhouse effects • Non-competition with food sources • Comparatively simple production processing	• Enhancement of microalgae growth rate and synthesis of product • Algae culture dewatering for production of biomass, • Pre-treatment of biomass, and optimization of algal fermentation process to bioethanol • Insufficient information to run on trials	• Generation of information for commercial scale trials • Optimization of process stages for enhancement of product synthesis	[3, 12, 45]
4th generation	• Technologies are still at the early stages	• Environment friendly • carbon negative since carbon captured are more than the carbon produced	• Technologies are costly and less economical	• Development of more economical and sustainable technologies	[32, 33, 37]

7 Conclusion

Bioethanol production from various feedstocks has advanced from being novel to sustainable. This review has comprehensively discussed successes and lapses in applying various feedstocks for bioethanol production. Although many commercial-scale productions still adopt the first-generation feedstocks for ethanol production, the more promising second, third and fourth generation alternative that does not compromise food security should be embraced for sustainability and economic viability. More research should focus on optimization process stages and circumventing the challenges of these latter three generations for better improvements in future bioethanol production applications.

References

1. Abdullah B, Anuar S, Muhammad S et al (2019) Fourth generation biofuel: a review on risks and mitigation strategies. Renew Sustain Energy Rev 107:37–50. https://doi.org/10.1016/j.rser.2019.02.018
2. Abramson M, Shoseyov O, Hirsch S, Shani Z (2013) Genetic modifications of plant cell walls to increase biomass and bioethanol production. Introduction: the increasing importance, pp 315–338. https://doi.org/10.1007/978-1-4614-3348-4
3. Adeniyi OM, Azimov U, Burluka A (2018) Algae biofuel: current status and future applications. Renew Sustain Energy Rev 90:316–335. https://doi.org/10.1016/j.rser.2018.03.067
4. Adiya ZISG, Adamu SS, Ibrahim MA et al (2022) Comparative study of bioethanol produced from different agro-industrial biomass residues. Earthline J Chem Sci 7:143–152. https://doi.org/10.34198/ejcs.7222.143152
5. Al AQ, Nixon BT, Fortwendel JR (2016) The enzymatic conversion of major algal and cyanobacterial carbohydrates to bioethanol. Front Energy Res 4:1–15. https://doi.org/10.3389/fenrg.2016.00036
6. Ale S, Femeena PV, Mehan S, Cibin R (2019) Crop production and benefits of multifunctional bioenergy systems. 195–217. https://doi.org/10.1016/B978-0-12-816229-3.00010-7
7. Azad AK, Rasul MG, Khan MMK et al (2015) Prospect of biofuels as an alternative transport fuel in Australia. Renew Sustain Energy Rev 43:331–351. https://doi.org/10.1016/j.rser.2014.11.047
8. Azin M, Moravej R, Zareh D (2007) Production of xylanase by *Trichoderma longibrachiatum* on a mixture of wheat bran and wheat straw: optimization of culture condition by Taguchi method. 40:801–805. https://doi.org/10.1016/j.enzmictec.2006.06.013
9. Balat M, Balat H (2009) Recent trends in global production and utilization of bio-ethanol fuel. Appl Energy 86:2273–2282. https://doi.org/10.1016/j.apenergy.2009.03.015
10. Banka A, Komolwanich T (2014) Potential Thai grasses for bioethanol production. https://doi.org/10.1007/s10570-014-0501-2
11. Baxter L (2005) Biomass-coal co-combustion: opportunity for affordable renewable energy. 84:1295–1302. https://doi.org/10.1016/j.fuel.2004.09.023
12. Behera S, Singh R, Arora R et al (2015) Scope of algae as third generation biofuels. 2:1–13. https://doi.org/10.3389/fbioe.2014.00090
13. Borah AJ, Singh S, Goyal A, Moholkar VS (2016) An assessment of the potential of invasive weeds as multiple feedstocks for biofuel production. RSC Adv 6:47151–47163. https://doi.org/10.1039/c5ra27787f

14. Brennan L, Owende P (2010) Biofuels from microalgae—a review of technologies for production, processing, and extractions of biofuels and co-products. 14:557–577. https://doi.org/10.1016/j.rser.2009.10.009
15. Bušić A, Morzak G, Belskaya H, Šantek I (2018) Bioethanol production from renewable raw materials and its separation and purification: a review. 9818:0–3. https://doi.org/10.17113/ftb.56.03.18.5546
16. Camargos CV, Moraes VD, de Oliveira LM et al (2021) High Gravity and very high gravity fermentation of sugarcane molasses by flocculating *Saccharomyces cerevisiae*: experimental investigation and kinetic modeling. Appl Biochem Biotechnol 193:807–821. https://doi.org/10.1007/s12010-020-03466-9
17. Chisti Y (2007) Biodiesel from microalgae. Biotechnol Adv 25:294–306. https://doi.org/10.1016/j.biotechadv.2007.02.001
18. Cruz CHB, Souza GM, Cortez LAB (2014) Chapter 11-Biofuels for transport. In: Future energy, 2nd ed, pp 215–244. https://doi.org/10.1016/B978-0-08-099424-6.00011-9
19. Cuong DN, Tabil LG (2018) A review on techno-economic analysis and life-cycle assessment of second generation bioethanol production via biochemical processes
20. Dahman Y, Dignan C, Fiayaz A, Chaudhry A (2019) An introduction to biofuels, foods, livestock, and the environment. Elsevier Ltd
21. Demirbas A (2007) Progress and recent trends in biofuels. 33:1–18. https://doi.org/10.1016/j.pecs.2006.06.001
22. Demirel Y (2018) Biofuels. Comprehensive energy systems. Elsevier Inc., United States, pp 875–908
23. Derman E, Abdulla R, Marbawi H et al (2022) Simultaneous saccharification and fermentation of empty fruit bunches of palm for bioethanol production using a microbial consortium of S. cerevisiae and T. harzianum
24. Dey P, Pal P, Kevin JD, Das DB (2020) Lignocellulosic bioethanol production: prospects of emerging membrane technologies to improve the process—a critical review. 36:333–367
25. Duque A, Cristina Á, Dom P et al (2021) Advanced bioethanol production: from novel raw materials to integrated biorefineries. 1–30
26. Ethanol Industry Outlook (2021) Essential energy
27. Farrell AE, Plevin RJ, Turner BT et al (2006) Ethanol can contribute to energy and environmental goals. Science 311:506–508. https://doi.org/10.1126/science.1121416
28. Gandam PK, Chinta ML, Pabbathi NPP et al (2022) Second-generation bioethanol production from corncob—a comprehensive review on pretreatment and bioconversion strategies, including techno-economic and lifecycle perspective. Ind Crops Prod 186:1–9. https://doi.org/10.1016/j.indcrop.2022.115245
29. Ghosh SK (2020) Energy recovery processes from wastes. Springer Nature, Singapore Pte Ltd
30. Gomes D, Cruz M, de Resende M et al (2021) Very high gravity bioethanol revisited: main challenges and advances. Fermentation 7:1–18. https://doi.org/10.3390/fermentation7010038
31. Gupta M, Choudhary M, Singh A et al (2022) Meta-QTL analysis for mining of candidate genes and constitutive gene network development for fungal disease resistance in maize (*Zea mays* L.). Crop J. https://doi.org/10.1016/j.cj.2022.07.020
32. Halder P, Azad K, Shah S, Sarker E (2019) 8—Prospects and technological advancement of cellulosic bioethanol ecofuel production. Elsevier Ltd
33. He Y, Cassarini C, Lens PNL (2021) Bioethanol production from H_2/CO_2 by solventogenesis using anaerobic granular sludge: effect of process parameters. 12:1–13. https://doi.org/10.3389/fmicb.2021.647370
34. Hedegaard M, Henrik HN (2008) Sustainable bioethanol production combining biorefinery principles using combined raw materials from wheat undersown with clover-grass. 303–311. https://doi.org/10.1007/s10295-008-0334-9
35. Hoang T, Nghiem N (2021) Recent developments and current status of commercial production of fuel ethanol
36. Hossain ABMS (2015) Bio-solvent preparation from apple biomass for pharmaceutical, cosmetic and biomedical industrial application. 4:52–61

37. Huang Z, Grim G, Schaidle J, Tao L (2019) Using waste CO_2 to increase ethanol production from corn ethanol biorefineries: techno-economic analysis
38. Ishola MM (2014) Novel application of membrane bioreactors. University of Borås, Sweden
39. Ishola M, Brandberg T, Taherzadeh M (2014) Minimization of bacterial contamination with high solid loading during ethanol production from lignocellulosic materials. N Biotechnol 31:S93. https://doi.org/10.1016/j.nbt.2014.05.1828
40. Ishola MM, Brandberg T, Taherzadeh MJ (2015) Simultaneous glucose and xylose utilization for improved ethanol production from lignocellulosic biomass through SSFF with encapsulated yeast. Biomass Bioenerg 77:192–199. https://doi.org/10.1016/j.biombioe.2015.03.021
41. Ishola MM, Jahandideh A, Haidarian B et al (2013) Simultaneous saccharification, filtration and fermentation (SSFF): a novel method for bioethanol production from lignocellulosic biomass. Bioresour Technol 133:68–73. https://doi.org/10.1016/j.biortech.2013.01.130
42. Jin M, Gunawan C, Balan V, Dale BE (2012) Consolidated bioprocessing (CBP) of AFEX^{TM}-pretreated corn stover for ethanol production using *Clostridium phytofermentans* at a high solids loading. Biotechnol Bioeng 109:1929–1936. https://doi.org/10.1002/bit.24458
43. Jin M, Sarks C, Gunawan C et al (2013) Phenotypic selection of a wild *Saccharomyces cerevisiae* strain for simultaneous saccharification and co-fermentation of AFEX^{TM} pretreated corn stover. Biotechnol Biofuels 6:108. https://doi.org/10.1186/1754-6834-6-108
44. Józsa V (2011) Application of bioethanol in gas turbines. Period Polytech 55:91–94. https://doi.org/10.3311/pp.me.2011-2.05
45. Khan MI, Shin JH, Kim JD (2018) The promising future of microalgae: current status, challenges, and optimization of a sustainable and renewable industry for biofuels, feed, and other products. Microb Cell Fact:1–21. https://doi.org/10.1186/s12934-018-0879-x
46. Khan MMK, Sharma SC, Bhuiya MMK, Mofijur M (2016) A review on socio-economic aspects of sustainable biofuels. 10:32–54
47. Kim KH, Tucker M, Nguyen Q (2005) Conversion of bark-rich biomass mixture into fermentable sugar by two-stage dilute acid-catalyzed hydrolysis. 96:1249–1255. https://doi.org/10.1016/j.biortech.2004.10.017
48. Kumar SPJ, Kumar NSS, Devi A (2020) Bioethanol production from cereal crops and lignocelluloses rich agro-residues: prospects and challenges. SN Appl Sci 2:1–11. https://doi.org/10.1007/s42452-020-03471-x
49. Kumar V, Fdez-güelfo LA, Zhou Y et al (2018) Anaerobic co-digestion of organic fraction of municipal solid waste (OFMSW): progress and challenges. 93:380–399. https://doi.org/10.1016/j.rser.2018.05.051
50. Lee RA, Lavoie J (2012) From first- to third-generation biofuels: challenges of producing a commodity from a biomass of increasing complexity. 6–11. https://doi.org/10.2527/af.2013-0010
51. Lim W, Lee J (2013) Effects of pretreatment factors on fermentable sugar production and enzymatic hydrolysis of mixed hardwood. Bioresour Technol 130:97–101. https://doi.org/10.1016/j.biortech.2012.11.122
52. Liu LY, Chandra RP, Tang Y et al (2022) Instant catapult steam explosion: an efficient preprocessing step for the robust and cost-effective chemical pretreatment of lignocellulosic biomass. Ind Crops Prod 188:1–7. https://doi.org/10.1016/j.indcrop.2022.115664
53. Lopes ML, Cristina S, Paulillo DL et al (2016) Ethanol production in Brazil: a bridge between science and industry. Brazilian J Microbiol:1–13. https://doi.org/10.1016/j.bjm.2016.10.003
54. Malacara-becerra A, Melchor-mart EM, Eduardo J et al (2022) Bioconversion of corn crop residues: lactic acid production through simultaneous saccharification and fermentation
55. Manochio C, Andrade BR, Rodriguez RP, Moraes BS (2017) Ethanol from biomass: a comparative overview. Renew Sustain Energy Rev 80:743–755. https://doi.org/10.1016/j.rser.2017.05.063
56. Maulini-Duran C, Abraham J, Rodríguez-Pérez S et al (2015) Gaseous emissions during the solid state fermentation of different wastes for enzyme production at pilot scale. Bioresour Technol 179:211–218. https://doi.org/10.1016/j.biortech.2014.12.031

57. Moreno DA, Magdalena AJ, Oliva JM et al (2021) Sequential bioethanol and methane production from municipal solid waste: an integrated biorefinery strategy towards cost-effectiveness. Process Saf Environ Prot 146:424–431. https://doi.org/10.1016/j.psep.2020.09.022
58. Moutta RDO, Ferreira-leitão VS (2014) Enzymatic hydrolysis of sugarcane bagasse and straw mixtures pretreated with diluted acid. 32:93–100. https://doi.org/10.3109/10242422.2013.873795
59. Muregi MA, Abolarin MS, Okegbile OJ, Eterigho EJ (2021) Emission comparison of air-fuel mixtures for pure gasoline and bioethanol fuel blend (E20) combustion on sparking-ignition engine. Covenant J Eng Technol 5:46–55
60. Nigam PS, Singh A (2011) Production of liquid biofuels from renewable resources. Prog Energy Combust Sci 37:52–68. https://doi.org/10.1016/j.pecs.2010.01.003
61. Oke MA, Annuar MSM, Simarani K (2016) Mixed feedstock approach to lignocellulosic ethanol production—prospects and limitations. Bioenergy Res 9:1189–1203. https://doi.org/10.1007/s12155-016-9765-8
62. Paritosh K, Yadav M, Mathur S et al (2018) Organic fraction of municipal solid waste: overview of treatment methodologies to enhance anaerobic biodegradability. 6:1–17. https://doi.org/10.3389/fenrg.2018.00075
63. Patil V, Tran K, Giselrød HR (2008) Towards sustainable production of biofuels from microalgae. 1188–1195. https://doi.org/10.3390/ijms9071188
64. Patni N, Pillai SG, Dwivedi AH (2013) Wheat as a promising substitute of corn for bioethanol production. Procedia Eng 51:355–362. https://doi.org/10.1016/j.proeng.2013.01.049
65. Prasad S, Singh A, Joshi HC (2007) Ethanol as an alternative fuel from agricultural, industrial and urban residues. 50:1–39. https://doi.org/10.1016/j.resconrec.2006.05.007
66. Premjet S (2018) Potential of weed biomass for bioethanol production from sugarcane. In: Basso TP, Basso LC (eds) Fuel ethanol production from sugarcane. IntechOpen, pp 1–21
67. Rajesh Banu J, Poornima Devi T, Yukesh Kannah R et al (2021) A review on energy and cost effective phase separated pretreatment of biosolids. Water Res 198:117169. https://doi.org/10.1016/j.watres.2021.117169
68. Reijnders L (2008) Ethanol production from crop residues and soil organic carbon. 52:653–658. https://doi.org/10.1016/j.resconrec.2007.08.007
69. Sadh PK, Kumar S, Chawla P, Duhan JS (2018b) Fermentation: a boon for production of bioactive. Molecules 23:33. https://doi.org/10.3390/molecules23102560
70. Sadh PK, Duhan S, Duhan JS (2018a) Agro-industrial wastes and their utilization using solid state fermentation: a review. Bioresour Bioprocess:1–15. https://doi.org/10.1186/s40643-017-0187-z
71. Saha S, Sharma A, Purkayastha S et al (2019) 14 bio-plastics and biofuel: is it the way in future development for end users ? Elsevier Inc
72. Saqib A, Tabbssum MR, Rashid U et al (2013) Marine macro algae ulva: a potential feed-stock for bio-ethanol and biogas production. Asian J Agri Biol 1:155–163
73. Shahane A, Shivay YS (2016) Cereal residues—not a waste until we waste it: a review. Int J Bio-resource Stress Manag 7:162–173. https://doi.org/10.23910/ijbsm/2016.7.1.1401b
74. Sherief A, El-Tanash A, Atia N (2010) Cellulase production by *Aspergillus fumigatus* grown on mixed substrate of rice straw and wheat bran. Res J Microbiol 5:199–211
75. Shokravi Z, Shokravi H, Shokravi H (2019) The fourth-generation biofuel: a systematic review on nearly two decades of research from 2008 to 2019
76. Shukla A, Kumar D, Girdhar M et al (2023) Strategies of pretreatment of feedstocks for optimized bioethanol production: distinct and integrated approaches. Biotechnol Biofuels Bioprod"1–33. https://doi.org/10.1186/s13068-023-02295-2
77. Singh R, Shukla A, Tiwari S, Srivastava M (2014) A review on delignification of lignocellulosic biomass for enhancement of ethanol production potential. Renew Sustain Energy Rev 32:713–728. https://doi.org/10.1016/j.rser.2014.01.051
78. Singh AK, Garg N, Kumar A, Tyagi (2016) Viable feedstock options and technological challenges for ethanol production in India. https://doi.org/10.18520/cs/v111/i5/815-822

79. Smith AM (2008) Prospects for increasing starch and sucrose yields for bioethanol production. Plant J 54:546–558. https://doi.org/10.1111/j.1365-313X.2008.03468.x
80. Suriyachai N, Weerasaia K, Laosiripojana N et al (2013) Optimized simultaneous saccharification and co-fermentation of rice straw for ethanol production by *Saccharomyces cerevisiae* and *Scheffersomyces stipitis* co-culture using design of experiments. Bioresour Technol 142:171–178. https://doi.org/10.1016/j.biortech.2013.05.003
81. Swain MR, Mohanty SK (2018) Bioethanol production from corn and wheat: food, fuel, and future
82. Sánchez ÓJ, Cardona CA (2008) Trends in biotechnological production of fuel ethanol from different feedstocks. Bioresour Technol 99:5270–5295. https://doi.org/10.1016/j.biortech.2007.11.013
83. Tengborg C, Galbe M, Zacchi G (2001) Reduced inhibition of enzymatic hydrolysis of steam-pretreated softwood. Enzyme Microb Technol 28:835–844. https://doi.org/10.1016/S0141-0229(01)00342-8
84. Thangavelu SK, Ahmed AS, Ani FN (2016) Review on bioethanol as alternative fuel for spark ignition engines. Renew Sustain Energy Rev 56:820–835. https://doi.org/10.1016/j.rser.2015.11.089
85. Tiwari A, Kiran T, Pandey A (2019) Chapter 14. Algal cultivation for biofuel production. Elsevier Inc
86. Tomás-Pejó E, Oliva JM, Ballesteros M, Olsson L (2008) Comparison of SHF and SSF processes from steam-exploded wheat straw for ethanol production by xylose-fermenting and robust glucose-fermenting Saccharomyces cerevisiae strains. Biotechnol Bioeng 100:1122–1131. https://doi.org/10.1002/bit.21849
87. Tse TJ, Wiens DJ (2021) Production of bioethanol—a review of factors affecting ethanol yield. Fermentation 7:1–18
88. Verma D, Singla A, Lal B, Sarma PM (2016) Conversion of biomass-generated syngas into next-generation liquid transport fuels through microbial intervention: potential and current status. https://doi.org/10.18520/cs/v110/i3/329-336
89. Vohra M, Vinoba S, Civil B et al (2014) Bioethanol production: feedstock and current technologies. J Environ Chem Eng 2:573–584. https://doi.org/10.1016/j.jece.2013.10.013
90. Volynets B, Ein-mozaffari F, Dahman Y (2017) Biomass processing into ethanol: pretreatment, enzymatic hydrolysis, fermentation, rheology, and mixing. 1–22. https://doi.org/10.1515/gps-2016-0017
91. Vučurović VM, Puškaš VS, Miljić UD (2019) Bioethanol production from sugar beet molasses and thick juice by free and immobilised Saccharomyces cerevisiae. Inst Brew Distill:134–142. https://doi.org/10.1002/jib.536
92. Wingren A, Galbe M, Zacchi G (2003) Techno-economic evaluation of producing ethanol from softwood: comparison of SSF and SHF and identification of bottlenecks. Biotechnol Prog 19:1109–1117. https://doi.org/10.1021/bp0340180
93. Wirawan F, Cheng CL, Kao WC et al (2012) Cellulosic ethanol production performance with SSF and SHF processes using immobilized *Zymomonas mobilis*. Appl Energy 100:19–26. https://doi.org/10.1016/j.apenergy.2012.04.032
94. Yang Y, Sha M (2022) A beginner's guide to bioprocess modes—batch, fed-batch, and continuous fermentation. Hamburg, Germany. www.eppendorf.com, www.eppendorf.com
95. Yáñez-SM, Rojas J, Castro J et al (2013) Fuel ethanol production from Eucalyptus globulus wood by autocatalyzed organosolv pretreatment ethanol–water and SSF. J Chem Biotechnol 38:243–248. https://doi.org/10.1002/jctb.3895

Feedstock Conditioning and Pretreatment of Lignocellulose Biomass

Adeolu A. Awoyale, David Lokhat, Andrew C. Eloka-Eboka, and Adewale G. Adeniyi

Abstract This theoretical framework provides information on lignocellulosic biomass pretreatment and conditioning prior to their fermentation to produce bioethanol, which is a renewable source of energy. Because lignocellulosic biomass is so resistant and needs more intensive pretreatment before saccharification and fermentation, using it as a feedstock for the manufacture of bioethanol presents a significant difficulty. In order to establish a lignocellulosic feedstock-based biorefinery, the review seeks to give an insight into currently existing pre-treatment technologies for deconstruction and fractionation of lignocellulosic biomass. Feedstock conditioning is a very essential step in the bioethanol production loop because it helps to lessen toxicity and increase the sugars' accessibility to fermentation. This chapter will enable a good understanding of already used processes, assist in overcoming obstacles, and develop improvised technologies to simplify the pretreatment process.

Keywords Lignocellulosic biomass · Pretreatment · Conditioning · Feedstock · Bioethanol

---

A. A. Awoyale (✉)
Petroleum and Natural Gas Processing Department, Petroleum Training Institute, Effurun, Nigeria
e-mail: awoyale_aa@pti.edu.ng

D. Lokhat
Reactor Technology Research Group, School of Engineering, University of KwaZulu-Natal, Durban, South Africa

A. C. Eloka-Eboka
Centre of Excellence in Carbon based Fuels, School of Chemical and Mineral Engineering, North-West University, Potchefstroom Campus, South Africa

A. G. Adeniyi
Chemical Engineering Department, University of Ilorin, Ilorin, Nigeria

© The Author(s), under exclusive license to Springer Nature Switzerland AG 2023
E. Betiku and M. M. Ishola (eds.), *Bioethanol: A Green Energy Substitute for Fossil Fuels*, Green Energy and Technology,
https://doi.org/10.1007/978-3-031-36542-3_3

1 Introduction

Biomass is the residue left over after processing crops high in carbohydrates. The biomass is more frequently referred to as lignocellulosic biomass since it is abundant in lignin, cellulose, and hemicellulose. It has been proven that bioethanol made from lignocellulosic biomass is a good substitute for fuels made from fossil fuels [86]. Corn cobs, rice husks, cassava peels, yam peels, mango peels, sorghum straw, pineapple peels, potato peels, pawpaw peels, and sugar cane bagasse are some examples of lignocellulosic biomass from which bioethanol can be made [7]. Lignin, cellulose, and hemicellulose are the three main components of lignocellulosic biomass. Due to lignin and hemicellulose components of the biomass molecule blocking the enzyme's access to the cellulose and serving as an inhibitor, lignocellulosic biomass is by nature exceedingly resistant to being broken down to its component sugars [89]. Moreover, these macromolecular connections between cellulose, hemicellulose, and lignin are often very tight, which causes a reduction in the pore size of the molecules. This reduction in pore size improves the inhibitory properties of lignin and hemicelluloses [25]. The process by which cellulose ingredient is exposed to and made sensitive to enzymatic hydrolysis is known as "pretreatment" or "prehydrolysis." Since lignocellulosic biomass is refractory, it needs an appropriate and sufficient pretreatment in order for the subsequent hydrolysis to occur. The effectiveness of the pretreatment process affects the upstream selection of feedstock, recovery efficiency of cellulose, hemicellulose, and lignin components, and chemical and morphological characteristics of the resulting cellulosic component, which controls downstream hydrolysis and ultimately fermentation to bioethanol [45].

Given the variety of lignocellulosic biomass feedstocks, it is challenging to define a universal pretreatment procedure for all of them. In recent years, a number of pretreatment technologies have been proposed. The total cost of producing bioethanol is significantly impacted by the pretreatment technique chosen for the biomass [17]. Depending on the various forces or resources used in the pretreatment process, pretreatment technologies can be divided into physical, chemical, biological, and physio-chemical subcategories or a mix of these [4]. Different pre-treatment technologies in their distinctive classifications are presented in Fig. 1.

2 Role of Pretreatment

Pretreatment is the most time-consuming and expensive stage in the conversion of lignocellulosic biomass into ethanol. Hemicelluloses typically coat or sheath the cellulose in the biomass, generating a cellulose-hemicellulose complex that serves as a chemical barrier and blocks enzyme access when the complex is left in its natural state [1]. The lignin-encapsulated cellulose-hemicellulose complexes create a physical barrier to the hydrolysis of biomass to yield fermentable sugars [73]. Therefore, pretreatment is necessary to break up the lignin-cellulose-hemicellulose

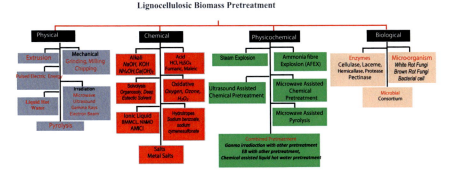

Fig. 1 Classification of lignocellulosic biomass pretreatment methods. Adapted from Kumar et al. [46]

complexes by changing the biomass' macroscopic, submicroscopic, and microscopic structures into a hydrolysable form. This will enable hydrolytic enzymes to easily break down cellulose and hemicellulose into fermentable sugars. The elimination of lignin, a decline in the crystallinity of cellulose, an increase in surface area and porosity of the biomass are the main changes that happened during pretreatment [81]. By hydrolysing the hemicellulose component of lignocellulose, some pretreatment procedures could also produce fermentable sugars.

3 Types of Pretreatments

Pretreatment is a crucial process in the conversion of biomass to bioethanol and other bio-based products. The chosen pretreatment method can have a significant impact on the final cost and design of a lignocellulosic-based biorefinery [12]. In recent years, several pretreatment techniques have been discovered, assessed, and tested in the laboratory, in pilot plants, and industrial operations [64]. Pretreatment uses energy, hence any pretreatment procedure should have its energy efficiency assessed as the basis or justification for its choice [86]. Generally, lignocellulosic biomass pretreatment can be classified into physical, chemical, biological and physicochemical.

3.1 Physical Pretreatment

Physical pretreatments include the use of mechanical and radiation-based techniques. The physical approach aids in increasing surface area, decreasing particle size, and altering cellulose crystallinity. Before pretreatment, the substrate is first prepared physically using a variety of techniques, some of which are combined with other pretreatment techniques [47].

3.1.1 Mechanical Pretreatment

The most often used mechanical size reduction techniques include wet milling, vibratory ball milling, compression milling, and dry milling [21]. Typically, mechanical processing aids in reducing crystallinity and increasing the material's contact surface area, which facilitates the hydrolysis of the polysaccharide polymers found in the cell wall. For low lignin-containing biomass, like an empty fruit bunch, mechanical pretreatment is typically advised. Because lignin cannot be removed normally with mechanical means, the method must also involve the hydrolysis of cellulose and hemicellulose, as well as an additional enzymatic delignification step. The cost of the entire process may rise as a result. However, in order to achieve the proper particle size prior to chemical or biological pretreatment, various mechanical techniques are necessary. But certain thorough mechanical treatments, like milling, can be used as a stand-alone pretreatment technique which helps reduce the crystallinity of cellulose and increase the digestibility of biomass [53].

The significance of mechanical pretreatment of biomass in bioenergy production characteristics was elucidated by Dell'Omo and Spena [27], using the giant reed and wheat straw in biogas production. With the two substrates treated to the dry milling before subjected to the mesophilic anaerobic digestion in 28 days, it was reported that a 137% biogas production gain as compared to the raw material was reported for the giant reed, while the similar condition yielded 49.1% biogas gain in favour of wheat straw. Apart from the biogas yield increase, no statistically significant differences were observed in physicochemical compositions of the treated following pretreatment when compared with the untreated.

3.1.2 Extrusion Pretreatment

Extrusion is the process of forcing materials through a die to produce desired cross-sectional profiles. When the material comes out of the die, it will expand. Due to the benefits of the process in sugar recovery from various biomass feedstocks, extrusion has been adopted as one of the physical continuous pretreatment methods for bioethanol production [88]. There are a few advantages of using extrusion pretreatment over other pretreatments, which include: (1) low cost and improved process monitoring and control of all variables; (2) no sugar degrading products; (3) good adaptability to diverse process modifications; and (4) high continuous throughput. In order to prepare lignocellulosic biomass for the synthesis of bioethanol, it appears that extrusion pretreatment is more practical [88].

For various lignocellulosic biomass, several extruder types, such as single-screw and twin-screw extruders, have been extensively investigated, leading to afterwards high enzymatic hydrolysis rates. With or without the use of chemical solutions, the extrusion pretreatment process can be utilized as a physical pretreatment for the bioconversion of biomass for production of ethanol in a wide range of systems.

Kupryaniuk et al. [48] sought to quantify the energy consumption of the extrusion-cooking of maize straw under various conditions (screw speed, moisture content), as well as to evaluate the efficiency of biogas production. The best results for methane and biogas production (51.63%) efficiency were obtained from pretreated corn straw moistened to 25% and processed at 110 rpm during the extrusion-cooking processes, as opposed to corn straw without pretreatment (49.57%).

3.2 Irradiations Methods

3.2.1 Microwave Pretreatment

Since the middle of the 1990s, there has been a surge in the use of microwave technology in the thermal treatment of biomass. This method not only speeds up the processing and uses less energy, but it also makes it possible to use new chemistry (a peculiar internal heating phenomenon linked to microwave energy) [58].

Microwave radiation is electromagnetic radiation with wavelengths between 0.01 and 1 m and a frequency between 0.3 and 300 GHz [91]. The wavelengths between 0.01 and 0.25 m are typically used for radar transmissions, and the remaining wavelengths are used for telecommunications [91]. Figure 2 shows the microwave region of the electromagnetic spectrum. The microwave region is situated between the infrared and radio frequencies as shown in the electromagnetic spectrum.

Lignin and cell wall constituents are disrupted by the thermal effect due to an increase in temperature and pressure. Under the action of an electromagnetic field,

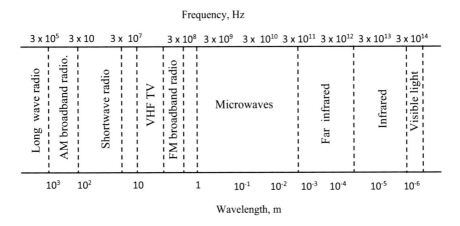

Fig. 2 The electromagnetic spectrum. Adapted from Motasemi and Afzal [58]

dielectric materials relax and polarize, producing a thermal or non-thermal effect forces [76].

Other elements that influence the performance of microwave heating include the following; dipole molecule dynamics, biomass type/composition, size, induction current of magnetic materials, ionic conduction of electrolytes, reaction time, heating rate, moisture content of biomass, microwave power output, and penetration depth [6].

Kłosowski et al. [43] assessed how microwave assisted pretreatment with sodium cumene sulphonate (NaCS) hydrotrope, used to make second-generation bioethanol, would affect wheat stillage. The best results for methane and biogas production (51.63%) efficiency were obtained from pretreated corn straw moistened to 25% and processed at 110 rpm during the extrusion-cooking processes, as opposed to corn straw without pretreatment (49.57%).

3.2.2 Ultrasound Pretreatment

The term "ultrasound" refers to sound waves with frequencies greater than the threshold of human hearing, or above 20 kHz [45]. Ultrasonic treatment of solutions has the potential to replace current pretreatment methods. Although ultrasound demands a large amount of additional energy input and generates radicals in a solution, however, it improves mass transfer through streaming within the solution [18].

To improve physical (mechanoacoustic) and chemical (sonochemical) processes, ultrasonic waves induce pressure variations inside a solution. This happens at frequencies that are higher than human hearing, usually between 20 and 1000 kHz. Either magnetostrictive or piezoelectric transducers produce ultrasound. The piezoelectric property of various ceramics is manipulated via piezoelectric transducers, which are more often utilized today. An alternating current will cause mechanical vibrations in the piezoelectric material, which will result in ultrasound with a specific frequency. A lot of current study into the properties of ultrasound is restricted as a result of the characteristic frequency of transducers. The mechanical vibrations of the piezoelectric material produce a pressure wave through the solution when the transducer is connected to a vessel containing a sonication solution [51]. An ultrasonic reactor typically consists of three basic parts: a reactor, a transducer, and an ultrasonic frequency generator. An electric current is used to create the ultrasonic frequency via a sinus wave generator, whose signal is then amplified by a power amplifier. An ultrasonic transducer is used to convert this signal into a pressure wave, which is then sent into the reactor space, which is already filled with the solution to be sonicated. The pressure wave is often transmitted to the solution using a horn, transducer plate system, or ultrasonic bath [18].

In order to achieve the enhanced anaerobic digestion of *Sida hermaphrodita*, in its combination with cattle manure, ultrasonic pretreatment option was explored by Marta et al. [56] with the specific energy input 25 to 550 kJ/kg volatile solids employed, it is confirmed that the positive net energy gain was only achieved in the range of 25–50 kJ/kg ultrasonic energy employed. However, within the specific energy input range of 200–550 kJ/kg, the pretreatment process was adjudged to significantly increased the biomass solubility by 21.9% chemical oxygen demand (COD) and enhanced biogas yield by 24.7% as TOC.

3.2.3 Gamma Ray Pretreatment

Gamma (γ) rays are created by the radioactive decay of radioisotopes like cobalt-60 or cesium-137. This ionizing radiation can enter lignocellulosic biomass, and when it does, its energy is transmitted to the atoms of the lignocellulosic components, which modifies the lignin and reduces the crystallinity of the cellulose [36]. These structural alterations are brought on by the production of free radicals, which aid in the breakdown of polysaccharides and the destruction of cell walls by the scission of glycosidic linkages [13].

Low amounts of sodium hydroxide can be added to lower the required irradiation dose. The lignocellulosic material begins to degrade more quickly, and the process becomes more economically viable. The weak chemical connections between the lignin, hemicellulose, and cellulose units and alkali swelling are responsible for the high breakdown yield of lignocellulosic compounds utilizing minimal irradiation doses [53, 75].

In contrast to other radioactive isotopes, ^{60}Co, with a half-life of 5.26 years, is frequently utilized in industrial irradiation operations. The radioactive disintegration of the isotopes ^{137}Cs and ^{60}Co produces gamma rays. A gamma chamber can be used in batch and closed-system irradiation systems, which constrain the volume of biomass that can be processed. The quantity of irradiated biomass would be helpful for additional large-scale processing in a batch-continuous process. Such systems have a lengthy history of use in other fields and include safety features. Radicals are created as a result of the gamma rays' energy being transmitted to the biomass's atoms [3].

Su et al. [72] presented a pioneering inquiry into the structural alterations of Miscanthus biomass exposed to various dosages of 60Co-ray radiation up to 1200 kGy. The findings demonstrate that biomass's intra- or intermolecular hydrogen bonds can be partially destroyed by irradiation treatment. Additionally, irradiation treatment can improve the specific surface area of biomasses while decreasing particle size and reducing the distribution range.

3.2.4 Electron Beam Irradiation Pretreatment

A linear accelerator is used to produce electron beam ionizing radiation. The goal of this pretreatment is to disrupt the structure of cell wall polymers (lignin, cellulose, and hemicellulose) by producing free radicals, inducing cross-link formation or chain scission, decrystallization, and/or reducing the degree of polymerization by irradiating lignocellulosic biomass with accelerated electron beams [33].

Because electron beam irradiation is most successful at depolymerizing cellulose, it should be used in conjunction with other pretreatments such steam explosion or alkali for hemicellulose and lignin hydrolyses for good pretreatment results [82].

Jusri et al. [39] explored the examples of the Electron beam irradiation pretreatment process for different LCBs in this work, *Gigantochloa albociliata* (GA), *Leucaena leucocephala* (LL), oil palm frond (OPF), acacia, and microcrystalline cellulose (MCC) were utilized as references. LCBs that were processed using a different dose of the combined electron beam irradiation and ionic liquid approach shows that the method can be improved. The lower transmittance at the wavelength of O–H bonds may suggest that more cellulose could be recovered following pretreatment, according to the FTIR spectra.

3.3 Hydrothermal/liquid Hot Water (LHW) Pretreatment

Bobleter and Pape [14] provided the first description of hydrothermolysis as a pretreatment before enzymatic hydrolysis in the fermentative synthesis of protein. The terms liquid hot water (LHW), autohydrolysis, aqueous liquefaction, aqueous extraction, aqueous pretreatment, aquasolv, hydrothermal pretreatment, and pressure cooking in water have all been used to describe this sort of hydrothermal procedure. Although LHW is the name given to the reaction most frequently, the name for the primary mechanism of reaction, autohydrolysis, is also frequently used [19]. Although higher temperatures can be employed, these processes typically operate at a temperature between 150 and 230 °C and require compressed hot water (pressure over saturation point). Depending on the temperature, the reaction time can range from a few seconds to several hours. Solid concentrations, also known as the liquid-to-solid ratio (LSR), typically range from 2 to 100 (w/w), with values around 10 being the most typical. Continuous reactors are typically associated with higher LSR. The raw material's ability to retain water has an impact on the LSR that can be used [60]. Due to the absence of specialist reactors or rigorous processing conditions, LHW pretreatment is regarded as more inexpensive and straightforward. Liquid hot water pretreatment of biomass is primarily used to change the ultrastructure of the cell wall, hemicellulose, and lignin content, and to break down lignocellulosic biomass into soluble components. Hemicellulose normally decomposes in liquid hot water in three steps that happen in succession: (1) radical products are produced from biomass surface interactions, (2) products dissolve in water, and (3) additional product breakdown [23].

Dimitrellos et al. [29] explored Three different types of lignocellulosic biomass (sunflower straw (SS), grass lawn (GL), and poplar sawdust (PS)) were used to produce hydrogen through dark fermentation (DF) and methane through anaerobic digestion (AD). The effects of liquid hot water (LHW) pretreatment with or without acid addition (A-LHW) on these processes were examined. Hemicellulose was degraded by both pretreatment techniques, although A-LHW released more potential inhibitors (furans and acids) than LHW pretreatment did. In comparison to untreated substrates, the cellulose-rich solid fractions generated following LHW and A-LHW pretreatment had increased biological hydrogen production (BHP).

3.4 Pyrolysis

The controlled anoxic thermal degradation of organic matter into solid/liquid residues and certain volatiles is known as pyrolysis. The potential of pyrolysis as a platform for the valorisation of homogenous and (un)processed lignocelluloses has been extensively explored [79]. Fast pyrolysis is the rapid breakdown of biomass to produce predominantly vapours and aerosols, along with some charcoal and gas. A dark brown homogenous mobile liquid with a heating value roughly half that of ordinary fuel oil is created after cooling and condensation. The majority of low-ash biomass diets produce a high yield of liquid. The following are the key components of a quick pyrolysis process for liquid production:

- because biomass normally has a low thermal conductivity, very rapid heating rates and very high heat transfer rates at the biomass particle reaction interface typically requires a finely ground biomass feed of typically less than 3 mm.
- pyrolysis reaction temperature is carefully monitored at 500 °C to maximize liquid output for most of the biomass,
- to minimize secondary reactions, short hot vapour residence periods of typically less than 2 s,
- swift elimination of product char to reduce vapour cracking,
- the bio-oil product is produced by quickly cooling the pyrolysis vapours [15].

According to Rollag et al. [66], Fast pyrolysis of lignocellulosic biomass provides little sugar or anhydrosugars in comparison to pyrolysis of pure polysaccharides because the pyranose and furanose rings are fragmented by naturally abundant alkali and alkaline earth metals (AAEM) in biomass. By pretreating the biomass with sulfuric acid before pyrolysis, which passivates the catalytic activity of the metals by changing them into thermally stable salts, sugar yields can be considerably enhanced.

3.5 Chemical Methods

The natural structures of lignocellulosic biomass are altered by chemical pretreatment carried out using different chemical agents. The following are some chemical pretreatment methods used in the pretreatment of lignocellulosic biomass:

3.5.1 Acid Pretreatment

Pretreatment techniques using acids have received the most research attention owing to their versatility [46]. By cleaving glucosidic linkages, the acid pretreatment approach has been shown to be very successful at disturbing the lignocellulosic matrix and turning polysaccharides into oligomeric and monomeric sugars [38]. The predominant component of the lignocellulosic biomass that is impacted by the acid pretreatment method is hemicellulose. However, part of the lignin and cellulose are also partially solubilized during this process, releasing oligomers and sugars [37]. Depending on how severe the pretreatment is, the temperature range for diluted acid pretreatment is typically 120–200 °C. It is basically recommended to combine low temperature with high acid concentrations or high temperature with low acid concentrations. For biomass deconstruction, several organic acids (acetic, formic, and oxalic acids) and inorganic acids (hydrochloric, maleic, nitric, nitrous, ortho-phosphoric, and sulfuric acid) are utilized [26]. However, sulfuric and hydrochloric acid are the two acids that are most frequently used in acid pretreatment. Several fermentation-inhibiting substances, including furfural and hydroxymethyl furfurals (HMF), are produced during the acid pretreatment process. As a result, choosing the right acid and adjusting its concentration, as well as controlling reaction time and temperature, may help reduce the production of inhibitors [49]. The main drawback of acid-based pretreatment is that it is corrosive, which raises the overall cost by necessitating pricey non-metallic reactors. Process complexity brought on by acid's toxicity, the necessity for neutralization, and wastewater treatment are additional factors [2].

Utilizing diluted acid pretreatment, Yildirim et al. [84] employed the use of a central composite design was to maximize sugar recovery from cotton straw and sunflower straw. Acid concentration, retention duration, temperature, and the response parameter of sugar production were all chosen input factors. For cotton straw, the ideal pretreatment conditions were 121.7 °C, 2.28% (vol/vol) acid concentration, and 36.82 min. For sunflower straw, the ideal pretreatment conditions were 87.03 °C, 3.68% (vol/vol) acid concentration, and 36.82 min. Under conditions for sunflower and cotton straw, the maximum sugar concentrations were 20 and 17.5 g L^{-1}, respectively.

3.5.2 Alkaline Pretreatment

The most advantageous pre-treatment method is an alkaline pre-treatment using NaOH, lime, and Na_2CO_3, which breaks up intra- and inter-unit linkages and successfully dissolves lignin and certain hemicelluloses [8, 32]. Ammonia can also be used for alkaline pretreatment but ammonia is relatively expensive and will need to be recovered, this will increase the overall cost of the process [9]. The most encouraging features of alkaline pre-treatment are that it preserves cellulose intact and significantly reduces delignification while solubilizing hemicelluloses without ever degrading them to furfural or HMF [20]. When rice husk is treated with thermally-assisted mild NaOH (2% w/w), the amount of lignin removed is highly effective, removing 54% (w/w), and the concentration of cellulose increases to 51.65% (w/w), with only 10.7–33.1% (w/w) of hemicellulose being solubilized [69].

Chen et al. [24] introduced a unique pretreatment technique called "Densifying Lignocellulosic Biomass with Alkaline Chemicals," which is straightforward but effective, uses less energy, and produces fewer hazardous compounds than more conventional pretreatment techniques. The findings demonstrated that DLC-CS has a high density and great stability, preventing contamination and entirely preserving sugar. This technique avoided the high temperature and high-pressure operations that are frequently utilized in conventional pretreatment techniques.

3.5.3 Oxidative Pretreatment

Oxidative pretreatment involves oxidation, which causes lignin to cleave, hemicelluloses to break down into their component sugars, and organic acids to partially degrade the cellulose [46]. A biomass suspension in water is subjected to an oxidative pretreatment by the addition of an oxidizing substance, such as hydrogen peroxide or peracetic acid. To make the cellulose more accessible, it is intended to eliminate the lignin and hemicellulose. Several events, including electrophilic substitution, side chain displacement, cleavage of alkyl aryl ether linkages, and oxidative cleavage of aromatic nuclei, may occur during oxidative pretreatment [34]. Because the oxidant being employed frequently is not selective, losses of cellulose and hemicellulose can happen. Once lignin is oxidized and soluble aromatic chemicals are produced, there is a considerable likelihood that inhibitors will develop [35].

To enhance biomass fractionation and boost enzymatic digestibility, Yuan et al. [85] looked into the use of oxygen (O_2) and hydrogen peroxide (H_2O_2) as co-oxidants during alkaline oxidative pretreatment. The two-stage alkaline pre-extraction/copper-catalysed alkaline hydrogen peroxide (Cu-AHP) pretreatment process demonstrated high overall sugar yields even at H_2O_2 loadings as low as 2% (w/w of original biomass) due to a significant improvement in delignification compared to using H_2O_2 alone during the second-stage Cu-AHP pretreatment. However, the presence of H_2O_2 was both essential and beneficial.

3.5.4 Solvent-Based Pretreatment

Organosolv Pretreatment

An effective pretreatment approach is organosolvation, which has shown promise for lignocellulosic biomass [4]. To solubilize lignin and produce treated cellulose appropriate for enzymatic hydrolysis, a variety of organic or aqueous solvent solutions, including methanol, ethanol, acetone, ethylene glycol, and tetrahydrofurfuryl alcohol can be used. The fundamental benefit of the organosolv procedure, when compared to other chemical pretreatments, is the recovery of relatively pure lignin as a by-product [87]. However, the chemical recovery problem hinders this approach from being used at an industrial scale due to the volatility of organic solvent. Meng et al. [57] explored the effects of two recent organosolv pretreatment techniques, co-solvent enhanced lignocellulosic fractionation (CELF) and -Valerolactone (GVL) pretreatments, on the physicochemical properties of poplar were examined in this study and contrasted with those of traditional ethanol organosolv pretreatment. When both pretreatments were compared side by side, it became clear that the CELF pretreatment removed lignin predominantly without appreciably altering the cellulose ultrastructure. On the other hand, crystalline cellulose was most severely harmed by GVL pretreatment, while EtOH pretreatment significantly altered cellulose DP.

Deep Eutectic Solvents (DES) Pretreatment

Deep eutectic solvents have recently attracted the attention of certain research groups as a green technique for utilizing biomass [46]. At a moderate temperature of between 60 and 80 °C, deep eutectic solvent (DES) mixes are created by combining hydrogen bond donors (alcohols, amides, and carboxylic acids) with acceptors (quaternary ammonium salts). These mixtures are made up of non-symmetric, low melting point, and lattice energy ions [68]. DESs are categorized based on the combinations of their chemical components. Due to the high melting points of the non-hydrated metal halides, Type I DESs have limited usage in the processing of biomass, but Type II DESs are more practical for industrial processes due to the relatively lower costs of the hydrated metal halides. Nonetheless, due to their quick and simple manufacture, non-reactivity with water, biodegradable nature, and cost-effectiveness, Type III DESs are the most investigated. Inorganic transition metals are used in Type IV DESs to create eutectic mixtures with urea, despite the fact that metal salts generally do not ionize in non-aqueous mediums [55].

Deep eutectic solvents (DESs), which are environmentally friendly substitutes for traditional solvents, were researched by Kohli et al. [44] as a way to delignify lignocellulosic biomass feedstocks in order to produce a more affordable and efficient technique. Miscanthus and birchwood feedstocks were delignified using six distinct DESs, including monocarboxylic acid/choline chloride (ChCl), dicarboxylic acid/ChCl, and polyalcohol/ChCl. ChCl.formic acid and ChCl.oxalic acid DES were used to extract the most lignin from the miscanthus and birchwood, respectively. The extracted lignin samples' TGA and 13C NMR characterization results showed that various lignin types were created utilizing various DESs.

3.5.5 Ionic Liquid (IL) Pretreatment

Large organic cations and small inorganic anions commonly make up ionic liquids (ILs), which usually have a melting point of 100 °C. By competing for hydrogen bonds, it can dissolve the lignocelluloses' biomass's carbohydrates and lignin polymer, uprooting the complex web of non-covalent connections between cellulose, hemicellulose, and lignin [4]. ILs are said to be "green" solvents because they do not produce any explosive or poisonous fumes. ILs with anion activity, such as the 1-butyl-3 methylimidazolium cation [C4mim]+, can dissolve both lignin and carbohydrates at the same time because they create hydrogen bonds in a 1:1 stoichiometry between their non-hydrated chloride ions and the sugar hydroxyl protons. As a result, the complex web of non-covalent connections between cellulose, hemicellulose, and lignin biomass polymers is successfully disrupted while the generation of breakdown products is minimized. More research is necessary to determine whether the majority of evidence demonstrating the efficacy of ILs can be applied to a more complicated mixture of components found in lignocellulosic biomass because most of the data have been obtained using pure crystalline cellulose [52].

Ziaei-Rad et al. [90] investigated the ionic liquid pretreatment of wheat straw in conditions characterized by a temperature of 130 °C, a high solid-to-solvent load ratio of 1:5 g/g, and 20 wt% water to separate lignin and carbohydrates from the source, which would then undergo enzymatic hydrolysis. The sample that had been processed with [TEA] [HSO$_4$] for three hours had the highest delignification rate, which was 80%, and the highest hemicellulose removal rate, which was 64.45%. The high glucose saccharification yield levels of two biomass samples pretreated for 0.5 h and 3 h were 50.36 and 87.19%, respectively, while the 3-h pretreatment sample produced 78.23% xylose yield in enzymatic hydrolysis using the commercial enzyme CelluMax in 72 h.

3.5.6 Hydrotropic Pretreatment

Compounds called hydrotropes, which have both hydrophilic and hydrophobic functional groups, let hydrophobic substances dissolve in aqueous solutions [5]. Sodium benzoate, sodium cymenesulfonate, sodium phenolsulfonate, sodium naphthalenesulfonate, sodium salicylate, sodium xylenesulfonate, and sodium toluenesulfonate are a few examples of hydrotrope agents. Amphiphilic salts called hydrotropes with aromatic anions can increase the solubility of lignin [28, 59]. In addition to being effective for lignin extraction, hydrotropic technology was also suitable for pretreating lignocellulosic biomass to increase the efficiency of enzymatic hydrolysis. For usage as a pretreatment technology, the treatment conditions employed in fractionation should be changed. High yields of sugar monomers should result from pretreatment settings that remove the lignin barrier while maintaining cellulose's enzyme accessibility [59].

In their study, Kłosowski et al. [43] assessed the impact of microwave-assisted pretreatment of wheat stillage while it was being produced into second-generation bioethanol using sodium cumene sulphonate (NaCS) hydrotrope. When compared to the raw material utilized before treatment, the composition of the wheat stillage biomass altered dramatically as a result of the microwave pretreatment. The lignin and hemicellulose content of the biomass were successfully reduced by microwave-assisted pretreatment with NaCS, leaving cellulose—which made up 42.91 0.10% of the biomass—as the predominant constituent.

3.5.7 Salts Pretreatment

Research has demonstrated that metal salts can stimulate the breakdown of lignocellulosic biomass. Limited study has been undertaken on pretreating lignocellulosic biomass using inorganic salts alone, without the addition of dilute acid. It is more economically feasible to employ inorganic salts in aqueous solutions directly in the processing of lignocellulosic biomass since they are less corrosive than acids and can be recycled. If realized, additional overliming to neutralize the hydrolysate would be avoided. However, more research is required to determine whether the inorganic salt pretreatment on hemicellulose degradation is as effective as the xylose monomer and xylotriose degradation [54]. The primary benefits of using metal salts in conjunction with other treatment methods include enhanced lignin removal, hemicellulose degradation, complete conversion of biomass, increased reaction rate, improved enzymatic hydrolysis, non-toxicity, environmental safety, and a lack of need for costly non-corrosive reactors [41].

In their work, Gao et al. [30] explored to determine how to use inorganic salts as adjuvants for ionic liquid–water pretreatment to increase the Ionic Liquids' tolerance to water and reusability. Following the processing of rice straw at 110 °C for 1 h with a solution of 40% 1-ethyl-3-methylimidazolium chloride ([C_2mim]Cl) + 53% water + 7% K_2CO_3, the residues became very amenable to enzymatic hydrolysis, removing 93.70% of the lignin and yielding 92.07% sugar. At ambient temperature, an aqueous biphasic [C_2mim]Cl–K_2CO_3 system was produced when the K_2CO_3 concentration above 30%, and a [C_2mim]Cl recovery of 94.32% was attained. The findings show that the cost of IL pretreatment can be significantly reduced by adding inorganic salts to Ionic Liquid aqueous solutions while maintaining an efficient enzymatic hydrolysis of lignocellulosic biomass.

3.6 Physicochemical Pretreatment

3.6.1 Steam Explosion (SE) Pretreatment/Autohydrolysis

An effective auto-hydrolysis technology for biomass pretreatment is known as steam explosion (SE), in which biomass particles are exposed to hot steam between 160 and 270 °C for a set amount of time. The residence period typically lasts between a few seconds and a few minutes. When the process is over, pressure is released, which causes moisture to evaporate and lignocellulosic biomass to break down. When pressure is suddenly released, the acetyl group in hemicelluloses undergoes autohydrolysis, which disrupts the structure of the cell wall and separates the individual biomass fibres [65]. By partially removing hemicellulose and lignin from pineapple leaves using SE pretreatment at a pressure of 20 kgfcm^2, cellulose accessibility was increased [46].

In their work, Balan et al. [10] looked into how the reaction parameters selected for the enzymatic hydrolysis affect the outcomes of a pretreatment optimization approach.

Beechwood was processed using a steam explosion, and the resulting biomass was then hydrolysed by enzymes using washed particles or the entire pretreatment slurry at glucan loadings of 1% and 5%. Beechwood was processed using a steam explosion, and the resulting biomass was then hydrolysed by enzymes using washed particles or the entire pretreatment slurry at glucan loadings of 1% and 5%.The glucose yields significantly increased with increasing severity and with increasing pretreatment temperature at identical severities for enzymatic hydrolysis in both reaction modes at a glucan loading of 1%, and maximum values were attained at a pretreatment temperature of 230 C. The ideal severity, however, was only 4.75 for entire slurry enzymatic hydrolysis and 5.0 for washed solids enzymatic hydrolysis.

3.6.2 Ammonia Fibre Expansion (AFEX)

A promising physicochemical procedure called ammonia fibre expansion (AFEX) pretreatment uses liquid ammonia applied at high pressure and low temperature. Under extreme pressure, the anhydrous ammonia splits into ammonium and hydroxide ions, which quickly raises the temperature [11]. As a result, the biomass swells, cellulose crystallinity decreases, hemicellulose is deacetylated, and the links between lignin and carbohydrates are broken [50]. The primary benefits of the AFEX pretreatment include its dry-to-dry nature, which eliminates the need for washing, low production of inhibitory compounds, lack of need for detoxification, retention of high cellulose and hemicellulose content, and lack of sugar loss with only a moderate temperature and brief residence time requirement [74]. However, the drawbacks of this pretreatment type are the high cost of ammonia, the need for a highly controlled atmosphere, and the consequent necessity for specialized equipment with greater capital and utility costs [16, 61].

Rhodococcus opacus NRRL B-3311 was tested by Wang et al. [80] as the only carbon source for microbial lipid synthesis using lignin from maize stover that had undergone Ammonia fibre expansion (AFEX) pretreatment and lignin model compounds. Without any prior preparation, R. opacus NRRL B-3311 could develop on the AFEX-lignin and collect lipids up to 32 mg/L in 72 h while only consuming 20% of the total lignin. The lipid made from lignin exhibited a fatty acid profile that was comparable to that of vegetable oil.

3.7 Biological Pretreatment

In biological pretreatment, lignin and hemicelluloses from the lignocellulosic biomass were degraded using microorganisms like brown, white, and soft rot fungi. Because they use less energy and minimal environmental harm, white rot fungi that can digest lignin are a promising biological pretreatment method [22, 71]. When the biological pretreatment is carried out under mild conditions, the by-products generated usually will not prevent further hydrolysis. Effective lignin degradation during biological pretreatment is dependent on lignolytic enzymes such as lignin peroxidase, manganese peroxidase, and laccase generated by basidiomycete. The white rot fungus aids in delignification during biological pretreatment, which boosts the rate of enzymatic saccharification [40]. Although biological pretreatment is an environmentally friendly method that does not produce any inhibitors, it takes a while to complete. The method can be optimized to be more effective by choosing the best strain and culture conditions, which will cut down on treatment time and carbohydrate loss [77]. The nature and composition of the biomass, as well as other factors including the type of microorganisms used, the incubation temperature, pH, incubation time, inoculums concentration, moisture content, and aeration rate, are significant process variables that affect biological pretreatment [71]. One of the efficient techniques for enzymatic saccharification is fungus pretreatment utilizing the wood rot fungus. *Gloeophyllum trabeum*, a brown rot fungus, produces enzymes that can depolymerize cellulose and hemicelluloses in wood, resulting in the production of modified lignin in the brown residue [31].

3.7.1 Enzymatic-Based Pretreatment

One fairly common method is the use of enzymes to remove specific biomass components [42]. Cellulosic and hemicellulosic component separation for ethanol production, when lignin breakdown is involved, it is aimed for enhanced carbohydrate polymer accessibility to hydrolysing enzyme. Enzymes made by microorganisms, such as cellulases and hemicellulases, typically aid in the biological breakdown of lignocellulosic material. The hydrolysis of cellulose and hemicellulose is the rate-limiting stage in the anaerobic digestion of lignocellulosic material. For biomass conversion, improving the process efficiency of anaerobic digestion requires increasing the hydrolysis rate [67].

In the work reported by Pascal et al. [62] the sugarcane bagasse structure was altered using mild acid-catalysed atmospheric glycerol organosolv (ac-AGO) pretreatment, which improved enzymatic hydrolysability. Ac-AGO pretreatment was optimized at 200 °C for 15 min with 0.06% H_2SO_4 addition using single factor and central composite design experiments, resulting in extraordinarily high cellulose retention of 98% and removal rates of hemicellulose and lignin of 82 and 52%, respectively. A modest cellulase loading and good enzymatic hydrolysability of the ac-AGO-pretreated substrate allowed for a 70% glucose production after 72 h.

3.7.2 Microbes-Based Pretreatment

The main factors influencing the microbe-based pretreatment are particle size, moisture content, incubation temperature, pH, and nutrition demand (carbon, nitrogen, phosphorus) [78]. Low cost, chemical consumption, energy requirement, water utilization, trash creation, inhibitor formation, and simple downstream processing are the main benefits of fungal pretreatment. However, fungal treatment is constrained since it uses the carbohydrate content of the biomass for growth, which reduces the overall yield of the process. The lengthy incubation period is the second significant disadvantage of fungal therapy [86, 88]. Using certain lignin-degrading bacteria can circumvent this. The employment of bacteria for microbe-assisted biomass treatment is restricted by their ineffectiveness at degrading lignin. Yet, bacteria can be employed for delignification because they are very flexible and genetically simple [83].

In their study, Shanmugam et al. [70] created a stable, xylan-using, anaerobic microbial consortium (MC1), and it was taxonomically determined that the genera Ruminococcus and Clostridium from this community played a significant part in the substrate utilization. Additionally, the bioaugmentation method was used to introduce the Clostridium sp. strain WST, which produces butanol. This greatly increased the butanol yield up to 0.54 g/g by 98-fold and led to the conversion of xylan to biobutanol up to 10.8 g/L. From 20 g/L of maize cob and other xylan-rich biomass, this system was able to produce 1.09 g/L of butanol.

4 Hydrolysate Conditioning

To lessen toxicity and increase the accessibility of sugar to fermentation, hydrolysate is subjected to processes known as conditioning. In general, it would be more cost-effective to alter the pretreatment processes to decrease the formation of toxic compounds rather than incurring the extra expense of a conditioning step, but as was already mentioned, less toxic pretreatment methods typically result in less sugar being released for fermentations, necessitating the addition of more hydrolytic enzymes [63]. Processes of conditioning can be divided into three categories: biological, physical, and chemical. The physical strategy is intended to remove harmful molecules from the hydrolysate, while the biological and chemical approaches are intended to transform harmful compounds into less toxic products. Recovery after toxic component sequestration will offer some understanding of hydrolysate fractionation and present opportunities for the production of chemical feedstocks for biorefineries [63].

5 Pretreatment Technology Challenges

The following can be used to outline the limitations of the lignocellulosic feedstock-based biorefinery [46]:

(i) Selection and perennial accessibility of appropriate biomass
(ii) Effective pretreatment method

(iii) Cost of the process, including equipment, reactors, chemicals, and process control
(iv) Toxic compounds produced during various pretreatment methods
(v) Post-pretreatment process, such as washing the substrate
(vi) Waste generated during the process and associated environmental risks
(vii) Recycling of the chemical or catalysts used during the process.

6 Conclusion

An emerging technology in the realm of renewable energy development is the generation of bioethanol from lignocellulosic biomass, such as maize cobs, rice husks, cassava peels, yam peels, sugar cane bagasse, and others. However, prior to the fermentation process, this biomass must be pretreated. The primary objective of pretreatment is to disrupt the crystalline structure of cellulose and break down the lignin structure in order to increase the accessibility of the cellulose to the enzymes during the hydrolysis step. This chapter therefore explicitly described some old and emerging pretreatment technologies as well as hydrolysate conditioning for effective lignocellulosic biomass conversion into bioethanol. It may also be inferred from this chapter that the pretreatment option to choose depends on the physicochemical makeup of the biomass. However, some biomass requires multiple pretreatments.

References

1. Agustini L, Efiyanti L, Faulina SA, Santoso E (2012) Isolation and characterization of cellulase- and xylanase-producing microbes isolated from tropical forests in Java and Sumatra. Int J Environ Bioenergy 3(3):154–167
2. Ajayi OA, Adefila SS (2012) Methanol production from cow dung. J Environ Earth Sci 2(7):2225–0948
3. Al-Masri M, Zarkawi M (1994) Effects of gamma irradiation on cell-wall constituents of some agricultural residues. Radiat Phys Chem 44(6):661–663
4. Alvira P, Tomás-Pejó E, Ballesteros M, Negro M (2010) Pretreatment technologies for an efficient bioethanol production process based on enzymatic hydrolysis: a review. Biores Technol 101(13):4851–4861
5. Andelin J, Niblock R, Curlin J (1989) Technologies for reducing dioxin in the manufacture of bleached wood pulp. Chap 2:17
6. Asomaning J, Haupt S, Chae M, Bressler DC (2018) Recent developments in microwave-assisted thermal conversion of biomass for fuels and chemicals. Renew Sustain Energy Rev 92:642–657
7. Awoyale AA, Lokhat D (2019) Harnessing the potential of bio-ethanol production from lignocellulosic biomass in Nigeria–a review. Biofuels, Bioprod Biorefin 13(1):192–207
8. Awoyale AA, Lokhat D (2021) Experimental determination of the effects of pretreatment on selected Nigerian lignocellulosic biomass in bioethanol production. Sci Rep 11(1):557
9. Baksi S, Saha D, Saha S, Sarkar U, Basu D, Kuniyal J (2023) Pre-treatment of lignocellulosic biomass: review of various physico-chemical and biological methods influencing the extent of biomass depolymerization. Int J Environ Sci Technol:1–28

10. Balan R, Antczak A, Brethauer S, Zielenkiewicz T, Studer MH (2020) Steam explosion pretreatment of beechwood. Part 1: Comparison of the enzymatic hydrolysis of washed solids and whole pretreatment slurry at different solid loadings. Energies 13(14):3653
11. Balan V, Bals B, da Costa Sousa L, Garlock R, Dale BE (2011) A short review on ammonia-based lignocellulosic biomass pretreatment. In: Chemical and biochemical catalysis for next generation biofuels. The Royal Society of Chemistry, London, pp 89–114
12. Banu JR, Kavitha S, Tyagi VK, Gunasekaran M, Karthikeyan OP, Kumar G (2021) Lignocellulosic biomass based biorefinery: a successful platform towards circular bioeconomy. Fuel 302:121086
13. Betiku E, Adetunji O, Ojumu TV, Solomon BO (2009) A comparative study of the hydrolysis of gamma irradiated lignocelluloses. Braz J Chem Eng 26:251–255
14. Bobleter O, Niesner R, Röhr M (1976) The hydrothermal degradation of cellulosic matter to sugars and their fermentative conversion to protein. J Appl Polym Sci 20(8):2083–2093
15. Bridgwater AV (2012) Review of fast pyrolysis of biomass and product upgrading. Biomass Bioenerg 38:68–94
16. Brodeur G, Yau E, Badal K, Collier J, Ramachandran K, Ramakrishnan S (2011) Chemical and physicochemical pretreatment of lignocellulosic biomass: a review. Enzyme Res 2011
17. Brodin M, Vallejos M, Opedal MT, Area MC, Chinga-Carrasco G (2017) Lignocellulosics as sustainable resources for production of bioplastics–a review. J Clean Prod 162:646–664
18. Bussemaker MJ, Zhang D (2013) Effect of ultrasound on lignocellulosic biomass as a pretreatment for biorefinery and biofuel applications. Ind Eng Chem Res 52(10):3563–3580
19. Carvalheiro F, Duarte L, Gírio F, Moniz P (2016) Hydrothermal/liquid hot water pretreatment (autohydrolysis): a multipurpose process for biomass upgrading. In: Biomass fractionation technologies for a lignocellulosic feedstock based biorefinery. Elsevier, pp 315–347
20. Chaturvedi V, Verma P (2013) An overview of key pretreatment processes employed for bioconversion of lignocellulosic biomass into biofuels and value added products. 3 Biotech 3:415–431
21. Che Kamarludin SN, Jainal MS, Azizan A, Safaai NSM, Mohamad Daud AR (2014) Mechanical pretreatment of lignocellulosic biomass for biofuel production. Applied mechanics and materials. Trans Tech Publ, pp 838–841
22. Chen S, Zhang X, Singh D, Yu H, Yang X (2010) Biological pretreatment of lignocellulosics: potential, progress and challenges. Biofuels 1(1):177–199
23. Chen W-H, Nižetić S, Sirohi R, Huang Z, Luque R, Papadopoulos AM, Sakthivel R, Nguyen XP, Hoang AT (2022) Liquid hot water as sustainable biomass pretreatment technique for bioenergy production: a review. Biores Technol 344:126207
24. Chen X, Yuan X, Chen S, Yu J, Zhai R, Xu Z, Jin M (2021) Densifying Lignocellulosic biomass with alkaline Chemicals (DLC) pretreatment unlocks highly fermentable sugars for bioethanol production from corn stover. Green Chem 23(13):4828–4839
25. Chu Q, Wang R, Tong W, Jin Y, Hu J, Song K (2020) Improving enzymatic saccharification and ethanol production from hardwood by deacetylation and steam pretreatment: insight into mitigating lignin inhibition. ACS Sustain Chem Eng 8(49):17967–17978
26. Czajczyńska D, Anguilano L, Ghazal H, Krzyżyńska R, Reynolds A, Spencer N, Jouhara H (2017) Potential of pyrolysis processes in the waste management sector. Therm Sci Eng Prog 3:171–197
27. Dell'Omo PP, Spena VA (2020) Mechanical pretreatment of lignocellulosic biomass to improve biogas production: comparison of results for giant reed and wheat straw. Energy 203:117798
28. Devendra LP, Kumar MK, Pandey A (2016) Evaluation of hydrotropic pretreatment on lignocellulosic biomass. Biores Technol 213:350–358
29. Dimitrellos G, Lyberatos G, Antonopoulou G (2020) Does acid addition improve liquid hot water pretreatment of lignocellulosic biomass towards biohydrogen and biogas production? Sustainability 12(21):8935
30. Gao J, Chen C, Wang L, Lei Y, Ji H, Liu S (2019) Utilization of inorganic salts as adjuvants for ionic liquid–water pretreatment of lignocellulosic biomass: enzymatic hydrolysis and ionic liquid recycle. 3 Biotech 9:1–10

31. Gao Z, Mori T, Kondo R (2012) The pretreatment of corn stover with *Gloeophyllum trabeum* KU-41 for enzymatic hydrolysis. Biotechnol Biofuels 5(1):1–11
32. Gottumukkala LD, Mathew AK, Abraham A, Sukumaran RK (2019) Biobutanol production: microbes, feedstock, and strategies. In: Biofuels: alternative feedstocks and conversion processes for the production of liquid and gaseous biofuels. Elsevier, pp 355–377
33. Grabowski C (2015) The impact of electron beam pretreatment on the fermentation of wood-based sugars. State University of New York. College of Environmental Science and Forestry
34. Hendriks A, Zeeman G (2009) Pretreatments to enhance the digestibility of lignocellulosic biomass. Biores Technol 100(1):10–18
35. Hon D, Minemura N (2001) Color and discoloration. In: Hon DN-S, Shiraishi N (eds) Wood and cellulosic chemistry. Marcel Dekker, New York
36. Hong SH, Lee JT, Lee S, Wi SG, Cho EJ, Singh S, Lee SS, Chung BY (2014) Improved enzymatic hydrolysis of wheat straw by combined use of gamma ray and dilute acid for bioethanol production. Radiat Phys Chem 94:231–235
37. Jönsson LJ, Martín C (2016) Pretreatment of lignocellulose: formation of inhibitory by-products and strategies for minimizing their effects. Biores Technol 199:103–112
38. Jung YH, Kim KH (2015) Acidic pretreatment. In: Pretreatment of biomass. Elsevier, pp 27–50
39. Jusri NAA, Azizan A, Zain ZSZ, Rahman AMF (2019) Effect of electron beam irradiation and ionic liquid combined pretreatment method on various lignocellulosic biomass. Key engineering materials. Trans Tech Publ, pp 351–358
40. Kalyani D, Lee K-M, Kim T-S, Li J, Dhiman SS, Kang YC, Lee J-K (2013) Microbial consortia for saccharification of woody biomass and ethanol fermentation. Fuel 107:815–822
41. Kang KE, Park D-H, Jeong G-T (2013) Effects of NH_4Cl and $MgCl_2$ on pretreatment and xylan hydrolysis of Miscanthus straw. Carbohyd Polym 92(2):1321–1326
42. Karthika K, Arun A, Rekha P (2012) Enzymatic hydrolysis and characterization of lignocellulosic biomass exposed to electron beam irradiation. Carbohyd Polym 90(2):1038–1045
43. Kłosowski G, Mikulski D, Bhagwat P, Pillai S (2022) Cellulosic ethanol production using waste wheat stillage after microwave-assisted hydrotropic pretreatment. Molecules 27(18):6097
44. Kohli K, Katuwal S, Biswas A, Sharma BK (2020) Effective delignification of lignocellulosic biomass by microwave assisted deep eutectic solvents. Biores Technol 303:122897
45. Kumar A (2013) Pretreatment methods of lignocellulosic materials for biofuel production: a review. J Emerg Trends Eng Appl Sci 4(2):181–193
46. Kumar B, Bhardwaj N, Agrawal K, Chaturvedi V, Verma P (2020) Current perspective on pretreatment technologies using lignocellulosic biomass: an emerging biorefinery concept. Fuel Process Technol 199:106244
47. Kumari D, Singh R (2018) Pretreatment of lignocellulosic wastes for biofuel production: a critical review. Renew Sustain Energy Rev 90:877–891
48. Kupryaniuk K, Oniszczuk T, Combrzyński M, Czekała W, Matwijczuk A (2020) The influence of corn straw extrusion pretreatment parameters on methane fermentation performance. Materials 13(13):3003
49. Larsson S, Palmqvist E, Hahn-Hägerdal B, Tengborg C, Stenberg K, Zacchi G, Nilvebrant N-O (1999) The generation of fermentation inhibitors during dilute acid hydrolysis of softwood. Enzyme Microb Technol 24(3–4):151–159
50. Laureano-Perez L, Teymouri F, Alizadeh H, Dale BE (2005) Understanding factors that limit enzymatic hydrolysis of biomass: characterization of pretreated corn stover. Appl Biochem Biotechnol 124:1081–1099
51. Leonelli C, Mason TJ (2010) Microwave and ultrasonic processing: now a realistic option for industry. Chem Eng Process 49(9):885–900
52. Li Q, He Y-C, Xian M, Jun G, Xu X, Yang J-M, Li L-Z (2009) Improving enzymatic hydrolysis of wheat straw using ionic liquid 1-ethyl-3-methyl imidazolium diethyl phosphate pretreatment. Biores Technol 100(14):3570–3575
53. Li W, Zheng P, Guo J, Ji J, Zhang M, Zhang Z, Zhan E, Abbas G (2014) Characteristics of self-alkalization in high-rate denitrifying automatic circulation (DAC) reactor fed with methanol and sodium acetate. Biores Technol 154:44–50

54. Liu L, Sun J, Cai C, Wang S, Pei H, Zhang J (2009) Corn stover pretreatment by inorganic salts and its effects on hemicellulose and cellulose degradation. Biores Technol 100(23):5865–5871
55. Loow Y-L, New EK, Yang GH, Ang LY, Foo LYW, Wu TY (2017) Potential use of deep eutectic solvents to facilitate lignocellulosic biomass utilization and conversion. Cellulose 24:3591–3618
56. Marta K, Paulina R, Magda D, Anna N, Aleksandra K, Marcin D, Kazimierowicz J (2020) Evaluation of ultrasound pretreatment for enhanced anaerobic digestion of Sida hermaphrodita. BioEnergy Res 13:824–832
57. Meng X, Bhagia S, Wang Y, Zhou Y, Pu Y, Dunlap JR, Shuai L, Ragauskas AJ, Yoo CG (2020) Effects of the advanced organosolv pretreatment strategies on structural properties of woody biomass. Ind Crops Prod 146:112144
58. Motasemi F, Afzal MT (2013) A review on the microwave-assisted pyrolysis technique. Renew Sustain Energy Rev 28:317–330
59. Mou H, Fardim P, Wu S (2016) A novel green biomass fractionation technology: hydrotropic pretreatment. In: Biomass fractionation technologies for a lignocellulosic feedstock based biorefinery. Elsevier, pp 281–313
60. Nabarlatz D, Farriol X, Montané D (2005) Autohydrolysis of almond shells for the production of xylo-oligosaccharides: product characteristics and reaction kinetics. Ind Eng Chem Res 44(20):7746–7755
61. Park J-y, Shiroma R, Al-Haq MI, Zhang Y, Ike M, Arai-Sanoh Y, Ida A, Kondo M, Tokuyasu K (2010) A novel lime pretreatment for subsequent bioethanol production from rice straw–calcium capturing by carbonation (CaCCO) process. Biores Technol 101(17):6805–6811
62. Pascal K, Ren H, Sun FF, Guo S, Hu J, He J (2019) Mild acid-catalyzed atmospheric glycerol organosolv pretreatment effectively improves enzymatic hydrolyzability of lignocellulosic biomass. ACS Omega 4(22):20015–20023
63. Pienkos PT, Zhang M (2009) Role of pretreatment and conditioning processes on toxicity of lignocellulosic biomass hydrolysates. Cellulose 16:743–762
64. Rabemanolontsoa H, Saka S (2016) Various pretreatments of lignocellulosics. Biores Technol 199:83–91
65. Rastogi M, Shrivastava S (2017) Recent advances in second generation bioethanol production: an insight to pretreatment, saccharification and fermentation processes. Renew Sustain Energy Rev 80:330–340
66. Rollag SA, Lindstrom JK, Brown RC (2020) Pretreatments for the continuous production of pyrolytic sugar from lignocellulosic biomass. Chem Eng J 385:123889
67. Romano RT, Zhang R, Teter S, McGarvey JA (2009) The effect of enzyme addition on anaerobic digestion of Jose Tall wheat grass. Biores Technol 100(20):4564–4571
68. Satlewal A, Agrawal R, Bhagia S, Sangoro J, Ragauskas AJ (2018) Natural deep eutectic solvents for lignocellulosic biomass pretreatment: recent developments, challenges and novel opportunities. Biotechnol Adv 36(8):2032–2050
69. Shahabazuddin M, Chandra TS, Meena S, Sukumaran R, Shetty N, Mudliar S (2018) Thermal assisted alkaline pretreatment of rice husk for enhanced biomass deconstruction and enzymatic saccharification: physico-chemical and structural characterization. Biores Technol 263:199–206
70. Shanmugam S, Sun C, Chen Z, Wu Y-R (2019) Enhanced bioconversion of hemicellulosic biomass by microbial consortium for biobutanol production with bioaugmentation strategy. Biores Technol 279:149–155
71. Sindhu R, Binod P, Pandey A (2016) Biological pretreatment of lignocellulosic biomass–an overview. Biores Technol 199:76–82
72. Su X-J, Zhang C-Y, Li W-J, Wang F, Wang K-Q, Liu Y, Li Q-M (2020) Radiation-induced structural changes of Miscanthus biomass. Appl Sci 10(3):1130
73. Taha M, Foda M, Shahsavari E, Aburto-Medina A, Adetutu E, Ball A (2016) Commercial feasibility of lignocellulose biodegradation: possibilities and challenges. Curr Opin Biotechnol 38:190–197

74. Teymouri F, Laureano-Pérez L, Alizadeh H, Dale BE (2004) Ammonia fiber explosion treatment of corn stover. In: Proceedings of the twenty-fifth symposium on biotechnology for fuels and chemicals held May 4–7, 2003, Breckenridge, CO. Springer, pp 951–963
75. Torun M (2017) Radiation pretreatment of biomass. In: Applications of ionizing radiation in materials processing, pp 447–460
76. Tsubaki S, Azuma J-I, Yoshimura T, Maitani M, Suzuki E, Fujii S, Wada Y (2016) Microwave-induced biomass fractionation. In: Biomass fractionation technologies for a lignocellulosic feedstock based biorefinery. Elsevier, pp 103–126
77. Van Kuijk S, Sonnenberg A, Baars J, Hendriks W, Cone J (2015) Fungal treated lignocellulosic biomass as ruminant feed ingredient: a review. Biotechnol Adv 33(1):191–202
78. Wan C, Li Y (2010) Microbial pretreatment of corn stover with *Ceriporiopsis subvermispora* for enzymatic hydrolysis and ethanol production. Biores Technol 101(16):6398–6403
79. Wang S, Dai G, Yang H, Luo Z (2017) Lignocellulosic biomass pyrolysis mechanism: a state-of-the-art review. Prog Energy Combust Sci 62:33–86
80. Wang Z, Li N, Pan X (2019) Transformation of ammonia fiber expansion (AFEX) corn stover lignin into microbial lipids by *Rhodococcus opacus*. Fuel 240:119–125
81. Wyman CE, Dale BE, Elander RT, Holtzapple M, Ladisch MR, Lee Y (2005) Coordinated development of leading biomass pretreatment technologies. Biores Technol 96(18):1959–1966
82. Xiang Y, Xiang Y, Wang L (2017) Electron beam irradiation to enhance enzymatic saccharification of alkali soaked *Artemisia ordosica* used for production of biofuels. J Environ Chem Eng 5(4):4093–4100
83. Yan X, Wang Z, Zhang K, Si M, Liu M, Chai L, Liu X, Shi Y (2017) Bacteria-enhanced dilute acid pretreatment of lignocellulosic biomass. Biores Technol 245:419–425
84. Yildirim O, Ozkaya B, Altinbas M, Demir A (2021) Statistical optimization of dilute acid pretreatment of lignocellulosic biomass by response surface methodology to obtain fermentable sugars for bioethanol production. Int J Energy Res 45(6):8882–8899
85. Yuan Z, Klinger GE, Nikafshar S, Cui Y, Fang Z, Alherech M, Goes S, Anson C, Singh SK, Bals B (2021) Effective biomass fractionation through oxygen-enhanced alkaline–oxidative pretreatment. ACS Sustain Chem Eng 9(3):1118–1127
86. Zabed H, Sahu J, Boyce AN, Faruq G (2016) Fuel ethanol production from lignocellulosic biomass: an overview on feedstocks and technological approaches. Renew Sustain Energy Rev 66:751–774
87. Zhao X, Cheng K, Liu D (2009) Organosolv pretreatment of lignocellulosic biomass for enzymatic hydrolysis. Appl Microbiol Biotechnol 82:815–827
88. Zheng J, Rehmann L (2014) Extrusion pretreatment of lignocellulosic biomass: a review. Int J Mol Sci 15(10):18967–18984
89. Zheng Y, Zhao J, Xu F, Li Y (2014) Pretreatment of lignocellulosic biomass for enhanced biogas production. Prog Energy Combust Sci 42:35–53
90. Ziaei-Rad Z, Fooladi J, Pazouki M, Gummadi SN (2021) Lignocellulosic biomass pre-treatment using low-cost ionic liquid for bioethanol production: an economically viable method for wheat straw fractionation. Biomass Bioenerg 151:106140
91. Zlotorzynski A (1995) The application of microwave radiation to analytical and environmental chemistry. Crit Rev Anal Chem 25(1):43–76. https://doi.org/10.1080/10408349508050557

Current Status of Substrate Hydrolysis to Fermentable Sugars

Olayomi Abiodun Falowo, Abiola E. Taiwo, Lekan M. Latinwo, and Eriola Betiku

Abstract This chapter examined different substrates used in the hydrolysis process to produce fermentable sugars. The substrates are classified under the generation biofuel produced to show a progressive substrate discovery and developmental shift over the years. Substrates under different generation biofuels are characteristically different in composition and have unique processing routes to fermentable sugars. The pretreatment methodologies applied before producing fermentable sugars are reviewed under physical, chemical, and biological techniques or a combination. Turning first-generation feedstocks into fermentable sugars is currently the most straightforward, advanced, and economical technology. Second-generation feedstocks (lignocellulose biomass) are abundantly available with no concern for food competition, but the associated high operation costs have limited their full application profitably. The third-generation feedstocks possessed huge potential, but the processing technology is still maturing and yet to be fully commercialized. Identifying microalgal strains capable of ensuring high fermentable sugars yield at low cost remains a major task. Hence, developing fourth-generation feedstocks (genetically engineered substrates) solves the challenges of utilizing algae biomass as substrate. Substrate combination is a promising strategy to improve the process efficiency, just as the integration of pretreatment techniques proves cost-effective. Thus, adopting this process development and modification would advance the large-scale operation of bioethanol production.

O. A. Falowo (✉)
Department of Chemical Engineering, Landmark University, P.M.B. 1001, Omu-Aran, Kwara, Nigeria
e-mail: falowo.olayomi@lmu.edu.ng

A. E. Taiwo
Faculty of Engineering, Mangosuthu University of Technology, Durban, South Africa

L. M. Latinwo · E. Betiku
Department of Biological Sciences, Florida Agricultural and Mechanical University, Tallahassee, FL 32307, USA

E. Betiku
Department of Chemical Engineering, Obafemi Awolowo University, Ile-Ife 220005, Osun, Nigeria

© The Author(s), under exclusive license to Springer Nature Switzerland AG 2023
E. Betiku and M. M. Ishola (eds.), *Bioethanol: A Green Energy Substitute for Fossil Fuels*, Green Energy and Technology,
https://doi.org/10.1007/978-3-031-36542-3_4

Keywords Feedstock · Pretreatment. · Assisted pretreatment · Substrate · Hydrolysis · Acid hydrolysis · Enzymatic hydrolysis · Fermentable sugars

1 Introduction

The shifting in the usage of fossil fuels to biomass fuel arises due to one or a combination of the factors such as rural development, climate change, and energy security. Fuels from biomass sources are renewable and environmentally friendly due to their carbon neutrality. It has been acknowledged that plant-based raw materials (biomass) have enormous potential to reduce dependence on fossil fuels as feedstock for transportation fuel. Biofuel, liquid fuel obtained from biomass, is classified based on the feedstock from which they were obtained. The bioethanol produced from starchy food, non-food crops, and lignocellulosic materials is known as the first, one-half, and second generation, respectively. In contrast, bioethanol from microalgae and genetically modified feedstock are called third- and fourth-generation fuels. There is strong opposition regarding substrate from first-generation feedstocks because food crops are essential for growing populations. The first-generation biofuel competition with food and feed raises ethical and environmental concerns [9]. The disadvantages of using food crops to generate biofuel led to the advancement of second-generation feedstock. The alternative to food crops with no competing interest is to consider non-food-based substrates and lignocellulosic materials. The main benefit of adopting the technology is that biofuel can be generated from different non-food crops. Lignocellulose biomass can be agricultural residues, designated lignocellulose crops, forestry, and industry leftovers. Currently, second-generation biofuel has not satisfactorily fulfilled the expectation to meet energy demand due to bottlenecks in its processing. The biorefining technologies for second-generation biofuel are still developing, and the whole process is capital-intensive on a large scale. Though second-generation fuel is a highly promising bioenergy, improving processing technology is vital before it can be adopted as a transportation fuel [34]. Algal biomass is another class of feedstock capable of generating bioethanol comparable in volume to second-generation feedstocks but with no attendant challenges limiting the large-scale operation of lignocellulosic biomass fuel. Biofuel from microalgae is known as a third-generation biofuel since it overcomes the shortcomings associated with the previous generations of biofuel in agricultural land and freshwater utilization. Microalgae have a rapid growth rate and fast harvesting cycle compared to classical lignocellulose biomass [54], which designates algae biomass as a sustainable feedstock for bioethanol. Similar to lignocellulosic fuel, the operation and processing cost of algae biofuel is a major barrier to its implementation to date. Moreover, improvement in third-generation fuel was made possible through metabolic engineering. The fourth-generation bioethanol is produced from genetically-modified algae or crops. The fourth-generation bioethanol is reported to be an improved algae fuel [80] because of the ability of metabolic technology to produce high carbohydrate concentrations through genetic manipulation. However, third- and fourth-generation biofuels are still in their infancy. Generally, each generation of bioethanol has benefits and drawbacks. The ideal bioethanol generation route should be able to integrate the advantages of each fuel generation from a viable economic standpoint. Sustainable development of bioethanol would require the utilization of non-controversial feedstocks, and all generation feedstocks apart from the first-generation feedstock met this criterion, with lignocellulose biomass taking the lead in this regard [32]. Due considerations in terms of economic, social, ethical, and environmental aspects are vital in ensuring sustainable bioethanol generation. Integrating these factors in biofuel development may position bioethanol to replace fossil fuels as a transportation fuel fully. All these feedstocks from the first to fourth generation must be broken down into fermentable sugars before bioethanol is produced through enzymatic fermentation. The process of producing fermentable sugars from a substrate is known as hydrolysis. However, the lignocellulose biomass and microalgae require initial processing, commonly known as pretreatment. The pretreatment sometimes converts polymeric carbohydrates into sugars that microbes or enzymes can easily digest for bioethanol production.

2 Feedstocks Classification

Substrates for bioethanol production can be classified based on the biofuel they produce. The need for food security and improvement in processing technology have resulted in using different substrates for bioethanol generation. Hence, the substrate can be categorized under first-, second-, third-, and fourth-generation feedstock. This classification is simple, and it avoids ambiguous categorization of a substrate.

2.1 First-Generation Feedstock

First-generation feedstocks are mainly agricultural crops. The substrates under this category are food crops consumed by man or used as animal feed. To date, first-generation bioethanol still accounts for most bioethanol produced globally, with the United States and Brazil contributing over 80% of the total production worldwide [93]. In first-generation bioethanol, food crops such as sugar, grains, and tuber roots serve as substrates. The chances of producing bioethanol from these substrates are high, and the processing technologies have advanced. However, the emergence of competition between food crops and energy, as well as its resultant effect on agricultural land, are the major obstacles [51]. It is widely acknowledged that biomass fuel from these substrates is more expensive than fossil-derived fuel, and the long-term effects are undesirable. The main advantages of these feedstocks are their high sugar concentration and ease of conversion into bioethanol.

2.1.1 Sugar

The first-generation plant utilizes starch-based feedstock from sugarcane to generate bioethanol. Generally, most bioethanol produced in Brazil comes from sugarcane. The major raw material for bioethanol production in India is sugarcane. Likewise, China generated a considerable amount of sugarcane-based bioethanol [40]. Sugarcane is a C4 carbohydrate with a photosynthetic pathway sufficiently high in converting light energy to biomass [52]. Sugarcane is from the grass family known as *Poaceae*, and it contains 57% of water, bagasse, straw, and sugar. Globally, sugarcane produces quality raw materials for sugars such as glucose, fructose, and sucrose. The composition of sugarcane juice as well as the production volume, certainly make it a desirable substrate for first-generation bioethanol. In addition to its rich carbohydrate content, it does not require a pretreatment step which further removes extra operational costs [93].

2.1.2 Grains

Starchy feedstocks such as corn, wheat, and sorghum have been used to produce first-generation bioethanol. As of 2022, a large percentage of bioethanol produced in the United States comes from corn. European countries and Canada utilize wheat as a raw material for bioethanol [78, 79]. Bioethanol is produced from corn grains either by wet milling or drying milling. The milled corn is transferred to a liquefaction tank to produce a mash, and enzymes are added at the stage to convert it to monomeric sugars. A direct relationship was established between the grain varieties and ethanol production. Grain corn for ethanol generation is efficient in high crops, producing 8.0 t/ha per hectare. Moreover, increased amounts of amylose in starch and granule size affected the enzymatic activity of amylase in hydrolyzing starch, subsequently influencing bioethanol production [37]. Another cereal-based bioethanol is from sweet sorghum, the substrate is a C4 plant, and it is highly rich in sugars and starch. Sweet sorghum is comparable to sugarcane in sugar content, yield per hectare, and ethanol production [2].

2.1.3 Tuber Roots

Carbohydrates, such as cassava, yam, and sweet potatoes, are another traditional substrate used in producing first-generation bioethanol. The renewable energy sector in Taiwan has sweet potatoes as the main raw material for bioethanol production. Root tubers contain a high concentration of carbohydrate starch and sugars in the case of sweet potatoes, and they can produce a large quantity of bioethanol [104]. Cassava is high in nutritional value because of its tuberous roots, and it is a staple food in Africa, Asia, and South America. Bioethanol production from cassava, when compared to corn-based ethanol, was reported to be higher than the three corn starches used for the fermentation [67]. The utilization of root biomass in producing bioethanol has been implemented to varying degrees, especially in Asia. Most root tubers used for bioethanol are not easily affected by drought and do not necessarily require huge farming practices to be maintained.

2.1.4 Non-food Industrial Products

(a) **Sugarcane molasses**

This substrate is rich in residual sugars since it is a liquid byproduct of sugar refining. Molasses contains a high concentration of fermentable sugars such as sucrose, fructose, and glucose. This substrate does not require an expensive pretreatment step since it is already in useable form. The substrate can be recycled from sugarcane processing and reused for bioethanol production [93].

(b) **Vinasse**

Vinasse is the liquid byproduct from biomass distillation of bioethanol production from sugarcane juice or molasses. This vinasse residue is rich in an organic compound, and this substrate can be recycled for fermentation purposes [32]. In sugarcane mills, a huge volume of vinasse is produced per one liter of bioethanol generated. Though bioenergy development on this substrate may not stand alone, it can still contribute to energy demand in the sugar mill industry.

2.2 Second-Generation Feedstocks

Lignocellulose biomass is a non-edible renewable feedstock that can serve as a substrate for chemicals and biofuel generation. Unlike food crops, these feedstocks are known as the dry matter of the crops due to their component. They are potential substrates for bioethanol production if the inherent lignin-cellulose-hemicellulose matrix is broken [42]. Lignocellulose is a naturally occurring biomass, and it is available in many regions. Apart from the availability of lignocellulose residues, other factors impacting potential bioethanol production are residue dryness, recovery, and theoretical ethanol yield [44]. Lignocellulose biomass can be categorized under the following divisions; agriculture residues, forestry products, industrial leftovers, and household solid wastes.

The advantages of using lignocellulosic biomass for bioethanol generation;

i. It contains a high level of cellulose and hemicellulose
ii. It is a renewable feedstock
iii. It is abundantly available
iv. It is relatively low cost since most are waste residues
v. It does not compete with food or animal feed.

Examples of cellulosic feedstock are wheat straw, rice husks, corn leaves, rice straw, corncobs, empty oil palm fruit, sorghum residue, and sugarcane bagasse. The lignocellulose biomass in residues or wastes is available in large quantities, meaning this feedstock class can contribute substantial biofuel to replace fossil fuel. Even though second-generation substrates are abundantly available, their processing technology is still developing [34]. The main characteristic of these feedstocks is in their composition, consisting of hemicellulose, cellulose, and lignin.

2.2.1 Agricultural Residues

The farming, harvesting, and processing of agricultural products generate a huge volume of waste that can be utilized in producing bioethanol. The current agricultural practices of disposing or treating agricultural residues are through burning and sometimes used for cooking. This waste management technique is unhealthy for the environment due to the various pollutants released. As a substrate, agricultural residues are available in various degrees, especially in rural communities due to high farming activities. With appropriate pretreatment, enormous fermentable sugars can be obtained from these substrates in the biorefinery. After harvesting crops, residues or wastes from jute, cotton, wheat, corn, sugarcane, and rice are potential energy-generated substrates. Figure 1 depicts different agricultural remains produced after the processing of food crops.

Fig. 1 Agricultural residues **a** Rice husks; **b** Corncobs; **c** Wheat straw; **d** Fruits wastes; **e** Cotton stalks; **f** Sugarcane bagasse

2.2.2 Sugarcane Stalks and Bagasse

Sugarcane is a crop cultivated in tropical and subtropical regions worldwide. The feedstock has a high sugar and lignocellulosic residues yield, making it an excellent bioethanol source. Sugarcane bagasse is the lignocellulosic residues left after the extraction of liquid sugar, and a huge amount is generated from every ton of processed sugarcane. Recently, cellulosic bioethanol has been produced from sugarcane bagasse in Brazil. On the contrary, China, one of the largest sugarcane producers, has yet to explore the enormous potential of sugarcane-based bioethanol fully [40]. The issues hindering commercial-scale bioethanol production from sugarcane in China are labor costs, limited farmland, infertile soil, and lack of new sugarcane varieties [40]. However, the prospect of financial return and instability in the energy market have caused substantial investment in the feedstock [93]. This results in a high volume of bagasse generation, a substrate capable of producing a large volume of bioethanol fuel. During sugar production, millions of tons of sugarcane bagasse are generated and converted to generate heat and energy, which can be applied in many industrial processes [44].

2.2.3 Rice Straw

Rice is one of the foods most consumed by humans globally, and rice straw is the most abundant lignocellulose biomass. Rice processing generates a huge waste that can serve as a substrate for bioethanol. Rice straw is a byproduct of rice processing. The fuel obtained from rice straw is considered non-threatening to the food supply and can potentially generate a high volume of bioethanol since it has relatively low lignin content [7]. An alkali-pretreated rice straw used in bioethanol generation indicates that biomass can be an ideal substrate for producing bioethanol [42].

2.2.4 Cotton Stalk

The cotton stalk is a major byproduct of cotton production, and it is generated in huge volumes annually. Biofuel such as bioethanol can be produced from this substrate since it is an inexpensive source of sugar. Cotton is an important crop that plays a major role in the economic and social activities of the world. Since cotton stalk is mainly lignocellulose and particularly high in holocellulose content, different pretreatment techniques or combined method is first applied on the substrate to release fermentable sugars [25].

2.2.5 Corncobs/Corn Stover

Corncob is another promising substrate for bioethanol production since it is highly rich in cellulose. Corn is the second most cultivated cereal plant globally, with the United States and China atop the maize-producing nations. This cereal crop is processed into livestock feed, food, and bioethanol, leaving a large quantity of corncob waste. Corncobs have unique lignocellulose composition with high cellulose and xylan and low lignin and ash contents [34], making them an excellent substrate for biorefinery. Corncob is first pretreated with suitable techniques to release fermentable sugars. Meanwhile, most corncobs-based bioethanol employs either acid or alkali pretreatment or a combination of both techniques. Corn stover is another feedstock in a biorefinery with tremendous potential for fermentable sugar production [1].

2.2.6 Wheat Straw

Wheat straw is the major substrate used in several European biorefineries to produce bioethanol. Countries such as Denmark, Spain, Germany, and the UK have already installed wheat straw facilities for bioethanol production [96]. The feedstock is abundantly available at a low cost compared with other lignocellulose biomass. Wheat straw contains almost 60% of residual sugars on a dry basis. It is a promising renewable substrate capable of generating several billion liters of ethanol since approximately 850 million metric tons are produced annually worldwide. Wheat straw has cellulose and hemicellulose composition similar to other agricultural residues such as bagasse, rice straw, and corn stover [34].

2.2.7 Fruits Residues/Peels

Fruits peels and residues are other categories of agricultural wastes that are potentially used in producing fermentable sugar for bioethanol. Though fruit peels/residues such as cashew nut, coconut shell, watermelon flesh, pineapple flesh, grapes seeds, coffee seeds, and palm kernel shell are agricultural residues [9], generally, they are not classified as lignocellulose biomass because of the absence of structured lignin-cellulose complex. After juice extraction, the semi-solid residues generated during processing are potential feedstocks for biofuel production. The management of fruit waste is another challenging task for the processing industries, and the disposal of this waste through landfilling method is dangerous as it can lead to pathogen proliferation and disturbance in environmental microbiota [30]. In several developing countries, fruit waste is generated at an alarming rate with low or no investment. This category of biomass constitutes part of the largest available agricultural wastes that can be processed to obtain bioethanol and other biofuels.

2.2.8 Wood and Forest Residues

Woods and forest residues such as sawdust, softwood, hardwood, wood chips, and bark residues (Fig. 2) are promising substrates in biorefinery industries to generate bioethanol. Due to forestry harvesting and processing from activities such as cutting,

Fig. 2 Forestry residues **a** Sawdust; **b** Wood log back; **c** Wood chips

edging, sizing, and re-sawing, many forestry waste products are generated, which can be converted to bioethanol [72]. Sawdust is the main residue of forestry wood processing, and between 45 and 65% of sawdust is generated in a sawmill per ton of wood [94]. The percent composition of lignin in rubberwood sawdust is almost equal to the hemicellulose and cellulose amount. The pretreatment of residues in this category is usually through thermochemical technique, and it possesses enormous fermentable sugar due to its high cellulose content.

2.2.9 Industrial Leftover

An industrial processing facility generates many sugar-containing waste products or side streams. These waste products are good candidates for fermentable sugar generation. Sugar, breweries, and wine-making facilities produce a significant amount of bagasse, spent grains, and vinasses which can serve as a substrate for bioethanol generation [32]. These substrates are highly rich in cellulose and, to some extent, contain a significant amount of lignin. In the processing or production of a product, the side streams and byproducts can be recycled and reused to generate fermentable sugars [6].

2.2.10 Brewer Spent Grains

Global beer production is in several billion hectoliters, and brewers' spent grains, the byproduct of beer-producing industries, account for 85% of the waste products. This substrate is a lignocellulose biomass rich in cellulose, hemicellulose, and lignin. Before it can be used for bioethanol, its moisture content must be reduced to < 55% and then pretreated by suitable techniques to release its sugar [32].

2.3 Third-Generation Feedstocks

Third-generation bioethanol is produced using macroalgae and microalgae (Fig. 3). This feedstock category is a promising substrate for biofuel production. Algae biomass can serve both as biofuel feedstock and as a food supplement. These substrates are widely believed to have distinct advantages over previous generational feedstocks. The high growth rate, all-year-round production, less pressure on freshwater, wastewater bioremediation, CO_2 sequestration, tolerance to harsh conditions, and high lipid and carbohydrate contents are the merits that make algae biomass a promising source of biofuel [27]. Like lignocellulose biomass, algae biomass requires pretreatment to obtain carbohydrate fraction. The cell disruption and extraction techniques used in the pretreatment stage can be any of the following; biological (enzymatic, autolysis), mechanical (bead mill, homogenizer), chemical (detergent, chelating agent, solvent-based, supercritical), and physicochemical (cell disrupted-assisted solvent extraction) [43].

2.3.1 Macroalgae (Seaweeds)

Seaweeds are multicellular marine macroalgae, broadly grouped as green, brown, and red, based on the photosynthetic pigment in their thallus. The marine macroalgae are fast-growing biomass with a high quantity of carbohydrates. As biofuel feedstock, it does not depend on the land, does not compete with food crops, or requires freshwater, which are the associated challenges of first- and second-generation feedstocks. Not all seaweeds can serve as a substrate for bioethanol. The bioethanol yield obtained from macroalgae biomass is proportional to the total carbohydrate present [73]. The higher the carbohydrate content in the seaweed cell wall, the higher the bioethanol production from the substrate. Seaweed consists of carbohydrates in the form of cellulose and starch, and they are converted to bioethanol by appropriate microorganisms, namely fungi, yeast, or bacteria. In converting macroalgae

Fig. 3 a Macroalgae and b Microalgae

to bioethanol, three common processes are usually involved: pretreatment, hydrolysis, and fermentation. Due to their distinctive features, green seaweeds are sustainable substrates for bioethanol production. Green seaweeds such as *Lactuca, Ulva intestinalis*, and *Enteromorpha* can grow under various environmental conditions, including salinities and temperature [54]. The cell wall is the primary source of carbohydrates, and the amount of carbohydrates in green seaweeds varies from one specie to another. A major drawback in utilizing macroalgae for ethanol production is the inability of microbes to ferment all sugars present in the seaweeds. A pretreatment of macroalga biomass releases various mixed sugars, including glucose, galactose, xylose mannitol, and rhamnose, some of which cannot be utilized by fermenting microbes.

2.3.2 Microalgae

Microalgae biomass is an attractive substrate for bioethanol production. This substrate contains three polymeric components in different proportions depending on their strains; carbohydrates, proteins, and lipids. Microalgae has higher lipid and protein content than macroalgae. After lipid extraction, starch hydrolysis is initiated by amylolytic enzymes to generate fermentable sugars, which are subsequently converted into bioethanol. Microalgae have a very short harvesting cycle and can be cultivated and harvested continuously within a year. Microalgae are photoautotrophic biomass that can grow rapidly and generate 20–300 times more oil than conventional biomass crops [80].

2.4 Fourth-Generation Feedstocks

There are a lot of technical limitations in the utilization of algae biomass for producing bioethanol, and issues such as low production output and high operational costs limit its expansion. These bottlenecks are resolved using metabolic engineering on photosynthetic microorganisms. Algae metabolic engineering procedure for bioethanol production uses recombinant DNA and other bioengineering techniques to modify the properties and cellular metabolism of algae biomass to improve biofuel generation [53]. The substrate under this category can be cultivated on a large scale to serve a commercial purpose. Examples of algae metabolic engineering substrates are Cyanobacteria (*Anacytis nidulans R2, Synechocystis* sp. *PCC 6803*), eukaryotic microalgae (*Chlorella vulgaris, Chlamydomonas reinhardtii*). The manipulation of the cellular activities of the photosynthetic algae makes them promising substrates for biofuel generation. These algae biomass are configured and constructed to be an improved substrate over third-generation feedstocks through engineering approaches. Though substrates under this category are still in the infancy stage, the enormous potential of generating a huge volume of bioethanol from them cannot be overlooked. Several microalgae strains have been improved through genome editing to potentially serve as bioethanol substrates, species such as *Chlorella vulgaris* UTEX395, *Tetraselmis* sp. *Phaeodactylum tricornutum, Chlamydomonas reinhardtii, Nannochloropsis Oceania, Anabaena* 7120, *Synechoccus* UTEX 2973, and *Synechocystis* 6803 have been modified for biofuel application [80].

3 Pretreatment Defined

In producing bioethanol from lignocellulosic materials, pretreatment is the first critical step and the most challenging stage. This step constitutes over 40% of the total production cost [81]. A pretreatment process includes mainly physical operations and thermochemical processes. All pretreatment operations can be classified under three categories: physical, chemical, and biological pretreatment. The physical step is mainly concerned with biomass-size reduction, while the thermochemical operation deals with the recalcitrant nature of the biomass through breakage and perturbation of bonds. This upstream operation enables the substrate to be highly porous, creates lignin redistribution, and ensures maximum exposure to cellulolytic enzymes. The pretreatment involving biological entities degrades lignin and hemicellulose from lignocellulose material without generating inhibitors. Even though lignocellulose biomass is naturally abundant, microbial and enzymatic conversion to bioethanol is impossible because of the recalcitrant structure of the substrate. The matrix of lignin and hemicellulose cements the cellulose in lignocellulose biomass. This arrangement prevents the accessibility of water, enzymes, and chemical through the substrate, thereby reducing the surface area available to enzymes and microbes.

The barriers around cellulose, including lignin and hemicellulose, are removed through the pretreatment step. This step alters the macroscopic, microscopic, and submicroscopic structure, size, and chemical composition of the lignocellulose biomass. A successful pretreatment facilitates rapid carbohydrate polymers breakdown to simple sugars and increases the number of fermentable sugars available for bioconversion (Fig. 4). For any pretreatment step to be commercially viable, the approach must be optimally effective with low operation costs [92]. The various factors responsible for the resistance of biomass to enzymatic hydrolysis include accessible surface area, cellulose crystallinity, lignin sheathing, and cellulose overlay.

3.1 Physical Pretreatment

Physical pretreatment involves high temperature, high pressure, irradiation, milling, and steaming to cause structural changes in the lignocellulosic biomass to reduce its recalcitrance nature [84]. Generally, substrate structure and particle size are usually reduced by grinding or milling lignocellulose biomass in most physical pretreatment stages. At times, a temperature between 180 and 240 °C is applied in the physical pretreatment of lignocellulose and mechanical shearing. This operation leads to the

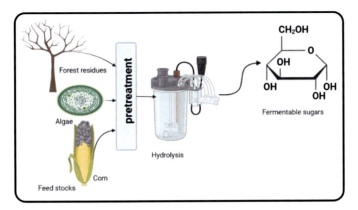

Fig. 4 Conversion of substrate to fermentable sugars

perforation of structure and increased pore surface. Generally, physical pretreatment causes changes in form, crystallinity index, specific surface area, polymerization degree, or biomass particle sizes. Residence time, feedstock type, operating pressure, and processing temperature affect physical pretreatment [92]. The common methods for biomass conversion under physical pretreatment are mechanical, microwave, and ultrasound.

3.1.1 Mechanical Pretreatment

Mechanical pretreatment of lignocellulose biomass usually involves size reduction by grinding, milling, and extrusion. The crushing mechanism alters the surface area, porosity, and degree of crystallinity of ground biomass. Different mechanical treatment of biomass can be achieved using various types of milling techniques, viz. ball milling, wet disk milling, two-roll milling, hammer milling, and rod milling, depending on the properties of the biomass [82]. Several works used ball and wet disk milling methods to cause physicochemical changes to the biomass prepared for bioethanol generation. The downsizing of biomass achieved during mechanical treatment will not separate lignin and hemicellulose from the cellulose. However, the structural alteration increases the contact surface area with the subsequent material, initiating hydrolysis. The demerit of mechanical pretreatment is the high-power consumption cost involved.

(a) **Ball milling**

Ball mills such as tumbler, planetary, attritor, and vibratory are designated based on the movement mode of the balls [82]. A planetary ball milling of oil palm empty fruit bunch and frond fiber was applied as a pretreatment method to evaluate the percentage of simple sugars generated for enzymatic hydrolysis. An extended oil palm biomass milling time resulted in reduced crystallinity index and particle sizes. The glucose and xylose amounts generated from the oil palm biomass pretreatment ranged from 70 to 87% and 81.6 to 82.3%, respectively [101]. The ball milling technique changes the microstructure of lignocellulose biomass by disrupting the compact structure and opening the substrate cell wall, thereby increasing the fermentable sugar yield. Furthermore, the biomass crystallinity is destroyed after ball milling it since the shade of lignin protecting the cellulose is ruptured. The huge disruption of this structure increases the biomass surface area and porosity, which is beneficial to reacting materials like enzymes or chemicals [102]. The physicochemical properties of the substrate using ball milling depend on the milling time, operation conditions, working temperature, milling modes, biomass properties, filling volume, and rotational speed of grinding pots [82].

(b) **Wet disk milling**

Wet disk milling is another technique used to pretreat lignocellulose biomass as it increases the degree of defibrillation. The biomass to be hydrolyzed is mixed with water and transferred into a grinder with revolving disks. The milled substrate slurry is transferred into the grinder again, and the process can be repeated up to 20 times. Wet disk milling pretreatment was applied to release fermentable sugars from rice straws. This operation leads to defiberization and decreased fiber thickness, hence increasing the surface area of the substrate for enzymatic fermentation [39]. The energy consumption of wet disk milling is low compared to ball milling pretreatment. However, higher sugar yield was released from ball milling. Besides the high effectiveness of wet disk milling, no fermentation inhibitors are generated during this operation.

3.1.2 Extrusion

Extrusion is a thermomechanical pretreatment technique to fractionate lignocellulose biomass. Extrusion is a physical technology involving mixing, heating, and shearing substrates to produce fermentable sugars. It works at high solid concentrations and can be combined with other pretreatment methods. Extrusion of biomass can be carried out with hot water, base, acid, steam, ammonia, and alkaline solution. Extrusion involves one or two screws spinning into a tight barrel coupled with temperature control. This action creates high shearing forces among the feedstock, the screw, and the barrel, accompanying pressure and temperature rise along the extruder [26]. The extrusion machine can be a single solid extruder or a twin-screw extruder. Extrusion pretreatment of biomass increases surface area and micro-fibrillation of substrates.

3.1.3 Pulse Electric Field

Pulse electric field pretreatment is the technology that causes electroporation of biomass by applying a very short eruption of high voltage. The substrate to be permeabilized is placed between two electrodes while a high-intensity electric field through induction is applied across its cell membrane [46]. The transfer of fast electric voltage leads to the structural changes of the cell as well as mechanical rupture of biomass tissue and subsequently increased membrane permeability. The benefit of the operation is the rapid pretreatment time compared with other methods. Besides the short duration, energy consumption is much lower than mechanical and thermal pretreatment [99].

3.1.4 Microwave-Assisted Pretreatment

Microwaves have historically been recognized in reactions, though their significance and positive effects have evolved with a typical working wavelength of 12.2–33.3 cm or 2.45–900 MHz [22, 63]. Under these settings, an electromagnetic field generated by microwave radiation results in energy-efficient internal homogenous heating. This thermal effect allows for a faster reaction rate because of dipole rotation in the presence of the oscillating electric field of ionic conduction and microwave radiation via swift translational movements. In contrast to conductive heating, the thermal mechanisms of rate acceleration include the creation of so-called molecular radiators, the superheating effect of solvents, selective heating using a microwave reagent or catalyst in a less polar reaction medium, and elimination of the wall effects brought on by convective heating. At times, enzyme hydrolysis is used extensively to produce fermentable sugar [89]. This method typically takes longer, costs more money, and uses more energy. Microwave-aided treatment is a potentially quicker method for thermally treating biomass at low pressure and temperature [19, 89]. The main benefits of microwaves are a regulated environment, quick and efficient heating, faster processing rates, and more than 80% quicker reaction times [45, 77]. Microwave pretreatment techniques can be used alone or combined with other pretreatment techniques. Acid hydrolysis combined with microwave irradiation is an energy-efficient, quick, and affordable alternative for producing bioethanol from various biomass.

3.1.5 Ultrasound-Assisted Pretreatment

Ultrasound-assisted irradiation is another step to support the biological conversion of waste materials and lignocellulosic materials to biofuels. Ultrasound is a term used to describe sound waves with frequencies > 20 kHz that move through a given medium in a series of compressions and rarefactions [16]. Gas and vapor bubbles form as the waves pass through areas of low pressure, and they then grow larger until they reach a critical diameter, at which point they implode.

In addition to its thermal effects, bubble implosion produces powerful shearing forces that may help in cellular disruption and membranes, including the fractionation of lignocelluloses. At the same time, oxidative radicals produced by the splitting of molecules due to cavitation may also aid in the oxidation of organics and their solubilization. The aggregate of these effects of ultrasonic irradiation adds to the breakdown of lignin structure, the cellulose structure dissolution, depolymerization, solubilization, the cell walls and membranes lysis, and improved organic matter solubilization [16]. It has been used as a stand-alone pretreatment procedure or in conjunction with other treatments. Sugarcane bagasse was hydrolyzed using sulfuric acid assisted with ultrasound, and it was reported that acid concentration rate and sonication time affected the sensitivity of the process [23]. The ideal processing settings were a liquid-to-solid mass ratio of 20:1, an acid content of 2% (g/100 mL), and a sonication pretreatment time of 45 min. The biomass was pretreated and hydrolyzed using ultrasound-mediated acid hydrolysis in another study. About 32 g of glucose were produced per 100 g of rice straw after 50 min of sonicated treatment at 80 °C with an acidic concentration of 10%. It has been demonstrated that sonication facilitates floc disintegration, cell lysis, and cellulose molecule rupture during cavitation, decreasing particle size and increasing surface area [16, 76].

Hay et al. [38] reported that about 19% of the lignin concentration in paper manufacturing mill effluent was reduced by sonication. Li et al. [49] discovered that sonication facilitated the breakdown of bamboo lignin when an ethanol medium was used. Mohapatra et al. [57] reported more than 80% lignin removal efficiency for annual tufted and Napier grass following ultrasound-assisted acid pretreatment. The authors discovered that increasing the sonication power boosted the efficacy of delignification. The breaking of hydrogen bonds in the cellulosic structure during cavitation weakens the cellulose morphology and causes the cellulose crystalline structure to rupture. This is due to very high temperature and pressure conditions during cavitation. Due to floc breakage and cell lysis brought on by heat effects, high shearing pressures, and the oxidative effect of radicals created during sonication, organic matter is dispersed [16, 59].

3.1.6 Hydrothermal or Autohydrolysis

Hydrothermal hydrolysis, also called autohydrolysis, is a low-energy and cost pretreatment method for deconstructing substrates and hydrolyzing internal compounds, enhancing fermentation biofuel output [20]. Nevertheless, fragile continuous reactors, variable rheological properties of substrate slurry, and complex reaction mechanisms impede its commercialization. Hydrothermal hydrolysis is a type of pretreatment technology that can use waste heat and operates under relatively mild conditions [50, 103]. During residence time, hydrothermal hydrolysis is typically carried out at pressures between 2 and 6 MPa and temperatures between 100 and 260 °C [56]. In hydrothermal hydrolysis of algae, cells in the slurry undergo several physicochemical changes, including cell wall deformation and organic matter dissolution from algal cells to the liquid phase, polymerization, hydrolysis, and accompanied side reactions.

The hydrothermal hydrolysis technique has a significant advantage in terms of energy efficiency because it immediately transforms wet algal biomass without the energy-intensive dehydration step. The flow properties of algal slurry, a non-Newtonian fluid type in hydrothermal hydrolysis reactors [20], are critical to selecting and controlling stirring paddle and pump in batch/continuous reactors, respectively.

The mixture of algae cells and water is subjected to high pressure and temperature conditions during hydrothermal hydrolysis. Under pressure, heated water in an algal slurry act as a reagent in different processes and a solvent for organic molecules. Hydrothermal fractionation systems are among the most promising lignocellulosic biomass-based biorefinery technologies [17]. This technique is promising because it does not require any chemicals, adheres to the sustainable biorefinery principles of green processing technology, effectively utilizes feedstocks to reduce waste generation, and minimizes energy use and environmental impact [103]. The hydrolysis is conducted via mechanisms catalyzed by hydronium in the autohydrolysis method, which utilizes lignocellulosic biomass and water. The cleavage of acetyl groups and the selective hydrolysis of heterocyclic ether bonds, both mediated by the hydronium ions created by water ionization, resulted in the depolymerization of the hemicellulose in the early stages of the reaction. When hemicellulose is hydrolyzed using autohydrolysis techniques, significant hemicellulose recovery occurs between 55 and 84%, and minimal inhibitory byproducts are observed [35]. It is necessary to use purifying procedures to increase the purity of the oligosaccharide-containing hydrolysate produced by autohydrolysis. Precipitation, extraction, surface-active material adsorption, membrane separation technologies, and chromatographic separation methods, including nano- and ultrafiltration, are some treatments that are used to accomplish this (or a combination of them).

3.1.7 Emerging and Combined Pretreatment Technology

Different approaches can be combined to pretreat the substrate, particularly lignocellulosic biomass, before hydrolysis and fermentation. The ideal pretreatment minimizes sugar loss, increases enzyme digestibility, and consumes minimum power [66]. The pretreatment is further boosted by new technology capable of producing higher efficiency than conventional heating. Applying ultrasound, hydrothermal, microwave, and pulsed electric field (PEF) to the substrate pretreatment have been adjudged to improve the hydrolysis stage [97, 98]. These technologies have been applied to the pretreatment of mostly lignocellulose materials and algae biomass, as shown in Table 1.

3.2 *Chemical Pretreatment*

Chemical pretreatment of biomass is the application of a chemical to deconstruct the crystalline structure of lignocellulose for easy release of fermentable sugars. In this technique, acid, alkali ozone, ionic liquid, and solvents (organosolv, deep eutectic) are applied to improve the digestibility of the biomass. This chemical pretreatment can be used alone or combined with other physical pretreatment techniques; even two chemical methods can be combined to degrade lignin from biomass.

Table 1 Pretreatment techniques and hydrolysates yield

Substrate	Pretreatment	Condition	Observation	Reference
Beech wood	Hydrothermal	High-pressure reactor, temp (130–220 °C), time (15–180 min)	70 wt% (glucose)	[62]
Chlorella sp.	Hydrothermal	High-pressure reactor, 1.5% sulfuric acid at 117 °C for 20 min, followed by α-amylase hydrolysis	51.06 g/L (total sugar)	[60]
Pleioblastus amarus (bamboo)	Hydrothermal	High-pressure microautoclave; temp of 160 °C for 60 min, followed by cellulase hydrolysis	52.7% (glucose)	[97]
Wheat straw	Hydrothermal	Pretreatment pilot reactor; 195 °C for 6–12 min	94% (cellulose)	[66]
Cotton stalk	Microwave-assisted	Alkaline/ethanol solvent at 160 °C for 15 min	82.41% (RS)	[20]
Rugulopterix okamurae (seaweed)	Microwave-assisted	Enzymatic hydrolysis at 220 °C for 20 min	35% (RS)	[31]
Kenaf powder	Microwave-assisted	Cholinium Ionic liquids solvent at 110 °C for 20 min, followed by enzymatic hydrolysis	90% (Cellulose conversion)	[61]
Corn straw	Microwave-assisted	95%v/v glycerol-NaOH solvent at 180 °C, enzymatic hydrolysis with Celluclast	18.28% (RS)	[24]
Chlorella pyrenoidosa NCIM 2738	Microwave-assisted	Time of 3 min, 700W and biomass concentration of 2 g/L	29.67 mg/L (Sugar)	[64]
Chlorella vulgaris ESP-31	Microwave-assisted acid hydrolysis	Wet torrefaction at 170 °C for 10 min, 0.2 M H_2SO_4	98.11 g/L (RS)	[98]

(continued)

Table 1 (continued)

Substrate	Pretreatment	Condition	Observation	Reference
Sweet sorghum bagasse	Ultrasound-assisted NaOH	Enzymatic hydrolysis (β-glucosidase) for 36 h at 120 rpm	92% (glucose)	[36]
Sweet sorghum bagasse	Ultrasound-assisted Phosphoric acid	Enzymatic hydrolysis (β-glucosidase) for 36 h at 120 rpm	81% (glucose)	[36]
Hybrid Napier grass	Ultrasonication-assisted NaOH	Time of 50 min at 40 °C, followed by Palkonal MBW Celluclast1.5 I + Xylanase hydrolysis	242.8 mg/g (RS)	[58]
Rice straw	Ultrasonication-assisted	Time of 15 min at 180 °C, followed by *Accellerase* 1500 hydrolysis	550 mg/g (RS)	[83]
Sugarcane bagasse	Sono-assisted alkaline	3% NaOH for 5 min, at 65 °C and 35 kHz	5.78 g/L (RS)	[28]
Spent coffee waste	Ultrasonication-assisted alkaline/KMnO$_4$	4% of KMnO$_4$ for 20 min, 47 kHz, followed by cellulase hydrolysis	35.15 mg/mL (RS)	[74]
Sugar beet tails	Pulsed electric field assisted	PEF intensity of 450 V/cm at the duration of 10 ms	8.9% (Sucrose)	[3]
Corncobs	PEF-assisted Alkaline (NaOH)	Field intensity of 9 kV/cm for 60 s	Increasing 40.59% (Cellulose)	[68]

RS—reducing sugar

A combined treatment of acid and alkali treatment of lignocellulose biomass can sometimes be used to facilitate microbial digestibility of substrate [42]. On the other hand, combining alkali and acid pretreatment techniques is found to have decreased overall water consumption by 10–15 folds [34].

3.2.1 Acid Pretreatment

Acid pretreatment of lignocellulose biomass involves using organic and inorganic acids to loosen the intertwined bond between lignin, hemicellulose, and cellulose.

The most common inorganic acids used are HCl, H_2SO_4, H_2SO_3, HNO_3, and H_3PO_4, while inorganic acids used in the chemical pretreatment include malic, oxalic, and acetic acids. At the mild condition of acid molarity, hemicellulose is easily solubilized from the substrate. However, a severe condition is required to unbound lignin and cellulose. Hemicellulose is easily solubilized from the substrate at the mild acid molarity condition [34]. Also, inhibitory and various side product formation is common in the acid-pretreated hydrolysate. Due to the negative impacts of inorganic acid, organic acids offer an improved route in producing fermentable sugar, however, with a lower yield.

3.2.2 Alkali Pretreatment

This involves using NaOH, $Ca(OH)_2$, Na_2CO_3, NaS, and aqueous ammonia to delignificate lignocellulose biomass. The mechanism of these chemicals involves cleavage of α, β-aryl ether linkages within the substrate, thereby causing detachment of lignin and cellulose swelling [34]. NaOH is the most applied alkali in the pretreating of the substrates. However, a further neutralizing process is required after hydrolysis. Applying $Ca(OH)_2$ in the pretreatment step is relatively cheaper and economical than NaOH since $Ca(OH)_2$ is easily recovered from the hydrolysate, although Ca salts may precipitate.

3.2.3 Organosolv Pretreatment

Using organic solvent is another chemical pretreatment method to fractionate internal bonds in lignocellulose. Organic solvents, including glycol, dioxane, ethylene, formaldehyde, acetone, ethanol, and formic acid, are used to separate polymeric lignin and depolymerize hemicellulose. In addition to organosolv, the organic acid is added to facilitate the dissolution of lignin. In this pretreatment, the organic acid serves as a catalyst so that the process can run at a lower temperature [41]. The benefit of the chemical method is the recyclability of the solvent to reuse it. However, the chance of forming a degradation product is high, and the pretreated substrate can be toxic to fermentation microorganisms [92].

3.2.4 Ionic Liquids Pretreatment

Ionic liquids are chemicals capable of solubilizing cellulose from lignin. Ionic liquid has inorganic anions and organic cations, making it an effective lignocellulose solvent. The ionic liquid can accept hydrogen bonds because of the interaction between its anions and hydrogen bond in lignin, thereby solubilizing cellulose from the lignocellulose structure. Ionic liquids include trifluoromethane sulfonate, 1-butyl-3-methylimidazolium acetate, 1-butyl-3-methylimidazolium chloride, and choline acetate have been utilized in biomass fractionation [41]. However, these chemicals are expensive and require multiple washing steps before enzymatic fermentation occurs.

3.3 Biological Pretreatment

Substrates pretreatment with metabolites of a microorganism for bioethanol production is a promising technology because of its various benefits, such as eco-friendliness, renewability, and economic viability. Besides being environmentally friendly, no inhibitors are generated for the following enzymatic fermentation process. Biological pretreatment can be applied to all feedstocks. For instance, sweet potato starch hydrolysis was reported to have decreased incubation time and increased sugar yield when mycelium concentration was increased [47]. For lignocellulosic biomass, pretreatment is carried out to remove lignin and hemicellulose bonded by covalent cross-linkages and non-covalent forces to the cellulose. Biological pretreatment involves using microbes such as soft rot, white and brown fungi to degrade the protective layers in lignocellulosic biomass. White rot fungi can efficiently degrade lignin in plant cell walls gaining access to holocellulose, consuming less energy, and generating less toxic compounds in the environment. The efficient degradation of lignin depends on the ligninolytic enzymes produced by basidiomycetes such as laccase, manganese peroxidase, and lignin peroxidase. The synergistic action of microbial consortiums, such as combined bacteria and fungi is highly desirable to improve the biodegradation of lignocellulosic materials. Factors including incubation time, temperature, aeration rate, moisture content, inoculum concentration, microorganism type, and nature of composition affect biological pretreatment [81].

Advantages of biological pretreatment

i. It is environmentally friendly.
ii. Biological pretreatment occurs in mild conditions.
iii. Economically viable for improving saccharification and inexpensive when compared with other pretreatment technologies.
iv. There is no need for recycling after pretreatment since chemicals are not used in the process.
v. It does not release toxic compounds or byproducts that interfere with the subsequent hydrolysis process.

Disadvantages of biological pretreatment

i. The process is very slow and not highly recommended for industrial application.
ii. There is a high operational cost associated with the large-scale implementation of the process since a sterile condition is a prerequisite.

3.4 Physicochemical Pretreatment

The challenges associated with separate chemical and mechanical pretreatment steps are tackled by coupling the mechanical size reduction method with a chemical treatment. Mechanical size reduction technique requires high energy, which increases its operational cost. The combination of these approaches helps to overcome the limitations peculiar to each method and thereby improve overall efficiency. The advanced chemical and mechanical pretreatment step is referred to as wet grinding. Similarly, the thermochemical technique is another method that combines the physical method with the chemical method. This involves thermal treatment in combination with acid or alkali treatment of a substrate. The acid or alkali usually acts as a swelling agent. This pretreatment approach was applied in deodar sawdust digestion [72]. A pretreatment study was conducted on rice straw using acid, alkali, ultrasound, and combination techniques. The combined method of acid treatment and ultrasonication techniques yielded the highest fermentable sugar of all other techniques before enzymatic fermentation [7].

4 Substrate Hydrolysis

Cellulose and hemicellulose form the bulk of the carbohydrates in lignocellulosic biomass. These components need to be degraded to produce the needed substrate, in which often extracted sugar may be transformed into ethanol or any other desired bioproduct [100]. Hydrolysis transforms biomass biopolymers into fermentable sugars [95]. At this stage, acids, enzymes, or an assisting medium coupled with an enzyme or an acid are used to hydrolyze polysaccharides in biomass resources into monosaccharide sugars. Cellulase or diluted or concentrated acid accelerates the conversion of the liberated cellulose to glucose. The biomass must be hydrolyzed to form simple sugars that can be converted by microbial activity into ethanol or any desired bioproduct [70].

4.1 Types of Hydrolysis

Acid and enzymatic techniques are commonly used to hydrolyze lignocellulosic biomass, with different performance levels based on biomass pretreatment factors, chosen feedstock, and hydrolyzing agent characteristics. The processing of different generation feedstocks to sugar is presented in Fig. 5. Acid hydrolysis is an established technology, but it has the disadvantages of producing dangerous acidic waste and recovering sugar from the acid. On the other hand, the enzymatic process is more efficient and operates under ambient circumstances (Table 2), producing no harmful

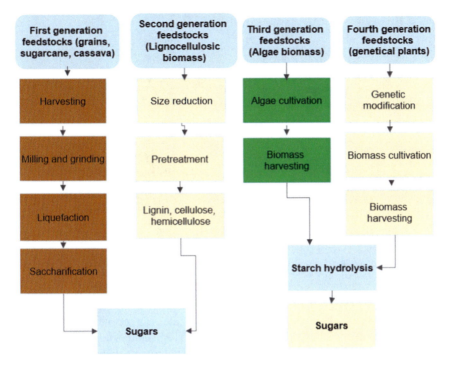

Fig. 5 Feedstocks processing stage to sugar

waste after hydrolysis. The latter strategy is rapidly evolving and has enormous cost and performance capacity [55].

Another hydrolysis method is through microbes. Some bacteria and filamentous fungi are cellulolytic microorganisms and can therefore produce cellulase and hemicellulose [88]. Further, different measures have been put in place to improve the rate of hydrolysis. This is usually referenced with the word "assisted." Acid and enzymatic hydrolysis have been combined with assisted measures like microwave, ultrasound, hydrothermal, and microbes.

4.1.1 Acid (Dilute or Concentrated) Hydrolysis

Hydrolysis with acid has a long industrial history, but they are expensive to operate and causes a variety of environmental and corrosion issues [100]. Acid hydrolysis uses an inorganic acid as a catalyst to break ether bonds [76]. In practice, dilute sulfuric acid is utilized because it is less corrosive to process equipment than hydrochloric acid. Dilute acid hydrolysis typically requires a temperature greater than 100 °C [70]. One of the simplest, oldest, and most efficient techniques for creating biomass ethanol is dilute acid hydrolysis. The biomass is hydrolyzed to sugars using dilute acid. The first step uses 0.7% sulfuric acid and a temperature of 190 °C to

Table 2 Hydrolysis type and total sugar recovery

Hydrolysis	Substrate	Condition	Sugar yield	References
Enzymatic	Cassava starch hydrolysate	α-amylase, glucoamylase, 24 h	45.01%	[13]
	Green seaweeds	Cellulase, 48 h	206.82 mg/g	[90]
	Bamboo	Cellulase, 50 °C, 150 rpm, pH 4.8, and 72 h	52.7%	[97]
Microbial	Rice straw	*Aspergillus fumigatus*	27.89 g/L	[42]
	Sweet sorghum	*Trichoderma harzianum*	10.34% (g/g)	[88]
Acid	*Chlorella vulgaris*	0.1 M sulfuric acid, microwave-assisted	98.11 g/L	[98]
	Soybean hull	2.7% H_2SO_4, 118 °C for 40 min	59%	[18]
	Gelidium amansii	2% (w/w) sulfuric acid	65 g/L (galactose)	[65]

hydrolyze the hemicelluloses available in the substrate. The second step is designed to produce a more resistant cellulose fraction [70]. The benefit of acid hydrolysis lies in its ability to penetrate the lignin in lignocellulosic biomass without pretreatment [23]. The hemicelluloses and cellulose are separated into different sugar molecules during the process.

The breakdown of crystalline cellulose is achieved through hydrolysis with either concentrated acid at low temperatures or dilute acid at high temperatures. Concentrated acid hydrolysis gives a high cellulose yield, but concentrated acid causes equipment damage. Conversely, hydrolysis at high temperatures with dilute acid accelerates the solubilization of hemicellulosic polysaccharides and the production of inhibitory substances such as furfural. Two-step dilute acid hydrolysis is used to overcome these downsides. In the first stage, the process is conducted at a moderate temperature (170–190 °C) to solubilize amorphous hemicellulose. In the second step, severe conditions (200–230 °C) were used to hydrolyze more crystalline cellulose. The use of acid in hydrolysis has significant downsides, including equipment corrosion, hazardous chemicals, and the additional neutralization step [23].

The earliest and most often used inorganic acids were HCl, H_2SO_4, HF, HNO_3, and H_3PO_4. Following this original finding, many dilute and concentrated organic and inorganic acids have been evaluated for their ability to hydrolyze lignocellulosic materials [71]. High sugar yields can be obtained from cellulose at low temperatures with concentrated acid. However, due to differences in intrinsic characteristics, the hydrolysis rate of crystalline cellulose is slower compared to amorphous hemicellulose. Therefore, pentoses and hexoses generated from hemicellulose are more vulnerable to breakdown to furfural and 5-hydroxymethyl furfural (5-HMF). These products are known to have inhibitory effects on the subsequent fermentation of the sugars when hydrolysis is conducted in one step. Concentrated acid also has other drawbacks, such as equipment deterioration, high acid consumption, environmental toxicity, and energy requirements for acid recovery. A two-stage sulfuric acid procedure has been developed to solve the problems with concentrated and diluted acids. Maleic acid has double pKa values that favor cellulose hydrolysis over glucose breakdown, which allows it to hydrolyze cellulose as effectively as sulfuric acid without promoting degradation processes. The primary benefit of acid hydrolysis is that it may dissolve lignin without first needing to treat biomass, dissolving hemicellulose and cellulose polymers to produce sugar molecules individually [95]. The two most popular catalysts for the hydrolysis of lignocellulosic biomass are sulfuric and hydrochloric acids [48, 69]. The amount of concentrated acid hydrolysis used in the procedure is between 10 and 30%. Low temperatures are used during the operation, which results in high cellulose hydrolysis yields (90% of theoretical glucose output) [95]. The primary problems of acid hydrolysis in the industry include the conversion of glucose into hazardous compounds, neutralizing hydrolysates before fermentation, and the high cost of equipment construction materials [87]. Fast reaction times, straightforward pretreatment of starchy biomass, easy accessibility, affordable acid catalyst, and relatively low reaction temperatures are all benefits of this method [89].

4.1.2 Enzymatic Hydrolysis

Like microbial breakdown of biomass, a critical component of the global carbon cycle, enzymatic hydrolysis is based on similar principles [100]. A conceptual framework for the enzymatic conversion of plants has been developed from the study of cellulose microbial degradation. [75] proposed that C1 enzyme first decrystallizes cellulose, followed by a Cx hydrolytic enzyme that converts the cellulose to sugar. Enzymes function in a more complex manner, according to further study. Enzymatic hydrolysis involves breaking down cellulose and hemicellulose using various enzymes isolated from microorganisms. Compared to cellulose, it is more effective at hydrolyzing hemicellulose [23].

Due to its higher stability and crystalline structure, cellulose cannot be easily depolymerized. The cellulase complex is used for hydrolyzing cellulose. Three key enzymes in the cellulase complex work together to hydrolyze cellulose into reducing sugar [21]. There are four types of glucanases: (i) endoglucanases, which act randomly on soluble and insoluble glucose chains; (ii) exoglucanases, which include glucan hydrolases that preferentially release glucose monomers from the end of the cellulose chain; (iii) cellobiohydrolases, which favorably release cellobiose; and (iv) glucosidases, which release soluble cellodextrins and cellobiose dimers produce D-glucose. These enzymes interact in a complicated interaction to efficiently decrystallize and hydrolyze native cellulose. Enzymatic hydrolysis is a technique for eliminating residual cellulose-enriched particles, yielding sugary liquids suitable for fermentation [14, 17]. This technique successfully modified the biomass composition and/or structure for it to work. The most easily accessible portion by the cellulase enzyme is cellulose. While hemicellulose solubilization should be linked to obtain the most efficient enzymatic hydrolysis process [23].

4.1.3 Microbial Hydrolysis

According to Tsegaye et al. [91], ligninolytic and cellulolytic microorganisms are typically used in microbial hydrolysis to produce active enzymes. The type of microorganism, substrate type, cultivation techniques, and hydrolysis conditions all affect the effectiveness of biological delignification [5]. Microbial hydrolysis is an inexpensive, effective, and environmentally friendly alternative to other hydrolyzing techniques. So many bacterial and fungal species are capable of degrading biomass made of lignocellulose [36, 42].

It is well known that fungi produce enzymes like laccase and peroxidases that break down lignin components found in lignocellulose biomass. Hydrolytic microorganisms can work synergistically to break down the polysaccharides in lignocellulosic biomass. The best examples of enzymes that degrade hemicellulose and cellulose entirely in biomass are xylanase and cellulase, respectively [87, 91]. The most effective microbial strains will generate more reducing sugars, increasing biofuel production [85]. The key variables influencing microbial delignification include the microbial strain utilized, the source and properties of the biomass, temperature, pH, moisture levels, inoculum concentration, culturing time, and air flow rate. The microbial strains are used to determine the required pretreatment period necessary for depolymerizing lignocellulose biomass. The critical challenge in microbial hydrolysis is the slow rate of removing lignin, which requires a longer pretreatment time.

5 Factors Affecting Substrate Hydrolysis to Fermentable Sugars

Temperature is a critical factor affecting hydrothermal hydrolysis reactions [33]. Temperature increase can cause algal cell deformation and release of organic materials. In addition, the temperature of the algal slurry is a major factor in the breakup and recombination of chemical bonds of organic macromolecules. Sonication aids in cell lysis, floc disintegration, and cellulose molecule breakdown [16]. Enzymatic hydrolysis and the production of sugars are consequently improved by the increased surface area available for microbial attack [4]. Several variables may substantially impact how quickly cellulose hydrolyzes in hot compressed water [100]. These elements can be divided into five categories: feedstock characteristics, hot compressed water characteristics, reactor designs, the time–temperature history that reacting particles have encountered, and catalysts and other additions [100]. In hot compressed water hydrolysis, reactor layouts also significantly affect hydrolysis. As a result of the hydrolysis products' prolonged residence in the batch-type system, glucose breaks down quickly into oil and char.

Reactor type is another factor affecting the hydrolysis yield. The degradation of sugar products can be decreased by using a flow-type system to accelerate the treating, heating, and cooling processes. Cellulose conversion can be examined using a batch- and flow-type reaction system. The flow-type system generated more hydrolysates. Due to the inability of cellulose to decrystallize under subcritical water conditions, the yield glucose could not be increased through subcritical treatment utilizing a batch-type method [29]. Aeration during biological pretreatment has a major effect on the generation of ligninolytic enzymes as well as their activity. It facilitates the movement of CO_2, heat, and oxygen while preserving humidity during metabolism [91]. Oxygen circulation is essential because lignin breakdown by bacteria is an oxidative process. The productivity of lignin peroxidase may rise if the rate of aeration is increased, and for biological delignification, it is crucial to regulate and optimize the rate of aeration [86].

pH considerably impacts on the biological delignification of lignocellulosic biomass since it changes the three-dimensional structure of enzymes, impacting the activities and functions of microbial enzymes. Laccase formation may be impacted by changing the pH of the medium [20]. A pH range of 4.0–5.0 is ideal for most white-rot fungi. Therefore, for microbial enzymes to operate properly during microbial hydrolysis, the pH must be at its ideal level. The enzymatic reaction and conditions influencing sugar yield are presented in Table 3.

Table 3 Techniques, substrate, and factors affecting hydrolysis to fermentable sugars

Hydrolysis techniques	Substrate source	Control factors	Product/Yield	References
Acid hydrolysis	Potato peel waste	Acid to substrate ratio 1:10 w/v 120 min; 100 °C	Ethanol (5.7 g/L)	[8]
Enzymatic	Sweet potato starch	pH 4.5; time 44 min; 52 °C, SPSH conc 241.92 g/L	Citric acid (86 g/L)	[10]
	Breadfruit starch	pH 4.5; time 24 h; hydrolysate; 120 g/L	Ethanol (4.10 v/v)	[14]
	Breadfruit starch	pH 5.5; time 72 h; 100 g/L hydrolysate	Gluconic acid (97.20 g/L)	[11]
	Sweet potato peel	Time 60 min; temp. 45 °C; glucoamylase 1% v/v	Hydrolysate (176.80 g/L)	[12]
	Ulva (green seaweeds)	Sugar 206.82 mg/g; pH 6.8; temp. 28 °C, 24 h	Ethanol 0.45 g/g	
	Cassava starch hydrolysate	Reducing sugar 27.54 g/L, 30 °C, pH 5, 24 h	Ethanol (44.47 g/L)	[13]
Pretreatment and steam explosion + Enzymatic	Starchy tuber	Sugar 21.60 g/L Heat power 24.64 MJ/kg; 120 min	Ethanol (8.68 g/L)	[15]

Temperature and residence time for autohydrolysis are key variables that affect the composition of the product [20]. The processes use compressed hot water (pressure above 0 saturation point), and the operating temperatures typically range from 150 to 230 °C, though higher temperatures could be used [17]. The reaction time can range from seconds to hours, according to the temperature. The liquid-to-solid ratio, which is the conventional name for solid concentrations, can range from 2 to 100 (w/w), but a value around 10 is ideal. Most frequently, continuous reactors are linked to a higher liquid-to-solid ratio. Depending on the experimental operating conditions, acetic acid, oligomers and monomers, and the end products of both pentoses and hexoses dehydration are the principal byproducts of autohydrolysis. Additionally, these compounds may go through breakdown processes that result in formic acid and levulinic acid. In addition, the partial solubilization of lignin may cause the release of phenolic chemicals into the reaction media [17].

6 Conclusion and Future Perspectives

The concepts of environmental sustainability, bioeconomy, and value chain are increasingly used in producing industrial value-added products. The biggest challenge for commercial bioethanol production is turning polymeric substrates and celluloses into fermentable sugars utilizing fast, cheap, and efficient methods. Though food-based substrates are still dominant in producing fermentable sugars, the discovery and development of lignocellulose, algae, and genetically engineering substrates indicate the direction for the current and future sustainability of bioethanol. Apart from first-generation feedstocks, substrates from other generational feedstocks can be combined to improve the process efficiency of fermentable sugar yield. Furthermore, unified technology to process substrates from different sources without major technical modification needs to be established to improve bioethanol production and economy. Pretreatment is a significant step in changing the structural properties of cellulosic biomass before it is then hydrolyzed to produce simple sugars, which are then processed to produce ethanol or any other desired value-added product. An ideal pretreatment technique must have distinct economic, environmental, technical, and ethical advantages. Consolidating the strength of the different pretreatment techniques by integrating two or more technologies would improve the wide acceptability of bioethanol by lowering the net operation costs. Using new technology such as ultrasound, pulsed electric field, and microwave activation to enhance the rate of biomass degradation is a very promising technique. This opens a more economical process that might achieve the scale-up objective of fast hydrolysis. Despite the numerous advances made in recent years that use any previously mentioned techniques for the pretreatment and hydrolysis of cellulosic biomass for bioethanol, many areas still need to be researched for industrial applications. These shortcomings include increased energy efficiency, process modeling and optimization, and control of hydrolysis operations. The problems with energy consumption, economics related to the process, and scale-up of the technology are some of the difficulties with using sonication as a pretreatment approach for enhanced biofuel production. Future research must concentrate on enhancing the pretreatment conditions and scaling up the technology used for hydrolysis. Likewise, future work on sonication as a pretreatment method must strongly focus on energy efficiency issues. Numerous studies using various bacterial and fungal species demonstrated the effectiveness and eco-friendliness of biological pretreatment of lignocellulosic biomass compared to other pretreatment techniques. Currently, the technology is not economically viable due to several challenges limiting the technology. Expanding the process to an industrial scale can be made possible if an efficient microbial consortium, reduced operational costs, and efficient reactor configuration can be established. Also, investigations on optimizing the process parameters and selecting the best microbial strains should be done to achieve the biological conversion of lignocellulosic materials to bioethanol at an industrial scale. Besides, using genetic engineering to manipulate the genetic makeup of microbial strains and the biosynthesis of plant biomass is another area of focus for future studies.

References

1. Aghaei S, Alavijeh MK, Shafiei M, Karimi K (2022) A comprehensive review on bioethanol production from corn stover: worldwide potential, environmental importance, and perspectives. Biomass Bioenerg 161:106447
2. Almodares A, Hadi M (2009) Production of bioethanol from sweet sorghum: a review. Afr J Agric Res 4(9):772–780
3. Almohammed F, Mhemdi H, Vorobiev E (2016) Pulsed electric field treatment of sugar beet tails as a sustainable feedstock for bioethanol production. Appl Energy 162:49–57
4. Amit K, Nakachew M, Yilkal B, Mukesh Y (2018) A review of factors affecting enzymatic hydrolysis of pretreated lignocellulosic biomass. Res J Chem Environ 22(7):62–67
5. Ananthi V, Ramesh U, Balaji P, Kumar P, Govarthanan M, Arun A (2022) A review on the impact of various factors on biohydrogen production. Int J Hydrogen Energy
6. Avanthi A, Mohan SV (2022) Emerging innovations for sustainable production of bioethanol and other mercantile products from circular economy perspective. Bioresour Technol:128013
7. Belal EB (2013) Bioethanol production from rice straw residues. Braz J Microbiol 44:225–234
8. Ben Taher I, Fickers P, Chniti S, Hassouna M (2017) Optimization of enzymatic hydrolysis and fermentation conditions for improved bioethanol production from potato peel residues. Biotechnol Prog 33(2):397–406
9. Bender LE, Lopes ST, Gomes KS, Devos RJB, Colla LM (2022) Challenges in bioethanol production from food residues. Bioresour Technol Reports:101171
10. Betiku E, Adesina OA (2013) Statistical approach to the optimization of citric acid production using filamentous fungus *Aspergillus niger* grown on sweet potato starch hydrolyzate. Biomass Bioenerg 55:350–354
11. Betiku E, Ajala O (2010) Enzymatic hydrolysis of breadfruit starch: case study with utilization for gluconic acid production. Ife J Technol 19(1):10–14
12. Betiku E, Akindolani O, Ismaila A (2013) Enzymatic hydrolysis optimization of sweet potato (*Ipomoea batatas*) peel using a statistical approach. Braz J Chem Eng 30:467–476
13. Betiku E, Alade O (2014) Media evaluation of bioethanol production from cassava starch hydrolysate using *Saccharomyces cerevisiae*. Energy Sources, Part A: Recov Util Environ Effects 36(18):1990–1998

14. Betiku E, Taiwo AE (2015) Modeling and optimization of bioethanol production from breadfruit starch hydrolyzate vis-à-vis response surface methodology and artificial neural network. Renew Energy 74:87–94
15. Bhuyar P, Trejo M, Mishra P, Unpaprom Y, Velu G, Ramaraj R (2022) Advancements of fermentable sugar yield by pretreatment and steam explosion during enzymatic saccharification of *Amorphophallus* sp. starchy tuber for bioethanol production. Fuel 323:124406
16. Bundhoo ZM, Mohee R (2018) Ultrasound-assisted biological conversion of biomass and waste materials to biofuels: a review. Ultrason Sonochem 40:298–313
17. Carvalheiro F, Duarte L, Gírio F, Moniz P (2016) Hydrothermal/liquid hot water pretreatment (autohydrolysis): a multipurpose process for biomass upgrading. In: Biomass fractionation technologies for a lignocellulosic feedstock based biorefinery. Elsevier, pp 315–347
18. Cassales A, de Souza-Cruz PB, Rech R, Ayub MAZ (2011) Optimization of soybean hull acid hydrolysis and its characterization as a potential substrate for bioprocessing. Biomass Bioenerg 35(11):4675–4683
19. Chen C, Boldor D, Aita G, Walker M (2012) Ethanol production from sorghum by a microwave-assisted dilute ammonia pretreatment. Biores Technol 110:190–197
20. Cheng J, Hu S-C, Geng Z-C, Zhu M-Q (2022) Effect of structural changes of lignin during the microwave-assisted alkaline/ethanol pretreatment on cotton stalk for an effective enzymatic hydrolysis. Energy 254:124402
21. de Souza TS, Kawaguti HY (2021) Cellulases, hemicellulases, and pectinases: applications in the food and beverage industry. Food Bioprocess Technol 14(8):1446–1477
22. Delbecq F, Len C (2018) Recent advances in the microwave-assisted production of hydroxymethylfurfural by hydrolysis of cellulose derivatives—a review. Molecules 23(8):1973
23. Devi A, Bajar S, Kour H, Kothari R, Pant D, Singh A (2022) Lignocellulosic biomass valorization for bioethanol production: a circular bioeconomy approach. Bioenergy Res:1–22
24. Diaz AB, de Souza Moretti MM, Bezerra-Bussoli C, Nunes CdCC, Blandino A, da Silva R, Gomes E (2015) Evaluation of microwave-assisted pretreatment of lignocellulosic biomass immersed in alkaline glycerol for fermentable sugars production. Biores Technol 185:316–323
25. Dimos K, Paschos T, Louloudi A, Kalogiannis KG, Lappas AA, Papayannakos N, Kekos D, Mamma D (2019) Effect of various pretreatment methods on bioethanol production from cotton stalks. Fermentation 5(1):5
26. Duque A, Manzanares P, Ballesteros M (2017) Extrusion as a pretreatment for lignocellulosic biomass: fundamentals and applications. Renew Energy 114:1427–1441
27. Dutta K, Daverey A, Lin J-G (2014) Evolution retrospective for alternative fuels: first to fourth generation. Renew Energy 69:114–122
28. Eblaghi M, Niakousari M, Sarshar M, Mesbahi GR (2016) Combining ultrasound with mild alkaline solutions as an effective pretreatment to boost the release of sugar trapped in sugarcane bagasse for bioethanol production. J Food Process Eng 39(3):273–282
29. Ehara K, Saka S (2002) A comparative study on chemical conversion of cellulose between the batch-type and flow-type systems in supercritical water. Cellulose 9:301–311
30. El Barnossi A, Moussaid F, Housseini AI (2021) Tangerine, banana and pomegranate peels valorisation for sustainable environment: a review. Biotechnol Reports 29:e00574
31. Fernández-Medina P, Álvarez-Gallego C, Caro I (2022) Yield evaluation of enzyme hydrolysis and dark fermentation of the brown seaweed Rugulopteryx okamurae hydrothermally pretreated by microwave irradiation. J Environ Chem Eng 10(6):108817
32. Ferreira JA, Brancoli P, Agnihotri S, Bolton K, Taherzadeh MJ (2018) A review of integration strategies of lignocelluloses and other wastes in 1st generation bioethanol processes. Process Biochem 75:173–186
33. Fu Q, Xiao C, Liao Q, Huang Y, Xia A, Zhu X (2021) Kinetics of hydrolysis of microalgae biomass during hydrothermal pretreatment. Biomass Bioenerg 149:106074
34. Gandam PK, Chinta ML, Pabbathi NPP, Baadhe RR, Sharma M, Thakur VK, Sharma GD, Ranjitha J, Gupta VK (2022) Second-generation bioethanol production from corncob– a comprehensive review on pretreatment and bioconversion strategies, including techno-economic and lifecycle perspective. Ind Crops Prod 186:115245

35. Gírio FM, Fonseca C, Carvalheiro F, Duarte LC, Marques S, Bogel-Łukasik R (2010) Hemicelluloses for fuel ethanol: a review. Biores Technol 101(13):4775–4800
36. Goshadrou A, Karimi K, Taherzadeh MJ (2011) Bioethanol production from sweet sorghum bagasse by *Mucor hiemalis*. Ind Crops Prod 34(1):1219–1225
37. Gumienna M, Szwengiel A, Lasik M, Szambelan K, Majchrzycki D, Adamczyk J, Nowak J, Czarnecki Z (2016) Effect of corn grain variety on the bioethanol production efficiency. Fuel 164:386–392
38. Hay JXW, Wu TY, Juan JC, Jahim JM (2015) Improved biohydrogen production and treatment of pulp and paper mill effluent through ultrasonication pretreatment of wastewater. Energy Convers Manage 106:576–583
39. Hideno A, Inoue H, Tsukahara K, Fujimoto S, Minowa T, Inoue S, Endo T, Sawayama S (2009) Wet disk milling pretreatment without sulfuric acid for enzymatic hydrolysis of rice straw. Biores Technol 100(10):2706–2711
40. Huang J, Khan MT, Perecin D, Coelho ST, Zhang M (2020) Sugarcane for bioethanol production: potential of bagasse in Chinese perspective. Renew Sustain Energy Rev 133:110296
41. Jędrzejczyk M, Soszka E, Czapnik M, Ruppert AM, Grams J (2019) Physical and chemical pretreatment of lignocellulosic biomass. In: Second and third generation of feedstocks. Elsevier, pp 143–196
42. Jin X, Song J, Liu G-Q (2020) Bioethanol production from rice straw through an enzymatic route mediated by enzymes developed in-house from *Aspergillus fumigatus*. Energy 190:116395
43. Jothibasu K, Dhar D, Rakesh S (2021) Recent developments in microalgal genome editing for enhancing lipid accumulation and biofuel recovery. Biomass Bioenerg 150:106093
44. Khan MU, ur Rehman MM, Sultan M, ur Rehman T, Sajjad U, Yousaf M, Ali HM, Bashir MA, Akram MW, Ahmad M (2022) Key prospects and major development of hydrogen and bioethanol production. Int J Hydrogen Energy 47(62):26265–26283
45. Kostas ET, Beneroso D, Robinson JP (2017) The application of microwave heating in bioenergy: a review on the microwave pre-treatment and upgrading technologies for biomass. Renew Sustain Energy Rev 77:12–27
46. Kumar P, Barrett DM, Delwiche MJ, Stroeve P (2011) Pulsed electric field pretreatment of switchgrass and wood chip species for biofuel production. Ind Eng Chem Res 50(19):10996–11001
47. Lee W-S, Chen I-C, Chang C-H, Yang S-S (2012) Bioethanol production from sweet potato by co-immobilization of saccharolytic molds and *Saccharomyces cerevisiae*. Renew Energy 39(1):216–222
48. Lenihan P, Orozco A, O'Neill E, Ahmad M, Rooney D, Walker G (2010) Dilute acid hydrolysis of lignocellulosic biomass. Chem Eng J 156(2):395–403
49. Li M-F, Sun S-N, Xu F, Sun R-C (2012) Ultrasound-enhanced extraction of lignin from bamboo (*Neosinocalamus affinis*): characterization of the ethanol-soluble fractions. Ultrason Sonochem 19(2):243–249
50. Liao Q, Sun C, Xia A, Fu Q, Huang Y, Zhu X, Feng D (2021) How can hydrothermal treatment impact the performance of continuous two-stage fermentation for hydrogen and methane co-generation? Int J Hydrogen Energy 46(27):14045–14062
51. Limayem A, Ricke SC (2012) Lignocellulosic biomass for bioethanol production: current perspectives, potential issues and future prospects. Prog Energy Combust Sci 38(4):449–467
52. Lisboa CC, Butterbach-Bahl K, Mauder M, Kiese R (2011) Bioethanol production from sugarcane and emissions of greenhouse gases–known and unknowns. Gcb Bioenergy 3(4):277–292
53. Lü J, Sheahan C, Fu P (2011) Metabolic engineering of algae for fourth generation biofuels production. Energy Environ Sci 4(7):2451–2466
54. Maity S, Mallick N (2022) Trends and advances in sustainable bioethanol production by marine microalgae: a critical review. J Clean Prod:131153

55. Mishima D, Tateda M, Ike M, Fujita M (2006) Comparative study on chemical pretreatments to accelerate enzymatic hydrolysis of aquatic macrophyte biomass used in water purification processes. Biores Technol 97(16):2166–2172
56. Mlonka-Mędrala A, Sieradzka M, Magdziarz A (2022) Thermal upgrading of hydrochar from anaerobic digestion of municipal solid waste organic fraction. Fuel 324:124435
57. Mohapatra S, Dandapat SJ, Thatoi H (2017) Physicochemical characterization, modelling and optimization of ultrasono-assisted acid pretreatment of two *Pennisetum* sp. using Taguchi and artificial neural networking for enhanced delignification. J Environ Manage 187:537–549
58. Mohapatra S, Mishra C, Merritt BB, Pattathil S, Thatoi H (2019) Evaluating the role of ultrasonication-assisted alkali pretreatment and enzymatic hydrolysis on cellwall polysaccharides of Pennisetum grass varieties as potential biofuel feedstock. ChemistrySelect 4(3):1042–1054
59. Mohd Ishak NA, Abdullah FZ, Muhd Julkapli N (2022) Production and characteristics of nanocellulose obtained with using of ionic liquid and ultrasonication. J Nanopart Res 24(8):1–22
60. Ngamsirisomsakul M, Reungsang A, Liao Q, Kongkeitkajorn MB (2019) Enhanced bioethanol production from *Chlorella* sp. biomass by hydrothermal pretreatment and enzymatic hydrolysis. Renew Energy 141:482–492
61. Ninomiya K, Yamauchi T, Ogino C, Shimizu N, Takahashi K (2014) Microwave pretreatment of lignocellulosic material in cholinium ionic liquid for efficient enzymatic saccharification. Biochem Eng J 90:90–95
62. Nitsos CK, Matis KA, Triantafyllidis KS (2013) Optimization of hydrothermal pretreatment of lignocellulosic biomass in the bioethanol production process. Chemsuschem 6(1):110–122
63. Olalere OA, Gan CY, Taiwo AE, Alenezi H, Maqsood S, Adeyi O (2022) Investigating the microwave parameters correlating effects on total recovery of bioactive alkaloids from sesame leaves using orthogonal matrix and artificial neural network integration. J Food Process Preserv:e16591
64. Onumaegbu C, Mooney J, Alaswad A, Olabi A (2018) Pre-treatment methods for production of biofuel from microalgae biomass. Renew Sustain Energy Rev 93:16–26
65. Park J-H, Hong J-Y, Jang HC, Oh SG, Kim S-H, Yoon J-J, Kim YJ (2012) Use of *Gelidium amansii* as a promising resource for bioethanol: a practical approach for continuous dilute-acid hydrolysis and fermentation. Biores Technol 108:83–88
66. Petersen MØ, Larsen J, Thomsen MH (2009) Optimization of hydrothermal pretreatment of wheat straw for production of bioethanol at low water consumption without addition of chemicals. Biomass Bioenerg 33(5):834–840
67. Pradyawong S, Juneja A, Sadiq MB, Noomhorm A, Singh V (2018) Comparison of cassava starch with corn as a feedstock for bioethanol production. Energies 11(12):3476
68. Putranto AW, Abida SH, Adrebi K (2021) Lignocellulosic analysis of corncob biomass by using non-thermal pulsed electric field-NaOH pretreatment. In: International conference on sustainable biomass (ICSB 2019), 2021. Atlantis Press, pp 273–280
69. Qin L, Li W-C, Zhu J-Q, Li B-Z, Yuan Y-J (2017) Hydrolysis of lignocellulosic biomass to sugars. In: Production of platform chemicals from sustainable resources. Springer, pp 3–41
70. Quintero JA, Rincón LE, Cardona CA (2011) Production of bioethanol from agroindustrial residues as feedstocks. In: Biofuels. Elsevier, pp 251–285
71. Rabemanolontsoa H, Saka S (2016) Various pretreatments of lignocellulosics. Biores Technol 199:83–91
72. Raina N, Slathia PS, Sharma P (2020) Response surface methodology (RSM) for optimization of thermochemical pretreatment method and enzymatic hydrolysis of deodar sawdust (DS) for bioethanol production using separate hydrolysis and co-fermentation (SHCF). Biomass Conv Bioref:1–21
73. Ramachandra T, Hebbale D (2020) Bioethanol from macroalgae: prospects and challenges. Renew Sustain Energy Rev 117:109479
74. Ravindran R, Jaiswal S, Abu-Ghannam N, Jaiswal AK (2017) Evaluation of ultrasound assisted potassium permanganate pre-treatment of spent coffee waste. Biores Technol 224:680–687

75. Reese ET, Siu RG, Levinson HS (1950) The biological degradation of soluble cellulose derivatives and its relationship to the mechanism of cellulose hydrolysis. J Bacteriol 59(4):485–497
76. Rehman MSU, Kim I, Chisti Y, Han J-I (2013) Use of ultrasound in the production of bioethanol from lignocellulosic biomass. Energy Educ Sci Technol Part A: Energy Sci Res 30(2):1931–1410
77. Rezania S, Oryani B, Cho J, Talaiekhozani A, Sabbagh F, Hashemi B, Rupani PF, Mohammadi AA (2020) Different pretreatment technologies of lignocellulosic biomass for bioethanol production: an overview. Energy 199:117457
78. Röder LS, Gröngröft A, Grünewald M, Riese J (2023) Assessing the demand side management potential in biofuel production; a theoretical study for biodiesel, bioethanol, and biomethane in Germany. Biofuels, Bioprod Biorefin 17(1):56–70
79. Saini R, Osorio-Gonzalez CS, Brar SK, Kwong R (2021) A critical insight into the development, regulation and future prospects of biofuels in Canada. Bioengineered 12(2):9847–9859
80. Shokravi H, Shokravi Z, Heidarrezaei M, Ong HC, Koloor SSR, Petrů M, Lau WJ, Ismail AF (2021) Fourth generation biofuel from genetically modified algal biomass: challenges and future directions. Chemosphere 285:131535
81. Sindhu R, Binod P, Pandey A (2016) Biological pretreatment of lignocellulosic biomass–an overview. Biores Technol 199:76–82
82. Sitotaw YW, Habtu NG, Gebreyohannes AY, Nunes SP, Van Gerven T (2021) Ball milling as an important pretreatment technique in lignocellulose biorefineries: a review. Biomass Conv Bioref:1–24
83. Song X, Zhang M, Pei Z (2013) Effects of ultrasonic vibration-assisted pelleting of cellulosic biomass on sugar yield for biofuel manufacturing. Biomass Conv Biorefinery 3:231–238
84. Swain MR, Singh A, Sharma AK, Tuli DK (2019) Bioethanol production from rice-and wheat straw: an overview. In: Bioethanol production from food crops, pp 213–231
85. Taiwo AE, Madzimbamuto TN, Ojumu TV (2018) Optimization of corn steep liquor dosage and other fermentation parameters for ethanol production by *Saccharomyces cerevisiae* type 1 and anchor instant yeast. Energies 11(7):1740
86. Taiwo AE, Madzimbamuto TN, Ojumu TV (2020) Optimization of process variables for acetoin production in a bioreactor using Taguchi orthogonal array design. Heliyon 6(10):e05103
87. Taiwo AE, Tom-James A, Falowo OA, Okoji A, Adeyi O, Olalere AO, Eloka-Eboka A (2022) Techno-economic analysis of cellulase production by *Trichoderma reesei* in submerged fermentation processes using a process simulator. S Afr J Chem Eng 42:98–105
88. Thanapimmetha A, Vuttibunchon K, Saisriyoot M, Srinophakun P (2011) Chemical and microbial hydrolysis of sweet sorghum bagasse for ethanol production. In: World renewable energy congress-sweden; 8–13 May; 2011, vol 057. Linköping University Electronic Press, Linköping, Sweden, pp 389–396
89. Thangavelu SK, Rajkumar T, Pandi DK, Ahmed AS, Ani FN (2019) Microwave assisted acid hydrolysis for bioethanol fuel production from sago pith waste. Waste Manage 86:80–86
90. Trivedi N, Gupta V, Reddy C, Jha B (2013) Enzymatic hydrolysis and production of bioethanol from common macrophytic green alga Ulva fasciata Delile. Biores Technol 150:106–112
91. Tsegaye B, Balomajumder C, Roy P (2019) Microbial delignification and hydrolysis of lignocellulosic biomass to enhance biofuel production: an overview and future prospect. Bull Natl Res Centre 43(1):1–16
92. Tu W-C, Hallett JP (2019) Recent advances in the pretreatment of lignocellulosic biomass. Curr Opin Green Sustain Chem 20:11–17
93. Vandenberghe L, Valladares-Diestra K, Bittencourt G, Torres LZ, Vieira S, Karp S, Sydney E, de Carvalho J, Soccol VT, Soccol C (2022) Beyond sugar and ethanol: the future of sugarcane biorefineries in Brazil. Renew Sustain Energy Rev 167:112721
94. Vega L, López L, Valdés CF, Chejne F (2019) Assessment of energy potential of wood industry wastes through thermochemical conversions. Waste Manage 87:108–118

95. Verardi A, De Bari I, Ricca E, Calabrò V (2012) Hydrolysis of lignocellulosic biomass: current status of processes and technologies and future perspectives. In: Bioethanol, vol 2012. InTech Rijeka, pp 95–122
96. Wang L, Littlewood J, Murphy RJ (2013) Environmental sustainability of bioethanol production from wheat straw in the UK. Renew Sustain Energy Rev 28:715–725
97. Yang H, Jin Y, Shi Z, Wang D, Zhao P, Yang J (2020) Effect of hydrothermal pretreated bamboo lignin on cellulose saccharification for bioethanol production. Ind Crops Prod 156:112865
98. Yu C-H, Sheen H-K, Chang J-S, Lin C-S, Ong HC, Show PL, Ng E-P, Ling TC (2020) Production of microalgal biochar and reducing sugar using wet torrefaction with microwave-assisted heating and acid hydrolysis pretreatment. Renew Energy 156:349–360
99. Yu GT, Grimi N, Bals O, Vorobiev E (2016) Pulsed electric field pretreatment of rapeseed green biomass (stems) to enhance pressing and extractives recovery. Biores Technol 199:194–201
100. Yu LX, Wu H (2008) Some recent advances in hydrolysis of biomass in hot-compressed water and its comparisons with other hydrolysis methods. Energy Fuels 22(1):46–60
101. Zakaria MR, Fujimoto S, Hirata S, Hassan MA (2014) Ball milling pretreatment of oil palm biomass for enhancing enzymatic hydrolysis. Appl Biochem Biotechnol 173(7):1778–1789
102. Zhang Z, Tahir N, Li Y, Zhang T, Zhu S, Zhang Q (2019) Tailoring of structural and optical parameters of corncobs through ball milling pretreatment. Renew Energy 141:298–304
103. Zhuang X, Liu J, Zhang Q, Wang C, Zhan H, Ma L (2022) A review on the utilization of industrial biowaste via hydrothermal carbonization. Renew Sustain Energy Rev 154:111877
104. Ziska LH, Runion GB, Tomecek M, Prior SA, Torbet HA, Sicher R (2009) An evaluation of cassava, sweet potato and field corn as potential carbohydrate sources for bioethanol production in Alabama and Maryland. Biomass Bioenergy 33(11):1503–1508

Bioethanol Production Using Novel Starch Sources

Gabriel S. Aruwajoye, Daneal C. S. Rorke, Isaac A. Sanusi, Yeshona Sewsynker-Sukai, and Evariste B. Gueguim Kana

Abstract Bioethanol production is a promising solution to the challenge of carbon emissions from global fuel consumption. However, to fully realize its potential, several bottlenecks must be overcome, including low product yield, food versus fuel factor, and feedstock logistics. One way to address these issues is to utilize highly dense and bioenergy-rich biomass for the fermentation process. Recent research has focused on novel starch sources, such as starch-based lignocelluloses, which have accessible starch in addition to their highly structured polysaccharides (hemicellulose, cellulose, and lignin). These have the potential to significantly increase the yield and efficiency of bioethanol production. Thus, this chapter highlights the properties of starch with their presence and usefulness as bioenergy together with the process of conversion of their well-known feedstocks to bioethanol. Additionally, the utilization of novel starch sources in the form of starch-based crops, agricultural wastes, and aquatic plants was elucidated. Lastly, innovative strategies towards improving the process of bioethanol generation from novel starch sources were discussed.

Keywords Bioethanol · Starch-based lignocellulose · Novel starch sources · Agricultural wastes · Bioenergy

1 Introduction

The need for environmentally benign energy sources as fossil fuel alternatives has escalated in recent years. This is attributed to the hazardous effect of combustive activities and the gradual depletion of fossil fuel reserves [60]. Bioethanol is one of the suitable alternative fuels that can replace fossil fuels as it is non-toxic and

G. S. Aruwajoye · D. C. S. Rorke · E. B. G. Kana (✉)
School of Life Sciences, University of KwaZulu-Natal, Pietermaritzburg, South Africa
e-mail: kanag@ukzn.ac.za

I. A. Sanusi · Y. Sewsynker-Sukai
Fort Hare Institute of Technology, University of Fort Hare, Private Bag X1314, Alice 5700, South Africa

renewable. Bioethanol can be produced from the fermentation of carbon sources in the presence of the fermentative organism. The global ethanol fuel market was valued at approximately 110 billion litres in 2018 [49, 70], reducing slightly to 98.6 billion litres in 2020 [31]. However, the implementation of large-scale ethanol production is hindered by the low-profit margins, low product yield, high process costs, feedstock logistics, and substrate intrinsic characteristics [54].

Most of the globally produced ethanol is made from first-generation feedstocks such as corn, sugarcane, sunflower oil, and soybeans. The largest ethanol producer is the United States (US), which produces approximately 55% of the global ethanol supply from corn starch as a feedstock. In 2021, the US produced about 57 billion liters of corn-based ethanol [78], while Brazil, which uses sugarcane juice as a feedstock, produced 30.4 billion liters [76]. A summary of the millions of gallons of bioethanol produced per country between 2015 and 2021 is shown in Fig. 1.

The choice of feedstock employed plays an integral role in bioethanol production. Although the most commonly applied technology for ethanol production from first-generation feedstocks is well-established, it is severely impacted by the high cost of the substrates and the ethical concerns of the food versus fuel debacle [7].

The use of abundantly available second-generation feedstocks such as lignocellulosic biomass (e.g. post-harvest agricultural waste, forestry waste) for ethanol production offers an alternative strategy to overcome the challenges of conventional first-generation ethanol production. However, the advanced processing required to break down the recalcitrant lignocellulosic material often renders these processes financially non-viable at the industrial scale level. Lignocellulose biomass has a general

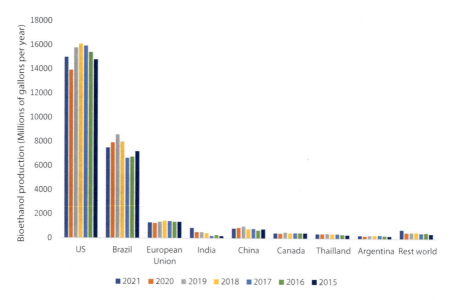

Fig. 1 Summary of the millions of gallons of bioethanol produced per country between 2015 and 2021 (*Data source* Renewable fuel association [63] and Statistica [73])

composition of 6–42% lignin, 4–50% cellulose, and 7–30% hemicellulose, and these can be harnessed for hydrolysis to fermentable sugars following relevant processing strategies [4]. Starch-based lignocellulosic biomass has greater potential for a higher yield of fermentable sugars and hence bioethanol. The presence of starch, in addition to lignin, cellulose, and hemicellulose, makes them more advantageous.

On the other hand, third-generation bioethanol production, which uses microalgae as a feedstock, has recently emerged as an efficient alternative to both first and second-generation sources, however, it is still in its infancy. Presently, starch-based bioethanol production is of particular interest due to the potential of high fermentable sugar recovery connected with the additional starch composition of the lignocellulose substrate utilized. Several novel starch substrates have appeared recently for bioethanol production and include but are not limited to breadfruits, potato peels, sweet potato peels, cassava peels, cassava wastes, and pumpkin waste. Most of these starch substrates have only emerged recently, while some are still underutilized.

Currently, on the global bioethanol scene, the fermentable glucose for bioethanol production is obtained from both sugar and starch substrates. The sugar substrates that are generally utilized are sugarcane, sugar beet, and sweet sorghum while the starch substrates include corn, wheat, barley, rye, triticale (a hybrid of wheat and rye), sorghum, potato, and cassava [27]. Global bioethanol production is based on approximately 60% starch and 40% sugar substrates [80].

This chapter focuses on bioethanol production from novel starch-based substrates. A general summary of the different bioethanol substrates is provided in addition to the structure and properties of starch. Furthermore, an overview of starch conversion to bioethanol was discussed. Subsequently, bioethanol fermentation was described with the outline of available novel starch sources. Lastly, the challenges encountered during the bioethanol fermentation from starch sources and the possible strategies for the improvement of the process were elucidated.

2 Starch Structure and Properties

Starch is a giant polymer of glucose molecules linked together by glycosidic bonds. Glucose is generated in green plants through photosynthetic activities, where they are utilized for growth, and the excess is stored as starch in the form of granules. The starch polymer exists as either amylose or amylopectin with α-1,4-glycosidic bonds and α-1,6-glycosidic bonds, respectively (Fig. 2).

The percentage of amylose and amylopectin varies widely in different plants (Table 1). About 20–25% have been reported for amylose, while amylopectin makes up the remaining 75–80% [27]. These heavily packed polymers (polysaccharide reservoirs) confer a major advantage on starch substrates over sugar substrates during bioethanol production. This is due to heavyweight chains that often generate more monomeric sugars, which fermentative microorganisms can subsequently utilize compared to readily available sugars in sugar substrates.

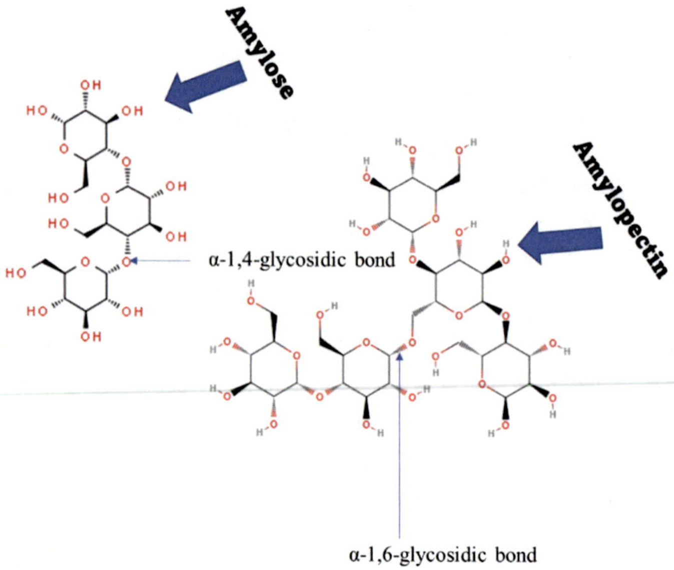

Fig. 2 The structure of amylose and amylopectin

Table 1 The percentage of amylose and amylopectin in starch sources

Starch source	Amylose (%)	Amylopectin (%)	References
Maize	25	75	[66]
Waxy corn starch	1	99	[75]
Cassava	18.6	81.4	[82]
Arrow root	20.5	79.5	[82]
Potato starch	29	71	[75]
Wheat	23	77	[82]
Rice starch	17	83	[75]
High amylose corn starch	50–85	15–50	[75]
Banana	20	80	[82]
Shoti	30	70	[66]

3 Bioethanol Feedstocks

Feedstocks for bioethanol production are categorized into four, namely, first, second, third and fourth generation [79]. The first-generation feedstock could be starchy materials such as corn, rice and barley or sugar-containing feedstock such as sugarcane. The major disadvantage of using first generation feedstocks for bioethanol production is the competition with food availability, which can generally hamper human survival. On the other hand, non-edible plants are utilized as second-generation feedstock. These are mostly lignocellulosic biomass such as agricultural wastes. Second generation feedstocks could also be starch-based (if there is presence of starch in their carbohydrate composition in addition to the structural polysaccharides—cellulose, hemicellulose and lignin) or non-starch based such as grass, wood, and other forest residues. Unlike the first-generation feedstocks, these are readily available, renewable and sustainable. The third-generation feedstocks are majorly aquatic/marine-based biomass such as microalgae which are also rich in starch and other polysaccharide components present in second generation feedstocks [46]. It has been estimated that land sizes subjected to microalgae cultivation has the potential of generating up to ten times bioethanol volume more than what to expect if corn is was the feedstock of choice.

In order to improve on the biofuel possibilities from first, second and third generation feedstocks, experts are using advanced technology to genetically engineer feedstocks with improved characteristics focused on biofuel production. These are categorized as fourth generation feedstocks [79]. Fourth generation feedstocks have the potential to overcome the limitations of previous generations of feedstocks. However, because of the huge costs associated with the processes involved, the utilization of fourth generation feedstocks for bioethanol production remains unpopular for the industrial space compared to the first three generations. Bioethanol yield from various generations of feedstock is shown in Table 2.

Table 2 Bioethanol yield from various generations of feedstock

Feedstock name	Feedstock generation	Bioethanol yield	References
Cassava	First	84%	[57]
Sugar beet	First	91.16–92.06 g/L	[40]
Corn	First	15.25–17.5% (v/v)	[83]
Sugarcane	First	70 L/t	[26]
Cassava peels	Second	143.31 g/Kg	[3]
Potato peels	Second	0.46 g/g	[41]
Wheat straw	Second	0.43 g/g	[77]
Sugarcane bagasse	Second	0.49 g/g	[38]
Microalgae	Third	18.57 g/L	[23]
Microalgae	Third	60%	[6]
Microalgae	Third	90%	[30]

4 Hydrolysis of Feedstocks for Bioethanol Production

The most significant components of interest present within feedstocks are the bioenergy compounds which include cellulose, hemicellulose, lignin and starch (for starch-based biomass). These carbohydrates become readily available and release fermentable sugar when subjected to various processes such as pretreatment followed by enzymatic hydrolysis. Subsequent to pretreatment activities using physical, chemical or biological methods, enzymatic hydrolysis of the biomass is implemented through the application of microbial enzymes, singly (e.g. amylase) or synergistically (e.g. cellulase and/or amyloglucosidase), for the deconstruction of their polysaccharide structures to simple, fermentable sugars [4, 74]. The carbohydrate polymers present in the biomass may exist as an intertwined holocellulose complex in the plant cell wall (which are cellulose, hemicellulose and lignin) or as granules (starch).

Hydrolysis of the holocellulose fraction involves the synergistic action of endoglucanases (EC 3.2.1.4), exoglucanase or cellobiohydrolases (EC 3.2.1.91), and β-glucosidases (EC 3.2.1.21) [47, 85]. The role of endoglucanases (endocellulase or endo-1-4-β-glucanase) is to cleave internal glycosidic linkages in the amorphous region of the cellulose, releasing shorter oligosaccharides with several degrees of polymerization (glucose units). This action opens up the cellulose molecule, producing reducing and non-reducing ends that are exposed to exoglucanase or cellobiohydrolase attack. On the other hand, cellobiohydrolases act on the crystalline portion of cellulose exposing more internal sites of the cellulose to endocellulose binding while attacking the already exposed reducing and non-reducing ends of the glucose chain converting them to cellobiose units. Finally, β-glucosidase cleaves the cellobiose units to glucose. Optimal temperature and pH of 40–50 °C and pH 4–5, have been reported for cellulase activities [3, 47].

For increased sugar yield during enzymatic hydrolysis of lignocelluloses, the enzyme cocktail can be supplemented with hemicellulases, lignin modifying enzymes and other accessory enzymes. Hemicellulases are complex group of enzymes that hydrolyses xylan, the major polymer in hemicelluloses. The enzymes are endo-β-1,4-xylanase (EC 3.2.1.8), and β-xylosidase (EC 3.2.1.37). Endo-β-1,4-xylanase randomly cleaves the internal bonds of xylan releasing xylooligosaccharides while β-xylosidase attacks the xylan from the non-reducing ends freeing xylose from the xylose chains [47]. The third composition of the holocellulose complex—lignin, can be degraded through the addition of lignin modifying enzymes such as lignin peroxidase, manganese peroxidase, and laccase [28]. However, the most preferred strategy for the removal of lignin from lignocelluloses is a pretreatment process consisting of a chemical and/or heating mechanism which is usually undertaken prior to the application of enzymes. In addition to the enzymes directed at the major polymers of lignocelluloses (cellulose, hemicellulose and lignin), other accessory enzymes specific for certain sugars such as arabinose and galactose have been reported [43, 22]. Hydrolysis of the granular content (starch) of bioethanol feedstocks if available can be achieved through the use of α-amylase and amyloglucosidase. These enzymes act by cleaving the α-1,4-D-glycosidic linkages and 1,4-linked maltooligosaccharides linkages of starch. α-amylase is specific for liquefaction activities where the starch present in the lignocellulose biomass is converted into dextrins and maltose, while the amyloglucosidase enzyme saccharifies the dextrins into glucose monomers. Optimal conditions for liquefaction and saccharification processes have been reported in the range of 90–105 °C (pH = 7.0) and 50–60 °C (pH = 4.5), respectively [3].

5 Bioethanol Production from Starch-Based Feedstocks

There are different fermentative processes for bioethanol production from starch-based feedstocks. These processes include separate hydrolysis and fermentation (SHF), simultaneous saccharification and fermentation (SSF), one-pot simultaneous saccharification and fermentation (one-pot SSF), and simultaneous saccharification and co-fermentation (SSCF). The most commonly used microbe for bioethanol fermentation is the fungal genera of *Saccharomyces* such *Saccharomyces cerevisiae* [16]. This fungus is preferred because it can easily assimilate and ferment hexose sugars to ethanol under facultative and acidic conditions. Additionally, *S. cerevisiae* has a wide operational temperature range of 15–35 °C [20]. Other microorganisms capable of ethanol fermentation include *Zymomonas mobilis* and *Thermoanaerobacterium* [20].

5.1 Separate Hydrolysis and Fermentation (SHF)

During separate hydrolysis and fermentation, enzymatic hydrolysis and subsequent fermentation are performed in separate reactors and sequentially at their optimal temperatures. Subsequently, the fermentation step is carried out, also under optimal physiochemical operation conditions and in the presence of fermenting microbes. Important factors during fermentation include; substrate type and concentration, pH, temperature, yeast strain, nutrient type and agitation [67]. The advantage of the SHF process is the possibility of independently optimizing the temperatures of the starch liquefaction, enzymatic hydrolysis, and fermentation processes. As a result, lesser enzyme concentration is required compared to the simultaneous saccharification and fermentation (SSF) process. Also, sterilizing the fermentable sugar in the saccharified hydrolysate is easier, thus reducing the risk of contamination. However, execution of the SHF process for starchy-lignocellulosic biomass requires three independent bioreactors for the starch liquefaction, saccharification, and fermentation stages, thus, heightening the capital cost compared to the SSF process [70].

5.2 Simultaneous Saccharification and Fermentation

SSF process combines enzymatic saccharification with fermentation in a single step after the initial pretreatment and starch liquefaction processes [8, 69]. The SSF process principle is based on the use of an enzymatic complex to hydrolyze cellulose and simultaneously ferment the released sugars [81]. This process has several advantages over other fermentation processes. These merits include the use of a single reactor for the saccharification and fermentation process thus, reducing processing time and capital costs [77]. In addition. reduction of inhibitory compounds from enzymatic hydrolysis can be achieved with SSF process, which in turn improves the overall performance of the process [8]. The disadvantages that have limited the use of SSF on an industrial level are the different pH and temperature requirements for the saccharification and fermentation processes since the optimum temperature of enzymatic saccharification is typically higher than the fermentation temperature [55].

5.3 One-Pot Simultaneous Saccharification and Fermentation

One-pot simultaneous saccharification and fermentation (one-pot SSF) is a bioprocess strategy developed to combine biomass pretreatment, enzymatic hydrolysis, and fermentation in a single fermenter [44]. Bioethanol production from lignocellulose substrates involves the pretreatment stage, enzymatic hydrolysis stage, and fermentation. Typically, in an SHF process, these three individual steps take place separately under different process conditions, while the SSF process combines enzymatic hydrolysis and fermentation in the same bioreactor, and simultaneously released sugars are fermented to ethanol. The pretreatment step was not included as a simultaneous process both in the SHF and SSF processes. One-pot SSF merges the pretreatment, enzymatic hydrolysis, and fermentation steps in a reactor [44].

5.4 Simultaneous Saccharification and Co-fermentation

Simultaneous saccharification and co-fermentation (SSCF) is another interesting approach that can be used in starch-based bioprocessing. Simultaneous assimilation of two different substrates by microorganisms could be difficult because substrates can compete for the transport systems [56]. For instance, in ethanol production (after enzymatic hydrolysis), the microorganism first consumes the glucose and when glucose is at a low concentration, it then consumes xylose [56]. Solving this problem involves gradually adding cellulases to the process or a pre-hydrolysis step to start the process with a low concentration of glucose, forcing the microorganism to consume the two substrates sequentially [36]. A drawback of the SSCF process is the differences in temperature, pH, and other conditions necessary for enzymatic hydrolysis and co-fermentation [36]. This indicates a need to employ microorganisms that can thrive under such conditions. This becomes a bottleneck using the SSCF process in lignocellulosic fermentation for bioethanol production. Genetic engineering could help to solve this challenge. The benefits of SSCF include less reaction time, and high process efficiencies [35].

6 Novel Starch Sources for Bioethanol Production

Bioethanol can be produced from several starch-based crops. Most of these crops and agricultural wastes have been innovatively explored recently and are herein elucidated.

6.1 Starch-Based Crops

One of the starch-based crops explored for bioethanol production is breadfruit. Breadfruit (*Artocarpus communis*) is a tropical fruit crop that has a starch composition of 77–89% [1, 25]. Bioethanol production from breadfruit starts with extraction of the starch component through milling followed by filtration, and subsequently subjecting the settled filtrate to sun drying until constant weight is achieved [12, 14]. Farida et al. [25] produced bioethanol from breadfruit starch through simultaneous saccharification and fermentation using a microbial consortium which generated up to 12.5 g/L of bioethanol. Similarly, [14] produced bioethanol from breadfruit starch hydrolysate (BFSH). A BFSH of 120 g/L was able to generate a bioethanol yield of 4.22% volume fraction.

Another starch-based crop with bioethanol potential is cassava (*Manihot esculenta*). The use of cassava for bioethanol production has a major advantage over other starch-based crops because of its growth tolerance and round-the-year availability [29]. When subjected to novel eco-friendly enzymes, cassava starch slurry reportedly generated up to 558 g of bioethanol per kg of starch [71]. Similarly, a 100 g/L of cassava starch hydrolysate (CSH) was used to generate 48.16 g/L of bioethanol when subjected to batch fermentation using *Saccharomyces cerevisiae* [13]. Also, Moshi et al. [51] used a thermoanerobe, *Caloramator boliviensis*, to produce bioethanol of 33 g/L from inedible wild cassava under fed-batch conditions.

Potato, *Solanum tuberosum*, is another starch-based tuber crop that has been utilized for bioethanol production. The plant tuber contains 65–80% starch [37, 45] produced 60.18 g/L of bioethanol from a repeated batch of simultaneous saccharification and fermentation of 12% (w/v) potato starch using a vertical mass flow type bioreactor. Similarly, [34] generated 19 g/L of bioethanol from 50 g/L of potato starch using a mixed culture of *S. cerevisiae* and *Aspergillus niger* in an electrochemical bioreactor.

6.2 Starch-Based Agricultural Wastes

Starch-based agricultural wastes are generated as residues of harvest or processing of crops. Starch-based wastes contain starch left as remains from the edible portion of the crops during processing. This is obtained as a compositional residue in addition to their lignocellulose structure. The lignocelluloses are rich in carbohydrates that can be converted to bioethanol. The major carbohydrates in lignocelluloses are cellulose, hemicellulose, and lignin. Lignocelluloses are sub-classified as energy crops (such as perennial grasses), forest materials, municipal solid wastes, and agricultural wastes [84]. Examples of starch-based agricultural wastes include cassava peels, potato peels, pumpkin peels, yam peels, and sweet potato peels. The percentage of starch composition of starch-based agricultural wastes varies between 15 and 81% [4]. A summary of the percentage composition of starch-based agricultural wastes is shown in Table 2.

Many starch-based agricultural wastes have been explored for bioethanol production. Bioethanol from the starch-based wastes commences with substrate drying followed by milling, often to a particle size of 1–2 mm. After milling, the substrate is subjected to pretreatment followed by a liquefaction process using a liquefying enzyme (e.g. α-amylase). The enzymatic saccharification process is executed before fermentation proceeds (SHF) or simultaneously with the fermentation process (SSF). [60] subjected 20% (w/v) of cassava peels to SHF and obtained 24.86 g/L. Similarly, 25% (w/v) simultaneous saccharification and fermentation of pretreated cassava peels reportedly yielded 27 g/L of bioethanol [52]. Additionally, Aruwajoye et al. [5] obtained 21.26 g/L of bioethanol from 10.16% (w/v) pretreated cassava peels through an SSF process using *S. cerevisiae*.

On the other hand, Ben Taher et al. [11] obtained 5.7 g/L of bioethanol after pretreatment and SHF. Also, Sanusi et al. [68] obtained 0.26 g/g of bioethanol after the simultaneous saccharification and fermentation of pretreated potato peels with the inclusion of Fe_3O_4 as a nano-biocatalyst (Table 3).

6.3 Starch-Based Aquatic Plants

Aquatic plants are usually found in freshwater, saltwater, marine water, and open oceans. Some aquatic plants such as microalgae, giant water lilies, and water hyacinths contain starch which can be harnessed for bioethanol production [32]. The cultivation of microalgae as biomass for bioethanol production can be very advantageous because they can be grown with minimal inputs of water, nutrients,

Table 3 Percentage composition of starch sources utilized for bioethanol production

Starch sources	Starch source	Starch (%)	Cellulose (%)	Hemicellulose (%)	Lignin (%)	References
Corn flour	Starch-containing plant	84	NR	NR	NR	[15]
Raw cassava waste	Starch-based agricultural waste	52	13.4	9.35	11	[61]
Waste potato mash	Starch-based agricultural waste	17–24	NR	NR	NR	[33]
Cassava stem	Starch-based agricultural waste	15	22.8	28.8	22.1	[59]
Cassava leaves	Starch-based agricultural waste	2.1	17.3	27.7	20.1	[59]
Potato peels	Starch-based agricultural waste	48.46	NR	NR	NR	[41]
Cassava peels	Starch-based agricultural waste	81.4	NR	NR	1.5	[50]
Cassava pulp	Starch-containing plant	67.5	40	18	9	[2]
Potato peels	Starch-based agricultural waste	20	4.03	10	6.07	[18]
Pumpkin waste	Starch-based agricultural waste	65.3	NR	NR	NR	[19]

and energy. The cultivation of microalgae can be executed in open raceway ponds or closed photobioreactors. In the process of microalgae growth, they produce starch as stored energy through photosynthesis. The starch content of microalgae can vary (or be influenced) depending on the species and growing conditions. For instance, growing *Chlorela vulagaris P12* with a limited concentration of nitrogen led to a 40% increase in starch accumulation [21]. This starch can be harnessed from the microalgae using several pretreatment methods, such as physical, chemical, and enzymatic application before subsequent fermentation to bioethanol. Other routes to produce bioethanol from microalgae are photofermentation and dark fermentation [24]. The bioethanol potential of several microalgae strains has been reported [6, 30, 48]. *Scenedesmus bijugatus* biomass (26% total carbohydrate content) was successfully converted to bioethanol (70% conversion) after pretreatment with H_2SO_4 [48]. Similarly, the subjection of *Chlamydomonas reinhardtii* biomass (59.7% total carbohydrate content) to enzymatic hydrolysis with α-amylase and glucoamylase, and subsequent fermentation process gave a 60% bioethanol yield [6]. Also, hydrolysing *C. vulgaris FSP-E* with sulfuric acid treatment gave a 90% bioethanol yield after fermentation [30].

Another starch-based aquatic plant that has been explored for bioethanol production is the giant water lily. The giant water lily, also known as the lotus, is a hardy plant that can grow in a variety of aquatic conditions and has a high starch content in its roots. Aside from the starch content, the major advantages of using giant water lilies as a feedstock for bioethanol production are their rapid growth and ease of harvesting. After harvesting, the plant must be rinsed thoroughly under sanitized conditions to remove algal, bacterial, and fungal contaminants. The biomass is then pretreated through chemical, physical, or biological methods for conversion of the inherent starch to glucose before the fermentation process commences. Junluthin et al. [39] reported a laboratory-scale bioethanol production from a giant water lily where the biogas effluent was used to obtain 4.2 g/L of bioethanol.

Water hyacinth (*Eichhornia crassipes*) is another aquatic plant that can be utilized as a starch-based feedstock for bioethanol production. Native to South America, it is invasive and known for its fast growth and ability to outcompete native plant species. Water hyacinth is rich in starch and cellulose which can be extracted for bioethanol production [62]. The process of converting water hyacinth biomass to bioethanol involves first harvesting the plant, then cleaning, drying, and washing it to remove any dirt or debris. The biomass is then prepared through chopping, blending, or grinding until a sample size of less than 1 mm is attained. Water hyacinth biomass can then be subjected to further routine steps of chemical, physical, or biological pretreatment before undergoing the fermentation process to bioethanol.

7 Bioethanol Production from Lignocelluloses

Starch-based lignocelluloses have been reported as an excellent substrate/feedstock for bioethanol production [3, 69]. Following milling of the dried substrate, it is subjected to pretreatment, enzymatic saccharification, fermentation, and distillation as shown in Fig. 3 [10]. Biomass pretreatment is an unavoidable step in the improvement of pre-processing of biomasses such as starch-based lignocellulosic for bioethanol production. Starch-based lignocellulosic biomasses are pretreated to pretreatment activities to reduce their sizes, weaken the recalcitrant lignin wall, and facilitate subsequent susceptibility to enzymatic attack. The subsequent enzymatic hydrolysis and fermentation are described in Sects. 4 and 5, respectively.

8 Strategies for Improvement of Bioethanol Production from Novel Starch Sources

Implementing novel strategies for bioethanol production from starch sources is pertinent for improved yield and an economically feasible bioprocess. These methods include the following:

Fig. 3 Process of bioethanol production from starch substrates

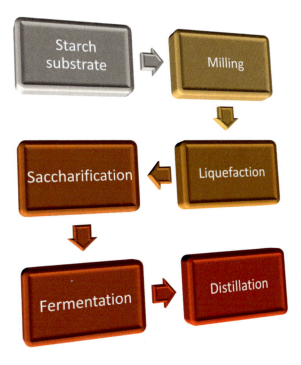

8.1 Engineered Simultaneous Saccharification and Fermentation (ESSF)

Engineered simultaneous saccharification and fermentation is a combination of SSF with bioprocess engineering. This process involves switching SSF from aerobic to fully anaerobic by stopping aeration activity in the reactor when the fermentative yeast approaches the end of the exponential growth phase [25]. This activity leads to the diversion of the yeast metabolic activity from respiration (and thus cellular multiplication) to fermentation, thus improving overall bioethanol yield. Farida et al. [25] obtained over a 17% increase in bioethanol yield from ESSF compared to the conventional SSF process using breadfruit), a starch-based crop.

8.2 Yeast-Modifying Activities and Immobilization

The introduction of yeast-modifying activities in starch-based bioethanol production is directed towards achieving bioethanol production in a single process instead of the routine three stages of liquefaction, enzymatic saccharification, and fermentation. This strategy will lower production costs, thereby making the bioprocess economically feasible. To achieve this, the bioethanol fermenting organism, *S. cerevisiae*, can be surface-engineered to display amylase and amyloglucosidase on the cell surface [72]. Thus, the modified yeast can liquefy the starch substrate, saccharify the slurry and ferment the sugar to bioethanol in a single process. Chen et al. [17] used *S. cerevisiae* co-displaying *Rhizopus oryzae* glucoamylase and *Streptococcus bovis* α-amylase on the cell surface for bioethanol production from uncooked raw starch. In seven days, the batch fermentation process generated 53 g/L of bioethanol from 50 g/L cells. Furthermore, immobilizing the cells within a loofa sponge in a packed bed reactor facilitated an additional 55% increase in repeated batch production [17]. Immobilizing microorganisms on various carriers, such as gel or loofa sponge, facilitates long-term bioethanol production for batch and continuous fermentation processes [58]. This is further demonstrated by Liu and Lien [45] using potato starch as a substrate. Three cycles of batch-SSF generating a stable 95.97 g/L bioethanol were achieved in a vertical mass flow bioreactor with the use of co-immobilized *Aspergillus. awamori*, *Rhizopus japonicas* (saccharification strains), and *Z. mobilis* (fermentation strain).

8.3 Mixed Starch-Based Substrate Concept

One of the limitations of commercial biomass-based biofuel is feedstock logistics [9, 64]. This is because of the challenges of seasonal availability of the plants from which the biomass is generated, coupled with the costs associated with collecting and transporting the substrates to the biorefinery. However, the mixture of more than one starch-based substrate during bioethanol production reduces the challenge of extensive storage of single biomass, which is usually done to make the substrate available for a prolonged season [4]. Furthermore, this strategy reduces the challenges associated with specialized personnel, resources, and equipment usually needed when strictly adhering to a single feedstock [65]. Moreover, it also meets the necessary conditions for multiple lignocellulosic biomass use in biofuel production. These include similar characteristics, the concession for enormous fermentable sugar yield potential (same composition of starch and lignocelluloses) as well as cheapness and abundance [42, 53].

9 Conclusion

Bioethanol production is a chemical product useful as a biofuel and precursor to other industrial bioproducts. The generation of bioethanol from cheap and affordable sources offers a promising solution to the challenge of carbon emissions from global fuel consumption. However, to overcome the bottlenecks of low product yield, food versus fuel factor, and feedstock logistics, highly dense and bioenergy-rich biomass should be utilized for the fermentation process. Novel starch sources have been the focus of intensive research lately. Many of them, such as starch-based lignocelluloses/

agricultural wastes, and aquatic plants have accessible starch in addition to their highly structured polysaccharides (hemicellulose, cellulose, and lignin). Upcoming challenges with using starch-based substrates for bioethanol can also be tackled with various strategies to increase substrate availability, yeast-modifying activities, and engineered fermentation processes. Thus, starch sources are viable for bioethanol generation in biomass-based biofuel processes.

References

1. Adewusi SR, Udio J, Osuntogun BA (1995) Studies on the carbohydrate content of breadfruit (*Artocarpus communis* Forst) from south-western Nigeria. Starch-Stärke 47(8):289–294
2. Akaracharanya A, Kesornsit J, Leepipatpiboon N, Srinorakutara T, Kitpreechavanich V, Tolieng V (2011) Evaluation of the waste from cassava starch production as a substrate for ethanol fermentation by *Saccharomyces cerevisiae*. Ann Microbiol 61(3):431–436
3. Aruwajoye GS, Faloye FD, Gueguim Kana E (2018) Process optimisation of enzymatic saccharification of soaking assisted and thermal pretreated cassava peels waste for bioethanol production. Waste Biomass Valorization. https://doi.org/10.1007/s12649-018-00562-0
4. Aruwajoye GS, Kassim A, Saha AK, Gueguim Kana EB (2020a) Prospects for the improvement of bioethanol and biohydrogen production from mixed starch-based agricultural wastes. Energies 13(24):6609
5. Aruwajoye GS, Sewsynker-Sukai Y, Kana EG (2020b) Valorisation of cassava peels through simultaneous saccharification and ethanol production: effect of prehydrolysis time, kinetic assessment and preliminary scale up. Fuel 278:118351
6. Ashokkumar V, Salam Z, Tiwari ON, Chinnasamy S, Mohammed S, Ani FN (2015) An integrated approach for biodiesel and bioethanol production from *Scenedesmus bijugatus* cultivated in a vertical tubular photobioreactor. Energy Convers Manage 101:778–786
7. Bajpai P (2013) Advances in bioethanol. Springer science & business media
8. Ballesteros M, Oliva JM, Negro MJ, Manzanares P, Ballesteros I (2004) Ethanol from lignocellulosic materials by a simultaneous saccharification and fermentation process (SFS) with *Kluyveromyces marxianus* CECT 10875. Process Biochem 39(12):1843–1848
9. Banerjee S, Mudliar S, Sen R, Giri B, Satpute D, Chakrabarti T, Pandey R (2010) Commercializing lignocellulosic bioethanol: technology bottlenecks and possible remedies. Biofuels, Bioprod Bioref: Inno Sustain Econ 4(1):77–93
10. Behera S, Kar S, Chandra R (2010) Comparative study of bio-ethanol production from mahula (*Madhuca latifolia* L.) flowers by *Saccharomyces cerevisiae* cells immobilized in agar agar and Ca-alginate matrices. Appl Energy 87:96–100
11. Ben Taher I, Fickers P, Chniti S, Hassouna M (2017) Optimization of enzymatic hydrolysis and fermentation conditions for improved bioethanol production from potato peel residues. Biotechnol Prog 33(2):397–406
12. Betiku E, Ajala O (2010) Enzymatic hydrolysis of breadfruit starch: case study with utilization for gluconic acid production. Ife J Technol 19(1):10–14
13. Betiku E, Alade O (2014) Media evaluation of bioethanol production from cassava starch hydrolysate using *Saccharomyces cerevisiae*. Energy Sources, Part A: Recov Util Environ Effects 36(18):1990–1998
14. Betiku E, Taiwo AE (2015) Modeling and optimization of bioethanol production from breadfruit starch hydrolyzate vis-à-vis response surface methodology and artificial neural network. Renew Energy 74:87–94
15. Białas W, Szymanowska D, Grajek W (2010) Fuel ethanol production from granular corn starch using *Saccharomyces cerevisiae* in a long term repeated SSF process with full stillage recycling. Biores Technol 101(9):3126–3131

16. Bourdichon F, Casaregola S, Farrokh C, Frisvad JC, Gerds ML, Hammes WP, Harnett J, Huys G, Lauland S, Ouwehand A, Powell IB, Prajapati JB, Seto Y, Schure ET, Boven AV, Vankerckhoven V, Zgoda A, Tuijtelaars S, Hansen EB (2012) Food fermentations: 241 microorganisms with technological beneficial use. Int J Food Microbiol 154:87–97
17. Chen JP, Wu KW, Fukuda H (2007) Bioethanol production from uncooked raw starch by immobilized surface-engineered yeast cells. In: Biotechnology for Fuels and Chemicals Springer, pp 59–67
18. Chohan NA, Aruwajoye G, Sewsynker-Sukai Y, Kana EG (2020) Valorisation of potato peel wastes for bioethanol production using simultaneous saccharification and fermentation: process optimization and kinetic assessment. Renew Energy 146:1031–1040
19. Chouaibi M, Daoued KB, Riguane K, Rouissi T, Ferrari G (2020) Production of bioethanol from pumpkin peel wastes: comparison between response surface methodology (RSM) and artificial neural networks (ANN). Ind Crops Prod 155:112822
20. Deenanath ED (2014) Production and characterization of bioethanol derived from cashew apple juice for use in internal combustion engine. PhD Thesis. University of the Witwatersrand, Johannesburg, South Africa
21. Dragone G, Fernandes BD, Abreu AP, Vicente AA, Teixeira JA (2011) Nutrient limitation as a strategy for increasing starch accumulation in microalgae. Appl Energy 88(10):3331–3335
22. Van Dyk J, Pletschke B (2012) A review of lignocellulose bioconversion using enzymatic hydrolysis and synergistic cooperation between enzymes—factors affecting enzymes, conversion and synergy. Biotechnol Adv 30(6):1458–1480
23. El-Mekkawi SA, Abdo SM, Samhan FA, Ali GH (2019) Optimization of some fermentation conditions for bioethanol production from microalgae using response surface method. Bull Nat Res Centre 43:1–8
24. de Farias Silva CE, Bertucco A (2016) Bioethanol from microalgae and cyanobacteria: a review and technological outlook. Process Biochem 51(11):1833–1842
25. Farida I, Syamsu K, Rahayuningsih M (2015) Direct ethanol production from breadfruit starch (*Artocarpus communis* Forst.) by Engineered simultaneous saccharification and fermentation (ESSF) using Microbes Consortium. Int J Renew Energy Dev 4(1):25
26. Food and Agriculture Organization (FAO) (2008) The state of food and agriculture 2008. Biofuels: prospects, risks and opportunities. http://www.fao.org/3/i0100e/i0100e.pdf. Accessed 16 March 2023
27. Friedl A (2017) Bioethanol from sugar and starch. In: Meyers RA (ed) Encyclopedia of sustainability science and technology. Springer New York, New York, NY, pp 1–21
28. Gunjal AB, Patil NN, Shinde SS (2020) Ligninase in degradation of lignocellulosic wastes. In: Enzymes in degradation of the lignocellulosic wastes. Springer, pp 55–70
29. Hanif M, Mahlia T, Aditiya H, Chong W (2016) Techno-economic and environmental assessment of bioethanol production from high starch and root yield Sri Kanji 1 cassava in Malaysia. Energy Rep 2:246–253
30. Ho SH, Huang SW, Chen CY, Hasunuma T, Kondo A, Chang JS (2013) Bioethanol production using carbohydrate-rich microalgae biomass as feedstock. Biores Technol 135:191–198
31. Hoang TD, Nghiem N (2021) Recent developments and current status of commercial production of fuel ethanol. Ferment 7(4):314
32. Hochman G, Palatnik RR (2022) The economics of aquatic plants: The case of algae and duckweed. Ann Rev Resour Econ 14:555–577
33. Izmirlioglu G, Demirci A (2012) Ethanol production from waste potato mash by using *Saccharomyces cerevisiae*. Appl Sci 2(4):738–753
34. Jeon B-Y, Kim D-H, Na B-K, Ahn D-H, Park D-H (2008) Production of ethanol directly from potato starch by mixed culture of *Saccharomyces cerevisiae* and *Aspergillus niger* using electrochemical bioreactor. J Microbiol Biotechnol 18(3):545–551
35. Jin M, Lau MW, Balan V, Dale BE (2010) Two-step SSCF to convert AFEX-treated switchgrass to ethanol using commercial enzymes and *Saccharomyces cerevisiae* 424A (LNHST). Bioresour Technol 101(21):8171–8178

36. Jin M, Sarks C, Gunawan C, Bice BD, Simonett PS et al (2013) Phenotypic selection of a wild *Saccharomyces cerevisiae* strain for simultaneous saccharification and co-fermentation of AFEX pretreated corn stover. Biotechnol Biofuels 6(108):1–14
37. Joginder SD, Ashok K, Sunil KT (2013) Bioethanol production from starchy part of tuberous plant (potato) using *Saccharomyces cerevisiae* MTCC-170. African J Microbiol Res 7(46):5253–5260
38. Jugwanth Y, Sewsynker-Sukai Y, Kana EG (2020) Valorization of sugarcane bagasse for bioethanol production through simultaneous saccharification and fermentation: Optimization and kinetic studies. Fuel 262:116552
39. Junluthin P, Pimpimol T, Whangchai N (2021) Efficient conversion of night-blooming giant water lily into bioethanol and biogas. Maejo Int J Energy Environ Commun 3(2):38–44
40. Kawa-Rygielska J, Pietrzak W, Regiec P, Stencel P (2013) Utilization of concentrate after membrane filtration of sugar beet thin juice for ethanol production. Bioresour Technol 133:134–141
41. Khawla BJ, Sameh M, Imen G, Donyes F, Dhouha G, Raoudha EG, Oumèma N-E (2014) Potato peel as feedstock for bioethanol production: a comparison of acidic and enzymatic hydrolysis. Ind Crops Prod 52:144–149
42. Kim KH, Tucker M, Nguyen Q (2005) Conversion of bark-rich biomass mixture into fermentable sugar by two-stage dilute acid-catalyzed hydrolysis. Biores Technol 96(11):1249–1255
43. Kumar D, Murthy GS (2013) Stochastic molecular model of enzymatic hydrolysis of cellulose for ethanol production. Biotechnol Biofuels 6(1):1–20
44. Li J, Lin J, Zhou P, Wu K, Liu H, Xiong C, Gong Y, Xiao W, Liu Z (2014) One-pot simultaneous saccharification and fermentation: a preliminary study of a novel configuration for cellulosic ethanol production. Biores Technol 161:171–178
45. Liu Y-K, Lien P-M (2016) Bioethanol production from potato starch by a novel vertical mass-flow type bioreactor with a co-cultured-cell strategy. J Taiwan Inst Chem Eng 62:162–168
46. Mahapatra S, Manian RP (2017) Bioethanol from lignocellulosic feedstock: a review. Res J Pharm Technol 10(8):2750–2758
47. Maitan-Alfenas GP, Visser EM, Guimarães VM (2015) Enzymatic hydrolysis of lignocellulosic biomass: converting food waste in valuable products. Curr Opin Food Sci 1:44–49
48. Miranda JR, Passarinho PC, Gouveia L (2012) Pre-treatment optimization of *Scenedesmus obliquus* microalga for bioethanol production. Biores Technol 104:342–348
49. Moscoviz R, Kleerebezem R, Rombouts JL (2021) Directing carbohydrates toward ethanol using mesophilic microbial communities. Curr Opin Biotechnol 67:175–183
50. Moshi AP, Crespo CF, Badshah M, Hosea KM, Mshandete AM, Elisante E, Mattiasson B (2014) Characterisation and evaluation of a novel feedstock, *Manihot glaziovii*, Muell. Arg, for production of bioenergy carriers: bioethanol and biogas. Biores Technol 172:58–67
51. Moshi AP, Hosea KM, Elisante E, Mamo G, Mattiasson B (2015a) High temperature simultaneous saccharification and fermentation of starch from inedible wild cassava (*Manihot glaziovii*) to bioethanol using *Caloramator boliviensis*. Biores Technol 180:128–136
52. Moshi AP, Temu SG, Nges IA, Malmo G, Hosea KM, Elisante E, Mattiasson B (2015b) Combined production of bioethanol and biogas from peels of wild cassava *Manihot glaziovii*. Chem Eng J 279:297–306
53. Nilsson D, Hansson P-A (2001) Influence of various machinery combinations, fuel proportions and storage capacities on costs for co-handling of straw and reed canary grass to district heating plants. Biomass Bioenerg 20(4):247–260
54. Oke MA, Annuar MSM, Simarani K (2016) Mixed feedstock approach to lignocellulosic ethanol production—prospects and limitations. BioEnergy Res 9:1189–1203
55. Olofsson K, Bertilsson M, Lidén G (2008) A short review on SSF—an interesting process option for ethanol production from lignocellulosic feedstocks. Biotechnol Biofuels 1(7):1–14
56. Olofsson K, Wiman M, Lidén G (2010) Controlled feeding of cellulases improves conversion of xylose in simultaneous saccharification and co-fermentation for bioethanol production. J Biotechnol 145:168–175

57. Pervez S, Aman A, Iqbal S, Siddiqui NN, Qader SAU (2014) Saccharification and liquefaction of cassava starch: an alternative source for the production of bioethanol using amylolytic enzymes by double fermentation process. BMC Biotechnol 14:49
58. Phisalaphong M, Budiraharjo R, Bangrak P, Mongkolkajit J, Limtong S (2007) Alginate-loofa as carrier matrix for ethanol production. J Biosci Bioeng 104(3):214–217
59. Pooja N, Padmaja G (2015) Enhancing the enzymatic saccharification of agricultural and processing residues of cassava through pretreatment techniques. Waste Biomass Valorization 6(3):303–315
60. Pooja N, Sajeev M, Jeeva M, Padmaja G (2018) Bioethanol production from microwave-assisted acid or alkali-pretreated agricultural residues of cassava using separate hydrolysis and fermentation (SHF). 3 Biotech 8(1):1–12
61. Pothiraj C, Arun A, Eyini M (2015) Simultaneous saccharification and fermentation of cassava waste for ethanol production. Biofuel Res J 2(1):196–202
62. Pratama JH, Amalia A, Rohmah RL, Saraswati TE (2020) The extraction of cellulose powder of water hyacinth (*Eichhornia crassipes*) as reinforcing agents in bioplastic. In: AIP conference proceedings 2219, 1, 100003. AIP Publishing LLC
63. Renewable Fuel Association. https://ethanolrfa.org/markets-and-statistics/annual-ethanol-production. Accessed 16 Nov 2022
64. Rentizelas A, Tatsiopoulos I, Tolis A (2009a) An optimization model for multi-biomass trigeneration energy supply. Biomass Bioenerg 33(2):223–233
65. Rentizelas AA, Tolis AJ, Tatsiopoulos IP (2009b) Logistics issues of biomass: the storage problem and the multi-biomass supply chain. Renew Sustain Energy Rev 13(4):887–894
66. Robyt JF (2008) Starch: structure, properties, chemistry, and enzymology. In: Glycoscience. Springer, Berlin, Heidelberg, pp 1437–1472
67. Rorke DCS, Kana EBG (2017) Kinetics of bioethanol production from waste sorghum leaves using *Saccharomyces cerevisiae* BY4743. Fermentation 3:19
68. Sanusi IA, Faloye FD, Gueguim Kana E (2019) Impact of various metallic oxide nanoparticles on ethanol production by *Saccharomyces cerevisiae* BY4743: screening, kinetic study and validation on potato waste. Catal Lett 149(7):2015–2031
69. Sanusi AI, Suinyuy TN, Kana GEB (2021) Impact of nanoparticle inclusion on bioethanol production process kinetic and inhibitor profile. Biotechnol Reports 29:e00585
70. Sarkar N, Ghosh SK, Bannerjee S, Aikat K (2012) Bioethanol production from agricultural wastes: an overview. Renew Energy 37(1):19–27
71. Shanavas S, Padmaja G, Moorthy S, Sajeev M, Sheriff J (2011) Process optimization for bioethanol production from cassava starch using novel eco-friendly enzymes. Biomass Bioenerg 35(2):901–909
72. Shigechi H, Uyama K, Fujita Y, Matsumoto T, Ueda M, Tanaka A, Fukuda H, Kondo A (2002) Efficient ethanol production from starch through development of novel flocculent yeast strains displaying glucoamylase and co-displaying or secreting α-amylase. J Mol Catal B Enzym 17(3–5):179–187
73. Statistica. https://www.statista.com/statistics/281606/ethanol-production-in-selected-countries/. Accessed 16 Nov 2022
74. Sun Y, Cheng J (2002) Hydrolysis of lignocellulosic materials for ethanol production: a review. Biores Technol 83(1):1–11
75. Thakur R, Pristijono P, Scarlett CJ, Bowyer M, Singh SP, Vuong QV (2019) Starch-based films: major factors affecting their properties. Int J Biol Macromol 132:1079–1089
76. The Future of Biofuels: A Global Perspective. Available online: https://www.ers.usda.gov/amber-waves/2007/november/the-future-of-biofuels-a-global-perspective/. Accessed 16 Nov 2022
77. Tomás-Pejó E, Oliva J, Ballesteros M, Olsson L (2008) Comparison of SHF and SSF processes from steam-exploded wheat straw for ethanol production by xylose-fermenting and robust glucose-fermenting *Saccharomyces cerevisiae* strains. Biotechnol Bioeng 100(6):1122–1131
78. U.S. Energy Information Administration. Biofuels Explained, Ethanol. Available online: https://www.eia.gov/energyexplained/biofuels/ethanol.php. Accessed 16 Nov 2022

79. Vasić K, Knez Ž, Leitgeb M (2021) Bioethanol production by enzymatic hydrolysis from different lignocellulosic sources. Molecules 26(3):753
80. Vohra M, Manwar J, Manmode R, Padgilwar S, Patil S (2014) Bioethanol production: feedstock and current technologies. J Environ Chem Eng 2(1):573–584
81. Watanabe I, Miyata N, Ando A, Shiroma R, Tokuyasu K, Nakamura T (2012) Ethanol production by repeated-batch simultaneous saccharification and fermentation (SSF) of alkali treated rice straw using immobilized *Saccharomyces cerevisiae* cells. Bioresour Technol 123:695–698
82. Young AH (1984) Fractionation of starch. In: Starch: chemistry and technology, 2nd Ed. Elsevier, Amsterdam, The Netherland, pp 249–283
83. Zabed H, Faruq G, Sahu J, Boyce A, Ganesan PA (2016) Comparative study on normal and high sugary corn genotypes for evaluating enzyme consumption during dry-grind ethanol production. Chem Eng J 287:691–703
84. Zabed H, Sahu J, Suely A, Boyce A, Faruq G (2017) Bioethanol production from renewable sources: current perspectives and technological progress. Renew Sustain Energy Rev 71:475–501
85. Zhao X, Zhang L, Liu D (2012) Biomass recalcitrance. Part I: the chemical compositions and physical structures affecting the enzymatic hydrolysis of lignocellulose. Biofuels, Bioprod Bioref 6(4):465–482

Bioethanol Production from Lignocellulosic Wastes: Potentials and Challenges

Esra Meşe Erdoğan, Pınar Karagöz, and Melek Özkan

Abstract The need for new energy sources is becoming more evident due to the increase in global population, development of industrialization, and limitations in the recovery, supply, and use of fossil fuels. Second-generation bioethanol produced from lignocellulosic agricultural wastes by fermentation is regarded as a reliable alternative energy source in terms of the food-water-energy security nexus as it will enable the use of lignocellulosic wastes remaining in agricultural lands and have a low contribution to greenhouse gas emission. Although lignocellulosic bioethanol production technology has great potential, there are some challenges preventing the widespread use of this energy including difficulty in removing lignin, costly enzymes for efficient saccharification, insufficient fermentation of sugars other than glucose found in the structure of lignocellulose, and limited ethanol production due to product inhibition. Recent studies show that new methods and technologies can solve these obstacles. In this chapter, the potential of lignocellulosic bioethanol is evaluated by analyzing new processes or inventions in the 2G bioethanol industry and the complex dynamics of the sector.

Keyword Bioethanol · Cellulases · Lignocellulosic wastes · Fermentation · Saccharification

1 Introduction

Today, most of the energy requirement in the World is supplied by fossil fuels, including oil, natural gas, and coal. The population and Gross Domestic Product (GDP) growth rate of each country tend to increase each year, resulting in a considerable rise in energy demands. It is obvious that no nation will restrain its economic

---

E. M. Erdoğan · M. Özkan (✉)
Department of Environmental Engineering, Gebze Technical University, Gebze-Kocaeli, Turkey
e-mail: mozkan@gtu.edu.tr

P. Karagöz
Department of Biochemical Engineering, University College of London, London, UK

© The Author(s), under exclusive license to Springer Nature Switzerland AG 2023
E. Betiku and M. M. Ishola (eds.), *Bioethanol: A Green Energy Substitute for Fossil Fuels*, Green Energy and Technology,
https://doi.org/10.1007/978-3-031-36542-3_6

growth for slowing down the depletion of energy sources. Fossil fuel reserves are limited since it is impossible to regenerate them at the rate they are consumed. The effects of limited fossil fuel reserves are felt mainly in countries that must import oil from another country, especially during times of crisis such as the Covid-19 pandemic or war between nations. In the last two years, we have experienced a tremendous increase in oil prices. These facts bring the need for accelerating research and development studies on alternative energy sources and increasing their use. Bioethanol production from lignocellulosic waste (2G bioethanol) is a promising solution for countries whose economy is based on agriculture and whose oil reserves are limited [41].

Global renewable energy resources include various types of energy sources such as wind energy, solar energy, hydro energy, and biomass energy. Solar, hydro, and wind energies are primarily used for power generation [178]. Energy based on biomass consists of biogas and liquid biofuels such as bioethanol and biobutanol. First-generation (1G) bioethanol defines as bioethanol produced from edible parts of plants. 1G bioethanol production is well commercialized globally, and it is widely used as a fuel blend in Brazil and USA, which are leading countries in bioethanol production. 1G bioethanol production from cane sugar, corn, and beet sugar provides high yields since lignin-containing residues of the plants are not included in the process as raw material. Offering a low CO_2 emission, bioethanol is a near-carbon–neutral solution. It also helps to decrease other Greenhouse gases (GHG) in the atmosphere. However, due to its effects on vital natural resources like the climate, water, food, and land, 1G bioethanol has some serious drawbacks [2, 106]. Ethanol production from food sources needs high amounts of water due to the cultivation of crops. According to reports, the development of biofuels increased the price of oilseeds by 4% between 2000 and 2010 and cereals worldwide by 1–2% during that time [84]. Moreover, the demand for land for expanding the agriculture of energy crops can create a conflict in terms of land use for food or energy [173]. More sustainable ways of bioethanol production are needed due to the precarious state of the World's water and food supplies [4]. Today, fossil fuel derived from petroleum is the predominant energy source used in transportation. It is also the major source of CO_2 emissions in the atmosphere [40]. The updated version Renewable Energy Directive (RED II), released in December 2018, set a legally enforceable goal of using at least 32% renewable energy by 2030, especially in transportation, where the goal was set at 14%. Furthermore, RED II mandates that by 2030, all EU Member States must employ 3.5% advanced biofuel usage in transportation [95].

As biomass wastes are plentily available and have the potential to be utilized for energy production, they are regarded as valuable feedstocks. The utilization of lignocellulosic waste for energy generation will help to decrease environmental issues and prevent the loss of economic value of the waste itself [37]. Lignocellulosic ethanol (2G bioethanol) is a promising renewable energy source that will remarkably help create a sustainable environment and circular economy. However, from lignin removal to sugar recovery and fermentation, there are many challenging steps in the process of lignocellulosic bioethanol. There has been an enormous amount of research to increase the efficiency of bioethanol production from lignocelluloses, which has resulted in a slight but gradual decrease in the cost of production and opened the way for its competition with fossil fuels. It was also promising that cellulosic feedstocks have had lower purchase costs than petroleum in the last decade [98]. The unavoidable increase in oil prices forces countries to find alternative energy production ways. The use of renewable fuels should become widespread not only because of the increase in oil prices but also to protect nature.

The potential of secondary bioethanol cannot be underestimated due to the need for low-carbon fuel for the transportation sector. Several international organizations, including OECD and the European Union, are committed to guiding renewable energy policies [106]. However, the experiences of pioneer cellulosic ethanol facilities established in the last decade have indicated that lignocellulosic ethanol production with current methods and technologies is not cost-competitive enough compared to stand-alone fuel production [96]. Novel approaches are needed to be developed or currently used methods should be improved for more economical 2G bioethanol production. In this chapter, in addition to the conventional ways of 2G bioethanol production, recently applied methods and technologies are analyzed, and the potentials and challenges of the 2G bioethanol industry are presented.

2 Lignocellulosic Feedstocks for Secondary Bioethanol Production

Bioenergy, with around 14% of the current total energy share, contributes to the World's energy demand [145]. Bioethanol can play a critical role in developing countries energy and economic security if generated locally from native biomass [133]. Common glucose and starch-based carbon sources for bioethanol production, such as corn and sugarcane, have importance as primary food sources. So existing legislation prohibits using these food crops as an energy feedstock in some countries. After 2020, the European Union decided to promote and financially support biofuels generated from lignocellulosic or algae biomass instead of food or feed crops [115]. For these reasons, researchers focus on finding suitable biomass for production. Many researchers have widely investigated bioethanol production using beverage or paper industry residues, agricultural crop wastes, and forestry plant residues [66, 90, 113, 193]. Lignocellulosic biomass, with an annual production of 181.5 billion tons, is considered the most available and abundant feedstock for biofuel production. Also, seven billion tons of biomass are produced from agricultural, grass, and forest lands [36]. Currently, most of the lignocellulosic biomass is burnt in the field or deposited in open landfills without any benefit.

Using lignocellulosic wastes for energy production has two key benefits: it is a renewable source for energy production. It recycles carbon rather than removing it from long-term storage, as in the case of fossil fuels [35]. High cellulose-containing lignocellulosic feedstocks are being considered as (1) agricultural residues such as barley straw, corn stover, and sugarcane bagasse, (2) non-grain or non-food energy crops like poplar, willow, miscanthus, sorghum, energy cane, (3) forest residues and (4) industrial wastes [87]. It is important that not all biomass types are suitable for economically viable biofuel production because of the complex composition and conversion characteristics and the presence of unfavorable contaminants [26, 69, 157].

2.1 Forest Biomass

Currently, forest areas cover approximately 30% of the land on Earth [90]. Forest biomass consists mainly of woody substances like hardwoods and softwoods. Softwoods are low-density and quickly-growing trees like pine, cedar, spruce, cypress, fir, hemlock, and redwood, whereas hardwoods are poplar, willow, oak, poplar, and aspen [184].

Bioethanol production from forest biomass may incorporate into the current wood value chains without placing additional strain on terrestrial ecosystems and enhance a circular economy approach [29]. *Eucalyptus globulus*, which is used in pulp manufacturing, is a hardwood with a rotation period of 5–10 years. It is estimated that 0.2 tons of bark residues are generated from each ton of bleached kraft pulp products [134]. Bioethanol production from hydrothermally pretreated *E. globulus* bark residues was investigated by [57]. Due to their high lignin content, thermochemical processes are more suitable than biochemical processes for bioethanol production from woody species [87, 110]. Biorefineries have already started to be established for bioethanol production from woody biomass. The Canadian company, Lignol, is attempting to construct biorefineries to manufacture biochemicals and fuel-grade ethanol from forest biomass and also supplies other cellulosic intermediates [108].

2.2 Agricultural Crop Residues

Globally, agricultural activities such as farming and food processing result in billions of tons of organic waste, such as corn stover, straws of corn, rice and wheat, and sugarcane bagasse. It is reported that almost 12 billion tons of solid waste are generated from agricultural activities [11, 48]. Within the past decades, advanced agricultural methods have been applied to meet the rising food demands of the increased population (approximately 23.7 million tons of food per day) [104]. Agricultural lands have been expanded by deforestation, leading to a rise in agro-industrial wastes and crop residues [55]. Figure 1 shows the amount of agricultural residue production in the World. China is the World's leader in agricultural residue production (Fig. 1). The country produces 900 million tons of residues, including rice straw, wheat straw, and corn stover [169]. The USA, the largest corn-growing country, has 215 million tons of residues annually, which consists of 150 M tons of corn stover and 65 M tons of wheat

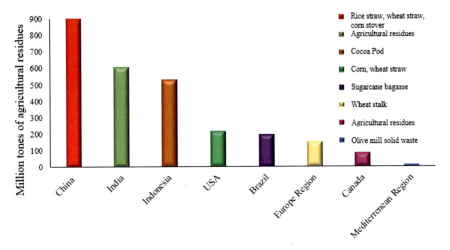

Fig. 1 Global agricultural residue production. The figure was adapted from the data given by Zhao et al. [190]

straw. It is estimated that 605 million tons of agricultural residue are produced annually [190] in India, one of the largest cereal-producer countries. Complete harvesting of all biomass residues from agricultural lands can reduce the soil organic matter and may cause soil erosion. Therefore, a proportion of these lignocellulosics should be remained in the field depending on the crop rotation, soil fertility, land slope, [73]. It is reported that approximately 1.7 Megagrams of wheat residue per hectare should be left in the field to prevent soil erosion [76].

Most lignocellulosic biomasses are mismanaged either by incineration or disposing of inappropriate landfills [59]. That results in a rise in environmental problems of groundwater and surface water pollution, greenhouse gas emissions, land pollution, and soil erosion. Burning fossil fuels and agricultural waste resulted in 36.73 tera-grams of CO_2 equivalent in total annual emissions [145]. Due to environmental concerns, the bioconversion of agro-waste into value-added products like biofuels, biogas, and biochar to reduce the dependence on fossil fuels has gained much more importance than before [81]. Recovery of leftovers from fruit and vegetable crops reduces pollution emissions and the investment in raw materials while generating new business models and local job opportunities [48].

The agricultural waste biomass consists of cellulose, hemicellulose, and lignin, which are important resources for value-added material production. A typical agricultural waste contains about 35–50% cellulose, 20–35% hemicellulose, and 15–20% lignin [13]. Cellulose and hemicelluloses can be used as a carbon source for microbial fermentation after hydrolyzing to monomeric sugars. Lignin is also important for producing polymer or thermal energy [12].

Bioethanol production from agricultural residues has been extensively investigated. For instance, Braide et al. [20] compared the bioethanol production efficiency of different kinds of agricultural residues, including sugarcane bagasse, sugarcane bark, cornstalk, corncob, and cornhusk and they found that these residues gave a maximum percentage ethanol yield of 6.72, 6.23, 6.17, 4.17, and 3.45, respectively. The use of a single type of biomass may not be economically feasible for biorefineries. Nguyen et al. [116] pretreated a mixture of agricultural biomass with popping at 1.47 MPa and 150 °C; they obtained an 88.1% maximum ethanol conversion yield.

2.3 Lignocellulosic Industrial Waste

Due to their high polysaccharide content, paper and pulp industry wastes may be a promising feedstock for advanced bioethanol production. Their use for bioethanol production also avoids landfill disposal, lowers the environmental effects, and provides economic benefits. Only about 45% of raw forest biomass can be turned into pulp, so the remaining portion has the potential to be used in biofuel production. Several types of residues produced from pulp and paper industries are primary and secondary sludges, bark, black and brown liquor, and lignin [139]. Sugars in pulp and paper industry sludges are known to be more accessible to enzymes due to the physicochemical processes in pulp and paper production. These characteristics of biomass prevent the costly pretreatment steps required for other unprocessed lignocellulosic feedstocks [139]. However, cost-effectiveness is still the biggest barrier to expanding these biorefineries.

There are studies on bioethanol production from various waste biomass such as livestock manure and municipal solid waste etc. For instance, more than 1100 Mton of livestock manure is produced annually in the European Union, primarily from pig and dairy farming [120]. Yan et al. used cow manure as a lignocellulosic carbon source to produce bioethanol. The ethanol yield with *Saccharomyces cerevisiae* LF1 was 0.19 g/g raw biomass after enzymatic hydrolysis, followed by alkali pretreated [174].

It is estimated that global municipal solid waste is approximately 2.01 billion tons annually, and 46% of urban waste effluent is biodegradable [159]. In a study by Mahmoodi et al. [99], bioethanol and biogas were produced from municipal waste, and 10,453 kJ energy, and 326.6 mL gasoline-equivalent per kg of dry biomass were generated.

2.4 Other Sectors in Need of Lignocellulosic Biomass

The need for lignocellulosic wastes in different areas may increase the economic value of these wastes, which brings the additional cost to the production of 2G bioethanol. Lignocellulosic biomass, with 1.3×10^{10} tons annual production, is a natural, renewable feedstock that can be converted to various high-value-added products. The utilization of lignocellulosic biomass can be divided into three categories, including thermochemical, biological, and chemical processes [91]. Combustion (to generate thermal energy or electricity), gasification (to produce H_2, CO, CO_2, and CH_4), and pyrolysis (to convert biomass to bio-oil, biochar, and gas) are thermochemical processes [75, 117]. Currently, lignocellulosic biomass, such as forest biomass, has been used for electricity generation on a commercial scale. Low efficiency and high energy demand of the thermochemical processes promote the investigation of the alternative high-value utilization of lignocellulosic biomass.

In the biological process, the lignocellulosic feedstock can be used to convert biomass to energy and biomaterials, e.g., bioethanol, biogas, biohydrogen, and bioproducts such as amino acids, organic acids, and vitamins [93]. Microorganisms can degrade and use cellulose and hemicellulose, the main sources of sugar and energy; some filamentous fungi have enzymes for breaking lignin, the most recalcitrant component of biomass. Unlike synthetic polymers, lignin is biodegradable and reported as the strongest biopolymer. It can also be used as a natural adhesive to produce composite materials [54]. It can be used for preparing synthetic aromatic polymer products such as polyimide, thermoplastics, phenolic resin, and composite films [165].

A high proportion of lignocellulosic biomass and industrial wastes have been studied to produce pulp, biofuel, nanofibrous materials, etc. Recently, the use of lignocellulosic biomass for the production of energy storage materials gained great attention [165]. Lignin-based chemicals can be supplied to the global market for application in different sectors, including construction, agrochemicals, animal feed, ceramics, lead acid batteries, soil containers, and gypsum boards [128].

The uses for lignocellulosic wastes vary depending on regional requirements such as the number of industries and diversity of industrial sectors, energy resources, and living conditions. According to a study conducted in China, wheat straw is used 72% for domestic fuel, food, fertilizer, and mushroom farming, 21% for energy production, and 7% for urban use. The uses of biomass can also vary depending on the location (urban and rural). In a city with substantial coal deposits, just 4% of straw is used as household fuel, with the remaining 90% being used as feed and fertilizer. However, the requirement for biomass in urban areas is bigger than in rural areas in terms of population density and traffic. For example, most of the lignocellulosic biomass power generation plants are located near industrially developed cities [88].

3 Lignocellulosic Bioethanol Production Process

Lignocellulosic bioethanol production is regarded as a sustainable way of energy generation due to waste utilization and reduction of CO_2 emission by the plants, which are the main raw material in the process. 2G bioethanol production starts with biomass harvesting and collection, and then transportation to the biorefinery (Fig. 2). After processing and biofuel production, biorefinery waste can be used for animal feed or electricity production. Figure 3 gives details of the process. Depending on the type of biomass, an appropriate size reduction (e.g., milling, grinding) is applied as a pre-processing step. Bioethanol production involves four steps: pretreatment,

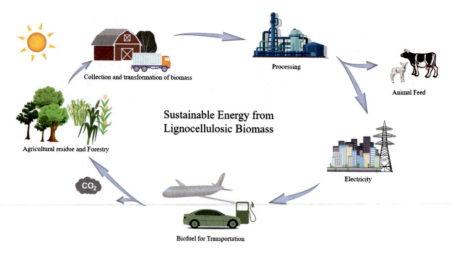

Fig. 2 Production and utilization of 2G bioethanol

hydrolysis, fermentation, and product upgrading [149]. In this part of the chapter, detailed information about 2G bioethanol production steps is presented. Examples of novel pretreatment methods applied to increase production efficiency are also included in related sections.

3.1 Pretreatment and Hydrolysis

Because of its complex structure, saccharides in lignocellulosic biomass are not easily available for fermentative microorganisms. Effective pretreatment techniques are required to make sugars accessible to enzymes and microorganisms [111]. Also, selecting the most appropriate pretreatment method is critical for economic bioethanol production. Two main targets of the pretreatment are, increasing the surface area of the lignocellulosic particles and separating the cellulose from hemicellulose and lignin fractions. These targets can be achieved by different types of pretreatment methods, including physical (hacking, grinding, milling, rolling, sonication, microwave radiation, etc.), chemical (acid pretreatment, alkali pretreatment, oxidation, organosolv pretreatment, etc.), physicochemical (ammonia fiber, steam or carbon dioxide explosions, and extrusion, etc.), biological (microbial and enzymatic treatment) and combined pretreatments [68]. Physicochemical pretreatment methods are applied to remove hemicellulose and so improve enzyme accessibility. Chemical pretreatment permits increasing the surface area of lignocellulose particles to provide a high reaction rate and efficiency and helps remove hemicellulose. Biological pretreatment can also degrade cellulose and hemicellulose to sugar monomers [149]. Current pretreatment methods are shown in Fig. 4. Among chemical pretreatments, the most studied and applied pretreatment methods include dilute acid and

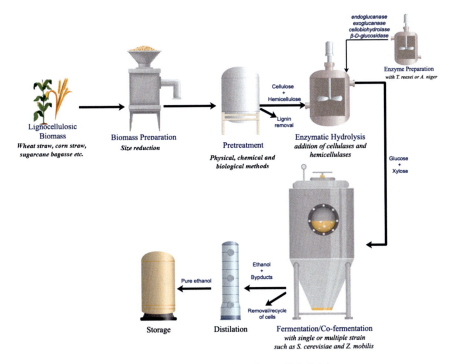

Fig. 3 The main steps of bioethanol production from lignocellulosic biomass

alkali pretreatments. Steam explosion is the main physical pretreatment. Related sections include several examples from literature to give an idea about the efficiency of the pretreatment methods.

(i) **Alkali pretreatment**: Alkaline solutions such as sodium hydroxide, ammonium hydroxide, lime, and sulfite have been used in alkaline pretreatment. Selective removal of lignin, simplicity of the process, mild reaction conditions (low chemical loading and low temperature), and low inhibitory by-product formation, which enhances enzymatic hydrolysis, are the most important advantages of alkaline pretreatment [89]. One disadvantage of the alkaline pretreatment is the long reaction time requirement ranging from several hours to one day. Sodium hydroxide is the most widely used alkaline solution, among others. The major effect of alkaline pretreatment is biomass delignification, as, in alkaline conditions, alkyl-aryl bonds in lignin can be easily broken [167]. Delignification via alkali pretreatment can increase the porosity and surface area of biomass and allows more enzymes to penetrate biomass and convert cellulose to monomeric sugars [188]. Liu et al. [89] performed alkali pretreatment for corn stover at 99 °C for 1 h and reported that the lignin content of biomass decreased to 20.49% in the presence of 0.06 g NaOH/g corn stover. However, they found the mechanical-alkali combined treatment more efficient than the alkali alone. In another study, Liu et al. [6] demonstrated that sodium hydroxide pretreatment effectively extracted lignin and hemicelluloses from

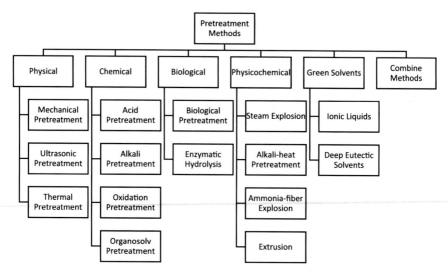

Fig. 4 Pretreatment methods of lignocellulosic biomass

Miscanthus, changed the compositions of the cell wall polymers, and significantly increased the biomass porosity, which led to the highest ethanol yield as compared to the other tested pretreatment techniques.

(ii) **Acid Pretreatment**: Acid pretreatment is used to break down the glycosidic bonds in lignocellulose by chemicals such as sulphuric acid, phosphoric acid, nitric acid, hydrochloric acid, etc. A high concentration of acid is not recommended because of its toxicity, corrosion, and high cost of acid [13, 130]. Solid loading, acid concentration, and reaction time are important parameters in acid pretreatments. The most important advantage of acid pretreatment is high monomeric sugar recovery from lignocellulosic biomass due to the polysaccharide-lignin bonds being broken easily under acidic conditions. The development of inhibitors and the high expense of acid recovery are significant drawbacks of this technology. It is important to note that the acid pretreatment procedure has improved regarding economic and environmental aspects.

Dilute acid pretreatment, as compared to concentrated acid treatment, has the advantages of lower acid consumption and lower cost of neutralizing hydrolysis mixture. However, higher reaction temperatures are needed to be applied to achieve a sufficient amount of glucose recovery from biomass [126]. Dilute sulphuric acid is generally used in lignocellulosic biomass pretreatment. A study by Dionísio et al. [45] applied dilute sulphuric acid pretreatment to sugarcane bagasse and obtained high xylose and arabinose recovery (82% of monomeric sugars from hemicelluloses). Lee and Yu [85] obtained 54.5% glucose recovery from acacia wood with 0.05% sulfuric acid at 200 °C for 5 min, and after mechanical refining and enzymatic hydrolysis, fermentation efficiency was increased to 94.9% ethanol yield.

(iii) **Organosolv Pretreatment**: Organosolv pretreatment is a physicochemical method that utilizes organic or aqueous-organic solvents at a temperature range between 100 and 250 °C [130]. For the processing of lignocellulosic biomass, a wide variety of solvents, including alcohol, phenol, esters, propionic acid, acetone, formaldehyde dioxane, and amines, are used [192]. The advantages of this method are easy solvent recovery and fractionation of the components with high purity. The major benefit of the method is high-value co-product formation due to the efficient fractionation of lignin [186]. Lignin recovery with a purity of 98.9% was achieved after acid-catalyzed ethanol organosolv treatment of *Eukalyptus pellita* biomass [33]. Ethanol and methanol are preferred organic solvents due to their low boiling points, easy recovery, and low cost [162, 192].

(iv) **Hot water Pretreatment**: Hydrothermal or liquid hot water pretreatment methods are based on lignocellulosic matrix degradation in the presence of water at high pressure and high temperature between 160 and 240°C. High temperature and pressure are needed because water can produce acidic (H_3O^+) and basic (OH^-) ions under these conditions that solubilize most of the hemicellulose and a small part of lignin [14]. It is accepted as a green technology due to chemical-free biomass degradation. However, low sugar yield, high energy and water requirements, and inhibitory by-product formation are drawbacks of this method [150].

Yang et al. [176] used alkali liquid hot water pretreatment and found that pretreatment partly removed and degraded hemicellulose and lignin from *Neosinocalamus affilis* (bamboo) biomass, the specific surface area of biomass increased after liquid hot water treatment, and 4.8 g/L ethanol was produced at 170 °C in the presence of 0.5% NaOH. In another study, sugarcane bagasse was pretreated with hot water, and 53.65 g total sugars per 100 g biomass were obtained after enzymatic hydrolysis [64]. The efficiency of pretreatment can be enhanced by combining two or more methods. For instance, hydrothermal pretreatment followed by mechanical extrusion is a favorable chemical-free process that can be used for bioethanol production from different kinds of biomass [158].

(v) **Steam Explosion**: As a thermophysical-chemical pretreatment method, steam explosion relies on vapor cracking and explosive decompression. The high-pressure saturated steam is applied for a few minutes to biomass, then pressure suddenly decreases. Due to the sudden pressure change, the structure of lignocellulosic biomass is disrupted and water-soluble (rich in hemicellulose), and water-insoluble (rich in cellulose and lignin) factions are separated [183]. Some of the most important advantages of the steam explosion method are depolymerization of lignin, hydrolyzation of hemicellulose fraction, and being an environmentally friendly and energy efficient process [7]. Steam explosion used for the pretreatment of sugarcane straw at 200 °C, 15 bar for 10 min increases the xylooligosaccharides yield of up to 35% and fermentable glucose yield of up to 78% after enzymatic hydrolysis [21].

In the ammonia fiber explosion method, similar to a steam explosion, the lignin-carbohydrate bond is broken, and lignin reacts with aqueous ammonia [151]. Although ammonia fiber explosion increases cellulase accessibility, ammonia recovery and the cost of pretreatment are the main barriers to commercial-scale applications [103].

(vi) *Green solvents*: Conventional pretreatment methods with alkali or acidic chemicals have drawbacks, such as corrosion to equipment, costly chemical recovery, and water treatment processes. Developing mild, environmentally friendly, and sustainable pretreatment procedures have recently received remarkable attention. The green solvents for biomass pretreatment have several benefits, including simplicity of preparation, reduced toxicity, high biodegradability, low melting points and volatility, high thermal stability, and non-flammability [122]. Ionic liquids and deep eutectic solvents are two common green solvents for biomass pretreatment [180].

Ionic Liquid Pretreatment: Because of the high solubility of biomass in ionic liquids and high sugar recovery, ionic liquid pretreatment has gained more attention. Numerous ionic liquids, based on imidazolium, pyridinium, pyrrolidinium, ammonium, phosphonium, and sulfonium, are available for pretreatment. Acidic ionic liquids, breaking the ether bonds, have a high potential for the delignification of biomass [10]. However, recycle of ionic liquids after pretreatment is difficult and consumes high energy. To decrease the cost of the chemicals, ionic liquids with high water content can be used for pretreatment [53]. There are many articles published about ionic liquid pretreatment of different kinds of lignocellulosic biomass. In one study, 80% total reducing sugar yield was achieved with the pretreatment of barley straw with 1-ethyl-3-methylimidazolium acetate [83]. It is reported that enzymatic saccharification of bamboo biomass was enhanced about 4.7-fold when pretreated with 1-ethyl-3- methylimidazolium acetate instead of dilute acid [109].

Deep Eutectic Solvents: Deep eutectic solvents (DES) are alternative green solvents used for pretreatment. They are similar to ionic liquids. DES are formed mainly with the mixture of two components: the donor and acceptor for hydrogen bonding. Intermolecular hydrogen bonds in DES break strong hydrogen bonds in the biomass structure. Thus, favorable degradation can be achieved [31]. The freezing point of the mixture is lower than its components [102]. Compared to ionic liquids, DESs are more cost-effective, less toxic, and more biodegradable [38, 168]. The use of triethyl benzyl ammonium chloride/lactic acid for the pretreatment of sugar cane bagasse at 120 °C for 4 h resulted in a remarkable delignification (88.7%) and cellulose recovery (95.9%) [92]. The HMF and furfural formation during benzyl trimethylammonium chloride pretreatment of corncob was investigated by Guo et al.; they showed that at the pretreatment temperature of 140 °C only a trace amount of byproduct formation occurred [60].

(vii) *Combined Pretreatment methods*: The compositional and structural change of various lignocellulosic biomass feedstocks can affect the efficiency of various pretreatment methods. An ideal pretreatment procedure should increase the amount of cellulose released and make the lignocellulose more accessible to the following procedures [166]. Due to the negative effects of using one type of pretreatment method as given above, researchers tried to combine two or more pretreatment methods to increase efficiency [105]. The combination of various pretreatment processes, such as hydrogen peroxide-steam, microbial-dilute acid, microbial-alkali, and bacteria-enhanced dilute acid, can be applied for the biomass [187]. It has been observed that fiber degradation and sugar yields can be effectively increased by combining energy-efficient biological treatment with a chemical method that shortens the pretreatment periods [105]. Single fungal pretreatment requires a long retention time for sufficient sugar yields, and when the combination of fungal and dilute acid hydrolysis is used for pretreatment, the process could be more economical, and chemical requirements can be reduced [148]. It was also stated that reducing sugar yield was increased, and inhibitory by-product formation was decreased when hydrogen peroxide was combined with a steam explosion [160]. One of the most important drawbacks of acid hydrolysis is inhibitory compound formation during the pretreatment process, when microbial pretreatment is used before acid hydrolysis delignification can be achieved and sugar recovery and so ethanol production yield can be increased [187].

Major barriers to the commercialization of lignocellulosic bioethanol production are economic obstacles and technological challenges, such as the cost of pretreatment and lower bioethanol production yields compared to sugar or starch-based biomass [39, 56]. Lignin removal is regarded to be the most challenging step of pretreatment. Various pretreatment methods, including physical, chemical, and biological ones, have been intensively investigated for lignin removal. Each pretreatment method has advantages and disadvantages, as mentioned before. Each method may not give the same yield for every plant biomass. Therefore, the suitability of the pretreatment method with the biomass structure should also be considered. Nevertheless, the efficiency of the pretreatment depends on several factors given in Fig. 5. The degree of depolymerization of lignocellulosic structures and the formation of inhibitors during pretreatment vary significantly for different feedstock types. Biomass with a hard structure is more difficult to degrade; fewer monosaccharides can be obtained and inhibitory compounds are released from biomass [65].

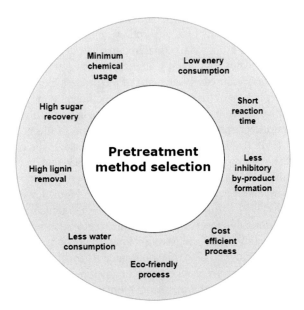

Fig. 5 Criteria for suitable pretreatment method selection

3.2 Detoxification

Another important advancement in cellulosic ethanol production is the development of efficient detoxification processes. Because of employing different chemicals and environmental conditions during the pretreatment processes, some inhibitory by-products which can severely affect the activity of microorganisms and the fermentation process can be present in the lignocellulosic hydrolysates [182]. Organic acids (acetic, formic, and levulinic acids), furan derivatives (furfural and 5-hydroxymethylfurfural (5-HMF)), phenolic compounds and metal ions can be listed among these inhibitory compounds. The combination of toxic compounds produced by various lignocellulosic feedstocks and pretreatment techniques, as well as the level of inherent resistance present in certain microbes, all influence how the fermentation process is affected by the toxic chemicals [123]. Hence, different physical and chemical methods, like dilution, adsorption, chemical precipitation, and biological detoxification, have been developed to remove or decrease the toxicity of these compounds. The most commonly used physical treatment methods are evaporation and membrane filtration. In the evaporation process, some volatile toxic compounds are removed from the hydrolysate by using vacuum evaporation. However, this method increases the amount of non-volatile toxic chemicals in the hydrolysate [24]. Membrane filtration offers several advantages over conventional liquid extraction, such as flexibility (use of different types of membranes) and less energy requirement. In this method, inhibitors that have a small molecular weight (such as phenolic acids) are allowed to pass through the membrane, while molecules with high molecular weight, like sugars are retained in the retentate [79]. Detoxification of lignocellulosic hydrolysate by activated charcoal (AC) and ion exchange resins are the chemical methods used by various authors [61, 86, 136]. The advantages of AC include its high adsorption capacity, high surface wide spectrum of surface functional groups, and compared to other detoxification methods AC is economical and does not significantly alter the number of recovered sugars [135]. The ratio of charcoal and hydrolysates, pH of the process, contact time, and temperature are the most important factors affecting the efficiency of this method [23]. Previously, ion exchange resins were used to treat corn cob hydrolysate and the study reported that resin treatment provided efficient removal of nitrate salt (by 70%), phenolics, and 5-HMF (by 70%) [78]. Overliming with $Ca(OH)_2$ is one of the earliest detoxification technologies specifically developed and used to detoxify dilute-acid hydrolysates. In this method, removing toxic products by precipitation and the instability of inhibitory compounds at high pH help increase the efficiency of fermentation by microorganisms [119]. Liquid–liquid extraction with organic solvents, such as ethyl acetate and trichloroethylene, is another chemical detoxification method that is especially used to remove small molecular weight toxic compounds such as acetic acid [5] and phenolics [30]. Biological detoxification methods involve specific microbial enzymes acting on inhibitors and changing the composition of enzymes, which are potential methods for eliminating inhibitors from biomass hydrolysates [182]. By genetic engineering, the tolerance of the fermenting microorganisms can be increased against a specific inhibitor, and this is an emerging approach in this field [79].

3.3 Enzymatic Hydrolysis

All major components of lignocellulosic biomass have specific and compositional characteristics, which ensure biological recalcitrance to the biomass [88]. Hence, in nature, organisms have developed complex enzyme systems for degrading this recalcitrant biomass. Different microorganisms growing on lignocellulosic biomass produce cellulosic inducible enzymes, cellulases. Cellulases depolymerize cellulose, the main polymer of the lignocellulosic biomass, to its building blocks [34]. Cellulase is a complex enzyme consisting of an endoglucanase, exoglucanase, cellobiohydrolase, and β-D-glucosidase. Endoglucanases work by breaking internal β-glycosidic bonds in the cellulose chain. This allows cellobiohydrolase access to the chain ends, and cellobiose, the byproduct, is then further broken down into glucose units by β-glucosidase [179].

Enzymes, especially cellulase, contribute to the high production costs of lignocellulosic ethanol production. The considerable decrease in its cost of production is the key solution to the commercialization of lignocellulosic biorefineries [51]. Wild-type microorganisms often do not produce enzymes at the levels and specifications required in the industry [80]. Therefore, developing novel enzymes has become an important challenge for biofuel production. Recent years have shown great progress in engineering cellulolytic and hemicellulolytic enzymes to improve the enzymatic processing step and reduce the cost of lignocellulosic bioethanol production. Research efforts have focused on three areas: (1) Improvement of catalytic efficiency, stability, and inhibitory tolerances of lignocellulosic enzymes by enzyme engineering; (2) Construction of enzyme cocktails and by using synergistic effects of multiple enzymes for the degradation of specific substrates; (3) Achieving high-level and low-cost enzyme production by the improvement of protein production process [88].

The filamentous fungus *Trichoderma reesei* has been widely used for industrial cellulase production. Increasing costs and depletion of fossil fuels provoke the demand for hyper-cellulase-producing *T. reesei* [175]. However, cellulases produced by this fungus are still too costly, and the enzymatic hydrolysis step is one of the bottlenecks for lignocellulosic bioethanol production [107]. For decades, the genetic engineering of *T. reesei* has provided detailed information about the regulators and transcription factors involved in enzyme expression has been gained, which has aided in the creation of novel cellulase-producing mutants [118]. In a recent study, *T. reesei* QS305 with high endoglucanase activity was developed from *T. reesei* Rut-C30 by replacing the transcription repressor gene *ace1* with the coding region of endoglucanase gene *egl1*. Compared to the parental strain, *T. reesei* QS305 showed a 90.0% and 132.7% increase in the activities of total cellulases and endoglucanases under flask culture conditions [107].

The introduction of hyperbranching is believed to increase protein secretion since most exocytosis is located at the hyphal apical tip [52]. Several genetic modifications can cause hyperbranching in filamentous fungi, for example, disruption of *gul1* gene, which encodes a putative mRNA-binding protein, in *T. reesei*. [191] developed a new strain, *T. reesei* QP4 Δgul1, to improve cellulase production and they reached 22% higher cellulase levels than the parent strain.

Increase in the cellulolytic activity of an isolated strain can be achieved by random mutagenesis. Silva et al. [143] obtained a new strain of *T. reesei* (teleomorph *Hypocrea jecorina*), which has up to twofold high cellulase production level as compared to the original strain, by exposing the spores from *T. reesei* QM9414 to UV light.

The selection of high-level producers among the variants in mutant libraries became a key issue in developing engineered hyper-cellulase-producing strains. He et al. [62] developed a droplet-based microfluidic high-throughput screening platform for selecting high-cellulase producers of filamentous fungus *T. reesei*. They used a fluorogenic assay to measure the amount and activity of cellulase. This high-throughput screening system could be applied to engineering *T. reesei* strains and other industrially valuable cellulase-producing filamentous fungi.

Although fungal and bacterial cellulases have completely different protein folds, they have evolved to acquire processivity through the same strategy of adding subsites to extend the substrate-binding site and forming a tunnel-like active site by increasing the number of loops covering the active site [155]. For efficient hydrolysis of lignocellulosic biomass, optimal synergistic cocktails of enzymes from different microorganisms are needed. Optimization of the composition of these mixtures is an effective strategy to improve hydrolysis efficiency and reduce protein demand during enzymatic degradation [46]. The hydrolytic efficiency of cellulase cocktails for lignocellulose degradation is affected by both the characteristics of the individual enzymes and their ratio in the cocktail [34]. The ideal cellulolytic cocktail must be highly active on the intended biomass feedstock, able to hydrolyze the biomass completely, operate well at mildly acidic pH, withstand the process stress, and be cost-effective [34]. According to a recent study, utilization of a novel two-thermostable enzyme-based cellulase cocktail (PersiCel1 and PersiCel2) implies a synergistic relationship and significantly increased the saccharification yield of lignocellulosic substrates up to 71.7% for sugar-beet pulp and 138.7% for rice-straw compared to maximum hydrolysis of Persicel1 or PersiCel2 separately at 55 °C [101].

Commercially available cellulolytic cocktails produced by leading biotech companies Novozymes (Cellic Ctec1, Cellic Ctec2) and Genencor (Accelerase 1000, Accelerase 1500, Accelerase XY, Accelerase DUET) are composed of cellulases from filamentous fungi such as *A. niger*, *T. longibrachiatum*, and *T. reesei* [34].

Even though it is produced by many industries worldwide due to widespread applications, the search for improved industrial-scale cellulase production, which can effectively hydrolyze lignocellulosic biomass, is still going on worldwide. Studies are focused on improving the thermal stability of the cellulase enzyme and its cost-effective production [144]. Enzymatic hydrolysis of lignocellulosic at high temperatures allows a higher reaction rate with reduced enzyme loading. Hence, thermostability has been the most desirable trait of cellulases for lignocellulosic bioethanol production as it will drastically reduce the dosage of enzymes resulting in its economic utilization [146].

3.4 Fermentation

The third step of lignocellulosic bioethanol production is fermentation, a biological process catalyzed by microorganisms that transform produced sugars into different products such as ethanol and acids. There are a large number of yeasts and bacteria that can ferment lignocellulose-based sugars. The ethanol fermentation of glucose is made according to the following reaction:

$$C_6H_{12}O_6 \rightarrow 2C_2H_5OH + 2CO_2 \quad (1)$$

The maximum theoretical yield of conversion of glucose into ethanol, calculated from this Eq. (1), is 0.511 g of ethanol/g glucose. This yield does not consider the sugar utilization for biomass production and the transformation of glucose into glycerol and other products such as lactic and acetic acid. Taking account of these losses, the maximum yield is estimated to be around 0.484 g of ethanol/g glucose [3]. In a cell, multiple enzymes are responsible for converting glucose to ethanol.

3.4.1 Cellulosic Ethanol-Producing Microorganisms

Although the bioethanol production industry has long been operational, finding an appropriate microbial strain for efficiently converting lignocelluloses is still an active field of study [72]. Cellulolytic bioethanol-producing organisms exist in many biological taxa and have been isolated from extremely diverse environments, and their evolutionary background provides a wide variety of metabolic conversion [58]. Table 1 shows the major metabolic characteristics of the most commonly studied microorganisms in cellulosic bioethanol production. Because of its high ethanol yield and tolerance, *S. cerevisiae* is the most commonly studied and industrially used yeast species for lignocellulosic bioethanol production. However, it is not natively capable of fermenting pentose sugars from the hydrolysis of various lignocellulosic biomass. *Zymomonas mobilis* is another facultatively anaerobic microorganism widely studied for ethanol production. It can metabolize sugars in a broad pH range and bring high ethanol yield and productivity [172]. *Z. mobilis* can also tolerate up to 120 g/L ethanol [72]. The availability of genome sequence information for multiple *S. cerevisiae* and *Z. mobilis* strains and ready-to-use advanced genetics tools have built up a stable basis for engineering these microorganisms and directed many researchers to develop pentose fermenting *S. cerevisiae* and *Z. mobilis* strains. Although *Z. mobilis* has a better yield and better productivity due to less biomass production, it is not commonly used because it is more susceptible to contamination and requires a nutrient-rich medium [3]. Natural pentose-fermenting yeasts, such as *Scheffersomyces stipitis*, *Candida shehatae*, and *Pachysolen tannophilus*, are studied for co-fermentation of pentoses and hexoses obtained from lignocellulosic biomass. Even though *S. stipitis* showed good potential in lab-scale experiments, it is difficult to accomplish high ethanol productivities with this strain on an industrial scale. This is because of its lower activity compared to traditional *S. cerevisiae* during the process, inhibition by glucose (catabolite repression), and the need for microaerophilic conditions [131]. Similar to *S. stipitis* the other xylose-fermenting yeasts show low tolerance to inhibitors; require a well-controlled oxygen supply for maximal ethanol production and are sensitive to high ethanol concentrations [63]. Hence, *S. cerevisiae* and *Z. mobilis* are still the strongest candidates for industrial-scale lignocellulosic bioethanol production.

3.4.2 Inventions and Developments in Fermentation Processes

According to the process configuration used for the fermentation of biomass, the ethanol production technologies are classified as separate hydrolysis and fermentation (SHF); simultaneous hydrolysis and fermentation (SSF); simultaneous saccharification and co-fermentation (SSCF); and consolidated bioprocess (CBP) (Fig. 6).

In SHF, hydrolysis and fermentation processes are carried out separately. Pretreated biomass is transferred into the hydrolysis reactor, where it is degraded into simple sugars via enzymatic activity. Then in the fermentation unit the produced sugars are converted to ethanol via microbial activity. The main advantage of this process is that each step is performed at its optimal temperature, which is usually different as most of the cellulase works at about 50 °C, and strains that produce ethanol require a temperature of 28–37 °C [42]. End-product inhibition is the main drawback of this process [141].

In SSF, hydrolysis and fermentation are carried out in one vessel simultaneously. This process eliminates the inhibition caused by the accumulation of the sugars because the sugars released are transformed into bioethanol instantly by using suitable microorganisms [42]. Other major advantages of this process are low investment cost [141], short reaction time, and low contamination risk [131]. The disadvantage of this process is the carbon source reduction due to glucose deficiency, as the enzymes are sensitive to ethanol. The optimal temperature difference between both processes limits the process and leads to lower productivity [42]. Using thermophilic or thermal tolerant strains in the SSF process can enhance ethanol productivity [94].

Simultaneous Saccharification and Co-Fermentation (SSCF) is a consolidated process used for increasing bioethanol production by hydrolyzing cellulose and co-fermenting hexose and pentose sugars (mainly glucose and xylose) in one step [142]. Genetically engineered strains or co-cultures can be used for the co-fermentation of glucose and xylose [67, 125, 137, 152, 189, 194]. SSCF is an advanced process and currently used, it is close-to-commercialization approach that is being continuously improved for high ethanol titter, pentose utilization, high mass transfer, reducing

Table 1 Characteristics of the most commonly used microorganisms for lignocellulosic ethanol production (Adapted from [58])

	Microorganism	Typical growth temperature (°C)	Oxygen requirement	Cellulase activity	Hemicellulase activity	Natural xylose utilization	Required mechanism for Xylose utilization
Prokaryotes	*Clostridium thermocellum*	60	Anaerobic	Yes	Yes	No	Xylose isomerase
	Clostridium cellulolyticum	34	Anaerobic	Yes	Yes	Yes	Xylose isomerase
	Zymomonas mobilis	40	Anaerobic	No	No	No	Xylose isomerase
	Escherichia coli	37	Facultative	No	No	Yes	Xylose isomerase
Eukaryotes	*Saccharomyces cerevisiae*	30	Facultative	No	No	No	Xylose reductase + Xylose dehydrogenase
	Shefferomyces sitipitis	30	Facultative	No	Yes	Yes	Xylose reductase + Xylose dehydrogenase
	Pachysolen tannophilus	30	Facultative	No	No	Yes	Xylose reductase + Xylose dehydrogenase

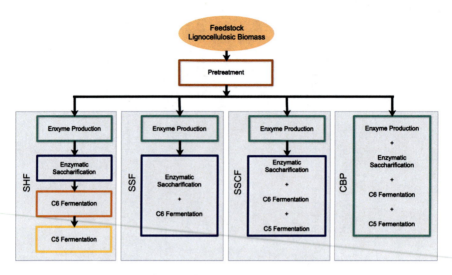

Fig. 6 Schematic process steps for SHF, SSF, SSCF, and CBP (Adapted from [124])

substrate inhibition on enzymes, and one-pot conversion strategies [142]. Like the SSF process, SSCF also helps reducing process time and investment costs.

Consolidated BioProcessing (CBP) is a single-step process that includes production of enzymes, hydrolysis of lignocellulosics, and fermentation of sugars. CBP has attracted close attention since it can provide an effective and potential solution for reducing investment and ethanol production costs and simplifying the operation process [93]. Cellulose-fermenting *Clostridium thermocellum* and the hemicellulose-fermenting *Thermoanaerobacterium saccharolyticum* are the microorganisms that have received attention for developing CBP processes [97]. These thermophilic organisms can tolerate high temperatures, which is important for reducing contamination risk and producing thermostable enzymes.

Immobilized microorganisms, including yeasts and bacteria, have been frequently used for batch or continuous ethanol production processes due to their feasibility for repeated use with high biomass retention during the operation [171]. Using immobilized cells also enables them to work with higher cell densities and protect cells from toxic compounds and harsh environmental conditions. Immobilized cells have been used for ethanol production in different reactor configurations, such as stirred tanks, flow-through columns, fixed-bed columns, and rotating-bed bioreactors [71]. The selection of a suitable matrix for immobilizing microorganisms is important for successfully applying the immobilized cells to bioprocess [154]. The physical structure of the carrier, such as pore size and distribution; chemical groups on the surface, have a significant effect on cell immobilization efficiency. The studies in the literature suggest that immobilized or co-immobilized cultures in continuous bioreactors can achieve efficient and rapid conversion of mixed sugars to ethanol [71]. The main challenge with multi-culture fermentation is simultaneously providing the optimal conditions for different strains. Co-immobilization of cells by entrapment in a matrix provides local niches where different environmental conditions can be maintained. This approach can help solve one of the most important issues, the fermentation of hexoses and pentoses, in cellulosic ethanol production.

3.5 Bioethanol Purification

After fermentation, product purification operations must be added to achieve fuel-grade ethanol. More than half of the total bioethanol production process costs comes from the bioethanol separation steps. Conventional purification methods consume high amounts of energy due to azeotrope, which limits the feasibility of bioethanol production on a commercial scale [114]. So, the costs of production and product separation need to be brought down to make it competitive with fossil fuels. Azeotropic distillation, liquid–liquid extraction, and extractive distillation are conventional methods for bioethanol purification. Azeotropic distillation is the most widely used method for bioethanol purification. This method is based on the difference in the relative volatility of the products in the fermentation broth. A higher concentration of ethanol can inhibit the growth of microorganisms. Still, a low concentration of ethanol leads to a large distillation unit and increases the energy requirement [77].

The liquid–liquid extraction method is based on ethanol's solubilities in an organic phase, significantly increasing ethanol concentration and reducing energy requirement[141]. Distillation technology can purify ethanol by a concentration of up to 95.6% [147]. Extractive distillation is used to increase the purity of bioethanol by adding separating agents with a high boiling point [9]. These agents are solvents, solid salts, ionic liquids [140] and recently added deep eutectic solvents [121]. However, an additional column for the recovery of the agents is required, and energy consumption and the capital cost of the process increased. These obstacles in distillation have triggered the search to find an alternative technology for bioethanol purification [74].

Pervaporation is one possible approach for separating fermentation products, depending on the diffusivity and solubility of the components to be separated. This method is more efficient than distillation for the separation of azeotropes [70, 114]. Moreover, inhibition of microbial activity is prevented by the continuous removal of products from broth, which increases the bioethanol production rate. Pervaporation has been recognized as a promising membrane-based product separation technology [74].

4 Pilot Plants and Biorefineries for Bioethanol Production from Lignocellulosic Biomass

It is estimated that plants can stabilize about 25% of carbon emissions produced by human activities [22]. Therefore, the utilization of plants for energy production is a reasonable strategy in terms of sustainability. Not only wood parts of plants but also forestry and agricultural residues, yard, animal, and aquatic wastes can be utilized in a cellulosic biorefinery [129]. To reduce the cost of 2G bioethanol production, an adaptation of the infrastructure of the agribusiness sector is a requirement. Adopting complex bio-production methods and other technological innovations should be considered for cost-effective processes [106]. Various modifications can be applied, and the potential of these modifications must be analyzed techno-economically before investment [18]. For example, establishing a plant close to the feedstocks can be a solution for cost reduction. Indeed, many refineries producing corn-based ethanol are located in regions where corn is grown. Effective coordination of transport and distribution of the product should also be considered for increasing the preferability of biofuels. This will require large infrastructure investments, which depend on the willingness of investors. The willingness of consumers to adopt changing technologies and fuels is one of the main triggers for investors [164]. The growing market size of ethanol in the transportation sector might be viewed as a sign that investors and consumers are open to using bioethanol as a fuel.

It is reported that the market size of lignocellulosic bioethanol increased from 4.5 billion dollars in 2021 to 4.8 billion dollars in 2022. The estimates will increase to 5.1 billion dollars in 2024 [156]. Until now numerous 2G bioethanol production pilot plants have been established in many countries, including the USA, Brazil, Sweden, Norway, Finland, and Italy; however, most of them have been put on hold today. Moreover, some demo plants have already been set up in Europe, such as Beta-renewables in Crescentino (Italy), Inbicon in Kalundborg (Denmark), and Clariant's in Straubing-Munich (Germany). BIOLYFE started to produce 2nd generation bioethanol production project and has a capacity of 75 million liters of annual bioethanol production in Crescentino, Italy [32].

Production capacities of the 2G bioethanol plants established up to now mostly range between 10 and 85 k tons per year. One of the major plants, Granbio Bioflex 1, with 65 k tons of bioethanol production capacity, has been operational since 2014 in Brazil. In this plant, bagasse and straw are used as lignocellulosic waste, which is pretreated with SO_2 and ethanol for the fragmentation of sugars and lignin. When the chemical-free process is preferred, liquid hot water or steam power and commercial thermochemical green technologies are used for monomeric sugar production [127]. In most 2G bioethanol refineries, chemical treatment and steam explosion are applied for pretreatment, organosolv treatment and ultrasonication are also preferred in some of them. Enzymatic hydrolysis is an important part of pretreatment affecting sugar recovery. Commercially available cellulase enzymes from different fungi species, such as *Aspergillus fumigatus*, *Aspergillus niger*, and *Penicilium purpurogenum* are preferred to obtain sugars after pretreatment. Dupond in the USA and Iogen in Canada apply enzymatic hydrolysis after the chemical and physical pretreatment processes [129]. The Renewable Fuel Association states that about 200 bioethanol refineries are established in the USA, 11 of which produce cellulosic bioethanol. There are a few 2G bioethanol refineries in Europe too. One of the biggest is Clariant, established in Germany and Romania and produces commercially available cellulosic bioethanol. The investment cost for the first plant of Clariant was reported to be 100 million euros.

Valuable information can also be gathered from news about biorefineries. It is promising that new plants are planned to be constructed, or the capacity of already established ones will be increased in different regions of the World, including Brazil and Japan. Sekisui Bio Refinery (SBR) completed the construction of a demonstration plant in Japan at the beginning of 2022. The first commercial-scale facility is targeted to begin production in 2025. Moreover, new methods or systems have been developed and patented in the 2G bioethanol sector. These developments help to decrease ethanol production costs and open the way for the commercialization of 2G bioethanol. Valmet (in Denmark) patented a force feed and a steam explosion technology. It is reported that the plant can handle five tons of dry biomass per hour.

In the Versalis plant at Crescentino, renewable electricity and steam can be produced from the thermoelectric power plant, so it is a self-sustaining facility. The facility has a water treatment plant enabling biogas production, which is used as the energy source for steam generation. Recycling used water in the plant drastically reduces consumption. The plant is capable of processing 200,000 tons of biomass annually. It is stated that up to 25,000 tons of bioethanol can be produced per year. Rayonier Advanced Materials' biorefinery (USA) provides a more sustainable operating model that captures residual sugars from its existing pulp process and uses them for 2G bioethanol production [17].

5 Food-Water-Energy Nexus

Many market dynamics impact supply and demand for essential resources like grains and oilseeds. Despite the reports that some nations' ethanol usage has decreased in the past two years since 2020, the world's ethanol demand has continued to rise (ethanol producer magazine, S and P global reports.). According to research on the Lignocellulosic Feedstock-based Biofuel Market by Conversion for 2020–2024, this market has the potential to increase in value by 7.83 billion USD over this period, and its growth speed will accelerate. Fluctuations in crude oil prices, stringent governmental regulations on environmental concerns, the negative impact of crude oil on the environment, and the World bank seizing funding for E&P activities have contributed to decreasing upstream expenditure for oil and gas exploration. These developments had positive reflections on the bioethanol market. However, global oilfield services and equipment are also witnessing a positive impact of factors such as increasing energy demand, exploration of unconventional resources, reduced drilling costs, etc. It is estimated that combined effects will cause a moderate growth in oilfield services and equipment [156].

Climate change, increasing populations, and economic instability threaten the sustainability of complex resource systems, such as the food, water, and energy nexus [163]. Problems of water scarcity, food security, and soil degradation are increasing, and these issues are not independent of bioenergy production [16]. The concept of circular bio-economy depends on the efficient integration of the substrates, products, and byproducts input/output streams to reduce waste release from a closed loop [15]. Bioethanol and biodiesel, which are mainly produced from starchy crops and vegetable oils, respectively, are two major liquid fuels used in the transportation sector. About 41.3 million ha, about 4% of the global arable area, was reported to be necessary for producing 65 and 21 million tons of bioethanol and biodiesel, respectively. It is estimated that 216 billion m3 of water, about 3% of the global water consumption for food production, is needed for this production [27, 28, 132]. Data shows that the need for energy crop production may necessitate deforestation and finding new water and land resources [127]. The agri-food sector is reportedly consuming 30% of the world's energy, and energy demand will double by 2050 [185]. Agriculture uses 12% of the world's land for producing crops [47, 185], and analyzes and projections reveal the necessity of using land primarily for food production.

Gruttola and Borello [43] predicted feedstock availability to produce advanced biofuel based on an autoregressive model. Their results indicate that several European countries can produce enough sustainable feedstocks in 2025 [43]. Some countries will be rich in agricultural feedstocks, while others will produce many forestry residues. Besides biomass availability, Technology Readiness Level (TRL), biofuel quality, and costs are important parameters affecting production capacity. It is estimated that 10–25 million tons of agricultural residues will be available, especially in France and Germany, for advanced biofuel production in 2025.

The World Bank estimates that complete agricultural market liberalization will result in average price increases of 5.5% for agricultural raw materials and 1.3% for food. The increase in demand for agricultural products like maize, wheat, sugar, and oilseeds for the production of biofuels, bioelectricity, and bioheat is regarded to be one of the main reasons for the increasing food prices. The International Food Policy Research Institute (IFPRI) reported that support for bioenergy production would cause an additional rise in food prices [138]. Commercialization of 2G bioethanol can be a solution for decreasing demand for agricultural commodities.

Optimization of the processes and strategic integration of technologies in biomass pretreatment will increase the efficiency of the processes by reducing energy usage and detrimental impacts on the environment, which will, in turn, help reduce the problems emerging from the energy-food-water supply chains [15].

6 Potential of 2G Bioethanol for the Transportation Sector and Future Projections

Bioethanol is regarded to be a promising energy source, especially for usage in the transportation sector. Gasoline blended with 5%—current automobile engines can burn 10% bioethanol without any modification [111]. Fuel gasoline engines are classified based on octane number as an indication of compression value. Although the heating value of ethanol-blended gasoline is lower, ethanol in the engine decreases detonation or knocking because ethanol booster, octane number, when blended with gasoline [170]. Reducing GHG emissions is regarded as one of the main advantages of blended fuel. It has been reported that corn ethanol can reduce emissions between 18 and 28%, while cellulosic ethanol can result in reductions of up to 87%. However, many factors must be considered, such as if coal is used as the source of electricity in its production, corn ethanol may not produce any carbon reduction benefits [153].

Compared to other liquid biofuels, bioethanol is produced in a higher quantity in the World and is used globally. Approximately 4 Gtons of agricultural waste in the World are produced from maize, sugarcane, rice, wheat, and cotton. Estimated theoretical biofuel production from these wastes reaches 1000 Gl in a year [49]. The agricultural waste potential of each country may vary depending on its agricultural economy. For example, in Turkey, it is estimated that 0.256 Mtons of ethanol is needed for 10% ethanol blended petroleum utilization in the transportation sector. Even the annual grain waste production in Turkey is about five folds the amount of agricultural waste required to produce that much 2G ethanol. However, the cost of production is a big challenge. According to the estimated 2G ethanol production [56], an annual cost of approximately 500 million dollars is required for 0.256 Mtons of lignocellulosic bioethanol production [181]. This cost should be reduced to reasonable levels to initiate lignocellulosic ethanol production and expand its use.

According to the report of S&P Global, the sector for renewable biofuels will grow by 85% over the next ten years. They also report that by 2050 oil will continue to be the top energy source in the World based on the data received from US EIA. However, it is emphasized that we will run out of crude oil reserves since current reserves can supply the demand until 2052 based on annual consumption rates of 4 billion tons [100]. Of course, limited oil reserves are the most important driving force for many countries to turn to alternative energy sources. However, for a dependable alternative energy strategy, the availability of oil reserves and the solar, wind, and water resources of the countries must be determined, and the extent to which they can benefit from these resources for energy production should be analyzed. Moreover, the abundance of potential agricultural lands that can be directed to bioethanol production is one of the most important indicators of any country's bioethanol production potential. In particular, a stable strategy and commitment of a country to bioethanol production and usage are necessary for the success of the bioethanol sector. For example, despite heavy criticism, the Brazilian Government made a significant effort to develop bioethanol research and build infrastructure between 1975 and 1989. Brazil, which produces the cheapest ethanol in the World, is now the second-largest country after the USA. It is a great achievement for a country not to rely on fossil fuels to meet its energy demands [161].

Projections on demands for different renewable energy types show changes over time and may not be so predictable in the far future. For example, the automotive industry is shifting toward producing electric vehicles, and it is expected that by 2030, 43% of all new cars sold of the US will be electric [19]. The European Commission Energy Department proposed a 100% reduction in CO_2 emissions from cars by 2035 by employing electric cars; however, there was a strong objection due to concerns about renewable fuel research [50]. Besides, the inadequacy of the infrastructure required to charge vehicles is a major challenge that will restrict the expansion of electric vehicle use [82, 112]. Although it is widely believed that electric vehicles have zero-emission, the upstream emissions associated with electricity generation and battery manufacturing are often overlooked and give this false impression [1]. According to a study, electric vehicle battery manufacturers produce up to 60% more CO_2 during fabrication than standard engine producers. Furthermore, previous studies showed that an electric vehicle charged on an exclusively coal-powered grid would reach similar carbon emissions performance as a conventional car [8]. Hence, the carbon benefits of electric vehicles depend on the manufacturing process and the electricity source required to charge them. We believe that diversifying energy sources could prevent the energy crisis that we experienced in the past. It is hard to guess what type of engines will be used in cars. Still, renewable energy sources will be needed for other sectors like aviation, shipping, or industry in the second quarter of this century. The European Commission's Energy Department indicated that fuels based on molecules would be difficult to produce; they should be used in those sectors where other alternatives are not so readily available.

Although lignocellulosic bioethanol is mainly considered a fuel for the transportation sector, biomass can also be directly combusted to produce heat and electricity. Combustible gases such as hydrogen or biogas and synthetic liquid fuels (biodiesel, bio-oil, methanol, ethanol, vegetable oil, etc.) can be produced from biomass, which is burned to produce heat and electricity [25]. Despite some problems, including R&D technology and policy standards, biofuels have a unique potential to serve the main energy demand sectors of heat and electricity [177]. Advanced biofuels like bioethanol and biodiesel are regarded to be advantageous over other renewable energy sources for electricity generation. Strong volatility, intermittent nature, and uncertainty of solar, wind, and hydroelectric power output have significantly impacted the power grid's reliability [177]. Policies and the financial environment of countries and organizations have promoted the extensive application of biofuel power generation in recent years, especially in China. Also, sugarcane bagasse is utilized for electricity generation in Brazil due to the high electricity prices. However, interest in more affordable renewable electricity sources such as wind power may change the economic scenario [44].

7 Conclusion

Lignocellulosic bioethanol is regarded to be a sustainable solution for decreasing GHG emissions and increasing the utilization of agricultural and municipal waste materials. There are several challenging steps related to the pretreatment of the biomass, especially for lignin removal and cellulose degradation. However, novel methods have been developed, and technological improvements have been achieved that bring the solution to the problems associated with the costly production of 2G bioethanol. There are many biorefineries with remarkably high production capacities in different regions of the World. Several biorefineries have already started commercial delivery of 2G bioethanol to the World. The achievements and developments in the 2G bioethanol industry are promising for the future of lignocellulosic ethanol, and it has great potential to be used as one of the main renewable energy sources in the World.

References

1. Abbott C (2022) Ethanol producers to electric car makers: we're greener than you are. In: Successful Farming. https://www.agriculture.com/news/business/ethanol-producers-to-electric-car-makers-we-re-greener-than-you-are. Accessed 17 Aug 2022
2. Abdali H, Sahebi H, Pishvaee M (2021) The water-energy-food-land nexus at the sugarcane-to-bioenergy supply chain: a sustainable network design model. Comput Chem Eng 145:107199. https://doi.org/10.1016/j.compchemeng.2020.107199
3. Abo BO, Gao M, Wang Y et al (2019) Lignocellulosic biomass for bioethanol: an overview on pretreatment, hydrolysis and fermentation processes. Rev Environ Health 34:57–68. https://doi.org/10.1515/reveh-2018-0054
4. Aghaei S, Karimi Alavijeh M, Shafiei M, Karimi K (2022) A comprehensive review on bioethanol production from corn stover: worldwide potential, environmental importance, and perspectives. Biomass Bioenergy 161:106447. https://doi.org/10.1016/j.biombioe.2022.106447
5. Aghazadeh M, Ladisch MR, Engelberth AS (2016) Acetic acid removal from corn stover hydrolysate using ethyl acetate and the impact on Saccharomyces cerevisiae bioethanol fermentation. Biotechnol Prog 32:929–937. https://doi.org/10.1002/btpr.2282
6. Alam A, Zhang R, Liu P et al (2019) A finalized determinant for complete lignocellulose enzymatic saccharification potential to maximize bioethanol production in bioenergy Miscanthus. Biotechnol Biofuels 12:99. https://doi.org/10.1186/S13068-019-1437-4
7. Alvira P, Tomás-Pejó E, Ballesteros M, Negro MJ (2010) Pretreatment technologies for an efficient bioethanol production process based on enzymatic hydrolysis: a review. Bioresour Technol 101:4851–4861. https://doi.org/10.1016/J.BIORTECH.2009.11.093
8. Anair D, Mahmassani A (2012) Electric vehicles' global warming emissions and fuel-cost savings across the United States. Cambridge
9. Arlt W (2014) New separating agents for distillation. distillation: operation and applications 403–428. https://doi.org/10.1016/B978-0-12-386876-3.00010-7
10. Asim AM, Uroos M, Naz S et al (2019) Acidic ionic liquids: promising and cost-effective solvents for processing of lignocellulosic biomass. J Mol Liq 287:110943. https://doi.org/10.1016/J.MOLLIQ.2019.110943

11. Atinkut HB, Yan T, Arega Y, Raza MH (2020) Farmers' willingness-to-pay for eco-friendly agricultural waste management in Ethiopia: a contingent valuation. J Clean Prod 261:121211. https://doi.org/10.1016/J.JCLEPRO.2020.121211
12. Awasthi MK, Sindhu R, Sirohi R et al (2022) Agricultural waste biorefinery development towards circular bioeconomy. Renew Sustain Energy Rev 158:112122. https://doi.org/10.1016/J.RSER.2022.112122
13. Awogbemi O, von Kallon DV (2022) Pretreatment techniques for agricultural waste. Case Stud Chem Environ Eng 6:100229. https://doi.org/10.1016/J.CSCEE.2022.100229
14. Barbier J, Charon N, Dupassieux N et al (2012) Hydrothermal conversion of lignin compounds. a detailed study of fragmentation and condensation reaction pathways. Biomass Bioenergy 46:479–491. https://doi.org/10.1016/J.BIOMBIOE.2012.07.011
15. Behera B, Mari Selvam S, Balasubramanian P (2022) Hydrothermal processing of microalgal biomass: circular bio-economy perspectives for addressing food-water-energy nexus. Bioresour Technol 359:127443. https://doi.org/10.1016/j.biortech.2022.127443
16. Benites-Lazaro LL, Giatti LL, Sousa Junior WC, Giarolla A (2020) Land-water-food nexus of biofuels: discourse and policy debates in Brazil. Environ Dev 33:100491. https://doi.org/10.1016/j.envdev.2019.100491
17. Bienergy International (2022) New 2G ethanol plans for Europe's growing biofuels market | Bioenergy International. In: Bioenergy International. https://bioenergyinternational.com/new-2g-ethanol-plans-for-europes-growing-biofuels-market/. Accessed 2 Jan 2023
18. Bondancia TJ, de Aguiar J, Batista G et al (2020) Production of nanocellulose using citric acid in a biorefinery concept: effect of the hydrolysis reaction time and techno-economic analysis. Ind Eng Chem Res 59:11505–11516. https://doi.org/10.1021/ACS.IECR.0C01359/SUPPL_FILE/IE0C01359_SI_002.XLSX
19. Boudway I (2022) More than half of us car sales will be electric by 2030—bloomberg. Bloomberg
20. Braide W, Kanu IA, Oranusi US, Adeleye SA (2016) Production of bioethanol from agricultural waste. J Fundam Appl Sci 8:372–386. https://doi.org/10.4314/jfas.v8i2.14
21. Brenelli LB, Bhatia R, Djajadi DT et al (2022) Xylo-oligosaccharides, fermentable sugars, and bioenergy production from sugarcane straw using steam explosion pretreatment at pilot-scale. Bioresour Technol 357:127093. https://doi.org/10.1016/J.BIORTECH.2022.127093
22. Canadell P (2014) Plants absorb more CO_2 than we thought, but ... The Conversation
23. Canilha L, Carvalho W, Felipe MDGDA et al (2010) Ethanol production from sugarcane bagasse hydrolysate using Pichia stipitis. Appl Biochem Biotechnol 161:84–92. https://doi.org/10.1007/s12010-009-8792-8
24. Canilha L, Chandel AK, dos Santos S, Milessi T et al (2012) Bioconversion of sugarcane biomass into ethanol: an overview about composition, pretreatment methods, detoxification of hydrolysates, enzymatic saccharification, and ethanol fermentation. J Biomed Biotechnol 2012:989572. https://doi.org/10.1155/2012/989572
25. Cao Y (2013) application status and development strategies of biomass energy in China. Adv Mat Res 608–609:261–264. https://doi.org/10.4028/WWW.SCIENTIFIC.NET/AMR.608-609.261
26. Carpenter D, Westover TL, Czernik S, Jablonski W (2014) Biomass feedstocks for renewable fuel production: a review of the impacts of feedstock and pretreatment on the yield and product distribution of fast pyrolysis bio-oils and vapors. Green Chem 16:384–406. https://doi.org/10.1039/C3GC41631C
27. Carr JA, D'Odorico P, Laio F, Ridolfi L (2013) Recent history and geography of virtual water trade. PLoS ONE 8:e55825. https://doi.org/10.1371/JOURNAL.PONE.0055825
28. Cassidy ES, West PC, Gerber JS, Foley JA (2013) Redefining agricultural yields: from tonnes to people nourished per hectare. Environ Res Lett 8:034015. https://doi.org/10.1088/1748-9326/8/3/034015
29. Cavalett O, Cherubini F (2018) Contribution of jet fuel from forest residues to multiple Sustainable Development Goals. Nat Sustain 1(12):799–807. https://doi.org/10.1038/s41893-018-0181-2

30. Chen K, Hao S, Lyu H et al (2017) Ion exchange separation for recovery of monosaccharides, organic acids and phenolic compounds from hydrolysates of lignocellulosic biomass. Sep Purif Technol 172:100–106. https://doi.org/10.1016/j.seppur.2016.08.004
31. Chen Y, Mu T (2019) Application of deep eutectic solvents in biomass pretreatment and conversion. Green Energy Enviro 4:95–115. https://doi.org/10.1016/J.GEE.2019.01.012
32. Chiaramonti D, Giovannini A, Janssen R, Mergner R (2013) Lignocellulosic ethanol production plant by Biochemtex in Italy
33. Choi JH, Jang SK, Kim JH et al (2019) Simultaneous production of glucose, furfural, and ethanol organosolv lignin for total utilization of high recalcitrant biomass by organosolv pretreatment. Renew Energy 130:952–960. https://doi.org/10.1016/J.RENENE.2018.05.052
34. Contreras F, Pramanik S, M. Rozhkova A et al (2020) Engineering robust cellulases for tailored lignocellulosic degradation cocktails. Int J Mol Sci 21
35. da Costa TP, Quinteiro P, Arroja L, Dias AC (2020) Environmental comparison of forest biomass residues application in Portugal: electricity, heat and biofuel. Renew Sustain Energy Rev 134:110302. https://doi.org/10.1016/J.RSER.2020.110302
36. Dahmen N, Lewandowski I, Zibek S, Weidtmann A (2019) Integrated lignocellulosic value chains in a growing bioeconomy: STATUS quo and perspectives. GCB Bioenergy 11:107–117. https://doi.org/10.1111/GCBB.12586
37. Das AK, Sahu SK, Panda AK (2022) Current status and prospects of alternate liquid transportation fuels in compression ignition engines: a critical review. Renew Sustain Energy Rev 161:112358. https://doi.org/10.1016/j.rser.2022.112358
38. Das L, Li M, Stevens J et al (2018) Characterization and catalytic transfer hydrogenolysis of deep eutectic solvent extracted sorghum lignin to phenolic compounds. ACS Sustain Chem Eng 6:10408–10420. https://doi.org/10.1021/ACSSUSCHEMENG.8B01763/SUPPL_FILE/SC8B01763_SI_001.PDF
39. Demichelis F, Laghezza M, Chiappero M, Fiore S (2020) Technical, economic and environmental assessement of bioethanol biorefinery from waste biomass. https://doi.org/10.1016/j.jclepro.2020.124111
40. Deshavath NN, Mogili NV, Dutta M et al (2022a) Role of lignocellulosic bioethanol in the transportation sector: limitations and advancements in bioethanol production from lignocellulosic biomass. In: Hussain CM, Singh S, Goswami L (eds) Waste-to-Energy Approaches Towards Zero Waste. Elsevier, pp 57–85
41. Deshavath NN, Vaibhav G v., Veeranki VD (2022b) Commercialization of 2G bioethanol as a transportation fuel for the sustainable energy, environment, and economic growth of India: theoretical and empirical assessment of bioethanol potential from agriculture crop residues. Biomass Conversion Biorefinery 1:1–13. https://doi.org/10.1007/S13399-022-03039-2
42. Devi A, Singh A, Bajar S et al (2021) Ethanol from lignocellulosic biomass: an in-depth analysis of pre-treatment methods, fermentation approaches and detoxification processes. J Environ Chem Eng 9:105798. https://doi.org/10.1016/j.jece.2021.105798
43. di Gruttola F, Borello D (2021) Analysis of the EU secondary biomass availability and conversion processes to produce advanced biofuels: use of existing databases for assessing a metric evaluation for the 2025 perspective. Sustainability (Switzerland) 13: https://doi.org/10.3390/su13147882
44. Dierks M, Cao Z, Rinaldi R (2021) Design of task-specific metal phosphides for the sustainable manufacture of advanced biofuels. Adv Inorg Chem 77:219–239. https://doi.org/10.1016/BS.ADIOCH.2021.02.002
45. Dionísio SR, Santoro DCJ, Bonan CIDG et al (2021) Second-generation ethanol process for integral use of hemicellulosic and cellulosic hydrolysates from diluted sulfuric acid pretreatment of sugarcane bagasse. Fuel 304:121290. https://doi.org/10.1016/J.FUEL.2021.121290
46. Du J, Liang J, Gao X et al (2020) Optimization of an artificial cellulase cocktail for high-solids enzymatic hydrolysis of cellulosic materials with different pretreatment methods. Bioresour Technol 295:122272. https://doi.org/10.1016/j.biortech.2019.122272

47. Dubois O (2011) The state of the world's land and water resources for food and agriculture: managing systems at risk
48. Duque-Acevedo M, Belmonte-Ureña LJ, Batlles-delaFuente A, Camacho-Ferre F (2022) Management of agricultural waste biomass: a case study of fruit and vegetable producer organizations in southeast Spain. J Clean Prod 359:131972. https://doi.org/10.1016/J.JCLEPRO.2022.131972
49. Erdoğan E, Karagöz P, Yılmaz E, Özkan M (2022) Biofuel production from agricultural waste. In: Kacprzak M, Attard E, Lyng K et al (eds) Biodegradable waste management in the circular economy: challenges and opportunities. John Wiley & Sons, pp 1–576
50. EURACTIV (2022) EU Parliament passes ban on new petrol, diesel cars by 2035—EURACTIV.com. https://www.euractiv.com/section/transport/news/eu-parliament-passes-ban-on-new-petrol-diesel-cars-by-2035/. Accessed 23 Aug 2022
51. Fang H, Li C, Zhao J, Zhao C (2021) Biotechnological advances and trends in engineering trichoderma reesei towards cellulase hyperproducer. Biotechnol Bioprocess Eng 26:517–528. https://doi.org/10.1007/s12257-020-0243-y
52. Fitz E, Gamauf C, Seiboth B, Wanka F (2019) Deletion of the small GTPase rac1 in Trichoderma reesei provokes hyperbranching and impacts growth and cellulase production. Fungal Biol Biotechnol 6:16. https://doi.org/10.1186/s40694-019-0078-5
53. Gao J, Xin S, Wang L et al (2019) Effect of ionic liquid/inorganic salt/water pretreatment on the composition, structure and enzymatic hydrolysis of rice straw. Bioresour Technol Rep 5:355–358. https://doi.org/10.1016/J.BITEB.2018.05.006
54. Ghaffar SH, Fan M, McVicar B (2015) Bioengineering for utilisation and bioconversion of straw biomass into bio-products. Ind Crops Prod 77:262–274. https://doi.org/10.1016/J.INDCROP.2015.08.060
55. Gibbs HK, Ruesch AS, Achard F et al (2010) Tropical forests were the primary sources of new agricultural land in the 1980s and 1990s. Proc Natl Acad Sci USA 107:16732–16737. https://doi.org/10.1073/PNAS.0910275107/SUPPL_FILE/PNAS.200910275SI.PDF
56. Gnansounou E, Dauriat A (2010) Techno-economic analysis of lignocellulosic ethanol: a review. Bioresour Technol 101:4980–4991. https://doi.org/10.1016/J.BIORTECH.2010.02.009
57. Gomes DG, Michelin M, Romaní A et al (2021) Co-production of biofuels and value-added compounds from industrial Eucalyptus globulus bark residues using hydrothermal treatment. Fuel 285:119265. https://doi.org/10.1016/J.FUEL.2020.119265
58. Gowen CM, Fong SS (2010) Exploring biodiversity for cellulosic biofuel production. Chem Biodivers 7:1086–1097. https://doi.org/10.1002/cbdv.200900314
59. Guo H, Chang Y, Lee DJ (2018) Enzymatic saccharification of lignocellulosic biorefinery: research focuses. Bioresour Technol 252:198–215. https://doi.org/10.1016/J.BIORTECH.2017.12.062
60. Guo Z, Zhang Q, You T et al (2019) Short-time deep eutectic solvent pretreatment for enhanced enzymatic saccharification and lignin valorization. Green Chem 21:3099–3108. https://doi.org/10.1039/C9GC00704K
61. Gupta R, Hemansi, Gautam S et al (2017) Study of charcoal detoxification of acid hydrolysate from corncob and its fermentation to xylitol. J Environ Chem Eng 5:4573–4582. https://doi.org/10.1016/j.jece.2017.07.073
62. He R, Ding R, Heyman JA et al (2019) Ultra-high-throughput picoliter-droplet microfluidics screening of the industrial cellulase-producing filamentous fungus Trichoderma reesei. J Ind Microbiol Biotechnol 46:1603–1610. https://doi.org/10.1007/S10295-019-02221-2/FIGURES/4
63. Hickert LR, da Cunha-Pereira F, de Souza-Cruz PB et al (2013) Ethanogenic fermentation of co-cultures of Candida shehatae HM 52.2 and Saccharomyces cerevisiae ICV D254 in synthetic medium and rice hull hydrolysate. Bioresour Technol 131:508–514. https://doi.org/10.1016/j.biortech.2012.12.135
64. Hongdan Z, Shaohua X, Shubin W (2013) Enhancement of enzymatic saccharification of sugarcane bagasse by liquid hot water pretreatment. Bioresour Technol 143:391–396. https://doi.org/10.1016/J.BIORTECH.2013.05.103

65. Hoppert L, Kölling R, Einfalt D (2022) Synergistic effects of inhibitors and osmotic stress during high gravity bioethanol production from steam-exploded lignocellulosic feedstocks. Biocatal Agric Biotechnol 43:102414. https://doi.org/10.1016/J.BCAB.2022.102414
66. Hu G, Heitmann JA, Rojas OJ (2008) Feedstock pretreatment strategies for producing ethanol from wood, bark, and forest residues. BioResources 3:270–294
67. Ire FS, Ezebuiro V, Ogugbue CJ (2016) Production of bioethanol by bacterial co-culture from agro-waste-impacted soil through simultaneous saccharification and co-fermentation of steam-exploded bagasse. Bioresour Bioprocess 3:1–12. https://doi.org/10.1186/S40643-016-0104-X/FIGURES/5
68. Jędrzejczyk M, Soszka E, Czapnik M et al (2019) Physical and chemical pretreatment of lignocellulosic biomass. Second and Third Generation of Feedstocks: The Evolution of Biofuels 143–196. https://doi.org/10.1016/B978-0-12-815162-4.00006-9
69. Jönsson LJ, Alriksson B, Nilvebrant NO (2013) Bioconversion of lignocellulose: inhibitors and detoxification. Biotechnol Biofuels 6:1–10. https://doi.org/10.1186/1754-6834-6-16
70. Kang Q, Appels L, Baeyens J et al (2014) Energy-efficient production of cassava-based bio-ethanol. Adv Biosci Biotechnol 05:925–939. https://doi.org/10.4236/ABB.2014.512107
71. Karagoz P, Bill RM, Ozkan M (2019) Lignocellulosic ethanol production: evaluation of new approaches, cell immobilization and reactor configurations. Renew Energy 143:741–752. https://doi.org/10.1016/j.renene.2019.05.045
72. Kazemi Shariat Panahi H, Dehhaghi M, Dehhaghi S et al (2022) Engineered bacteria for valorizing lignocellulosic biomass into bioethanol. Bioresour Technol 344:126212. https://doi.org/10.1016/j.biortech.2021.126212
73. Kerstetter JD, Lyons JK (2001). Logging and Agricultural Residue Supply Curves for the Pacific Northwest. https://doi.org/10.2172/900299
74. Khalid A, Aslam M, Qyyum MA et al (2019) Membrane separation processes for dehydration of bioethanol from fermentation broths: recent developments, challenges, and prospects. Renew Sustain Energy Rev 105:427–443. https://doi.org/10.1016/J.RSER.2019.02.002
75. Kim JS, Choi GG (2018) Pyrolysis of lignocellulosic biomass for biochemical production. Waste Bioref: Potent Perspect 323–348. https://doi.org/10.1016/B978-0-444-63992-9.00011-2
76. Kim S, Dale BE (2004) Global potential bioethanol production from wasted crops and crop residues. Biomass Bioenergy 26:361–375. https://doi.org/10.1016/J.BIOMBIOE.2003.08.002
77. Kiss AA, Suszwalak DJPC (2012) Enhanced bioethanol dehydration by extractive and azeotropic distillation in dividing-wall columns. Sep Purif Technol 86:70–78. https://doi.org/10.1016/J.SEPPUR.2011.10.022
78. Kumar V, Krishania M, Preet Sandhu P et al (2018) Efficient detoxification of corn cob hydrolysate with ion-exchange resins for enhanced xylitol production by Candida tropicalis MTCC 6192. Bioresour Technol 251:416–419. https://doi.org/10.1016/j.biortech.2017.11.039
79. Kumar V, Yadav SK, Kumar J, Ahluwalia V (2020) A critical review on current strategies and trends employed for removal of inhibitors and toxic materials generated during biomass pretreatment. Bioresour Technol 299:122633. https://doi.org/10.1016/j.biortech.2019.122633
80. Kun RS, Gomes ACS, Hildén KS et al (2019) Developments and opportunities in fungal strain engineering for the production of novel enzymes and enzyme cocktails for plant biomass degradation. Biotechnol Adv 37:107361. https://doi.org/10.1016/j.biotechadv.2019.02.017
81. Kuthiala T, Thakur K, Sharma D et al (2022) The eco-friendly approach of cocktail enzyme in agricultural waste treatment: a comprehensive review. Int J Biol Macromol 209:1956–1974. https://doi.org/10.1016/J.IJBIOMAC.2022.04.173
82. LaMonaca S, Ryan L (2022) The state of play in electric vehicle charging services—a review of infrastructure provision, players, and policies. Renew Sustain Energy Rev 154:111733. https://doi.org/10.1016/J.RSER.2021.111733
83. Lara-Serrano M, Morales-delaRosa S, Campos-Martín JM, Fierro JLG (2019) Fractionation of lignocellulosic biomass by selective precipitation from ionic liquid dissolution. Appl Sci 9:1862. https://doi.org/10.3390/APP9091862

84. Lászlok A, Takács-György K, Takács I (2020) Examination of first generation biofuel production in some selected biofuel producing countries in Europe: a case study. Agri Econ (Czech Republic) 66:469–476. https://doi.org/10.17221/237/2020-AGRICECON
85. Lee I, Yu JH (2020) The production of fermentable sugar and bioethanol from acacia wood by optimizing dilute sulfuric acid pretreatment and post treatment. Fuel 275:117943. https://doi.org/10.1016/J.FUEL.2020.117943
86. Lee JM, Venditti RA, Jameel H, Kenealy WR (2011) Detoxification of woody hydrolyzates with activated carbon for bioconversion to ethanol by the thermophilic anaerobic bacterium Thermoanaerobacterium saccharolyticum. Biomass Bioenergy 35:626–636. https://doi.org/10.1016/j.biombioe.2010.10.021
87. Li C, Aston JE, Lacey JA et al (2016) Impact of feedstock quality and variation on biochemical and thermochemical conversion. Renew Sustain Energy Rev 65:525–536. https://doi.org/10.1016/J.RSER.2016.06.063
88. Liu G, Qu Y (2021) Integrated engineering of enzymes and microorganisms for improving the efficiency of industrial lignocellulose deconstruction. Engineering Microbiology 1:100005. https://doi.org/10.1016/J.ENGMIC.2021.100005
89. Liu H, Pang B, Zhao Y et al (2018) Comparative study of two different alkali-mechanical pretreatments of corn stover for bioethanol production. Fuel 221:21–27. https://doi.org/10.1016/J.FUEL.2018.02.088
90. Liu WY, Lin CC, Yeh TL (2017) Supply chain optimization of forest biomass electricity and bioethanol coproduction. Energy 139:630–645. https://doi.org/10.1016/J.ENERGY.2017.08.018
91. Liu Y, Nie Y, Lu X et al (2019) Cascade utilization of lignocellulosic biomass to high-value products. Green Chem 21:3499–3535. https://doi.org/10.1039/C9GC00473D
92. Liu Y, Zheng X, Tao S et al (2021) Process optimization for deep eutectic solvent pretreatment and enzymatic hydrolysis of sugar cane bagasse for cellulosic ethanol fermentation. Renew Energy 177:259–267. https://doi.org/10.1016/J.RENENE.2021.05.131
93. Liu YJ, Li B, Feng Y, Cui Q (2020) Consolidated bio-saccharification: Leading lignocellulose bioconversion into the real world. Biotechnol Adv 40:107535. https://doi.org/10.1016/J.BIOTECHADV.2020.107535
94. Liu Z-H, Qin L, Zhu J-Q et al (2014) Simultaneous saccharification and fermentation of steam-exploded corn stover at high glucan loading and high temperature. Biotechnol Biofuels 7:167. https://doi.org/10.1186/s13068-014-0167-x
95. LIGNOFLAG (2021) Commercial flagship plant for bioethanol production involving a bio-based value chain built on lignocellulosic feedstock
96. Lynd LR, Beckham GT, Guss AM et al (2022) Toward low-cost biological and hybrid biological/catalytic conversion of cellulosic biomass to fuels†. Energy Environ Sci 15:938–990. https://doi.org/10.1039/d1ee02540f
97. Lynd LR, Guss AM, Himmel ME et al (2016) Advances in consolidated bioprocessing using Clostridium thermocellum and Thermoanaerobacter saccharolyticum. Indus Biotechnol Microorg 10:365–394
98. Lynd LR, Liang X, Biddy MJ et al (2017) Cellulosic ethanol: status and innovation. Curr Opin Biotechnol 45:202–211. https://doi.org/10.1016/j.copbio.2017.03.008
99. Mahmoodi P, Karimi K, Taherzadeh MJ (2018) Efficient conversion of municipal solid waste to biofuel by simultaneous dilute-acid hydrolysis of starch and pretreatment of lignocelluloses. Energy Convers Manag 166:569–578. https://doi.org/10.1016/J.ENCONMAN.2018.04.067
100. Mahmud S, Haider ASMR, Shahriar ST et al (2022) Bioethanol and biodiesel blended fuels—feasibility analysis of biofuel feedstocks in Bangladesh. Energy Rep 8:1741–1756. https://doi.org/10.1016/j.egyr.2022.01.001
101. Maleki M, Shahraki MF, Kavousi K et al (2020) A novel thermostable cellulase cocktail enhances lignocellulosic bioconversion and biorefining in a broad range of pH. Int J Biol Macromol 154:349–360. https://doi.org/10.1016/j.ijbiomac.2020.03.100
102. Martins MAR, Pinho SP, Coutinho JAP (2019) Insights into the nature of eutectic and deep eutectic mixtures. J Solution Chem 48:962–982. https://doi.org/10.1007/S10953-018-0793-1/FIGURES/9

103. Mathew AK, Parameshwaran B, Sukumaran RK, Pandey A (2016) An evaluation of dilute acid and ammonia fiber explosion pretreatment for cellulosic ethanol production. Bioresour Technol 199:13–20. https://doi.org/10.1016/J.BIORTECH.2015.08.121
104. Meade B, Thome K, Melton A (2018) USDA ERS—international food security expected to improve, but regional differences persist. In: Economic research service U.S. Department of agriculture. https://www.ers.usda.gov/amber-waves/2018/november/international-food-security-expected-to-improve-but-regional-differences-persist/. Accessed 4 Jan 2023
105. Meenakshisundaram S, Fayeulle A, Leonard E et al (2021) Fiber degradation and carbohydrate production by combined biological and chemical/physicochemical pretreatment methods of lignocellulosic biomass—a review. Bioresour Technol 331:125053. https://doi.org/10.1016/J.BIORTECH.2021.125053
106. Melendez JR, Mátyás B, Hena S et al (2022) Perspectives in the production of bioethanol: a review of sustainable methods, technologies, and bioprocesses. https://doi.org/10.1016/j.rser.2022.112260
107. Meng Q-S, Liu C-G, Zhao X-Q, Bai F-W (2018) Engineering Trichoderma reesei Rut-C30 with the overexpression of egl1 at the ace1 locus to relieve repression on cellulase production and to adjust the ratio of cellulolytic enzymes for more efficient hydrolysis of lignocellulosic biomass. J Biotechnol 285:56–63. https://doi.org/10.1016/j.jbiotec.2018.09.001
108. Menon V, Rao M (2012) Trends in bioconversion of lignocellulose: Biofuels, platform chemicals & biorefinery concept. Prog Energy Combust Sci 38:522–550. https://doi.org/10.1016/J.PECS.2012.02.002
109. Mohan M, Deshavath NN, Banerjee T et al (2018) Ionic liquid and sulfuric acid-based pretreatment of bamboo: biomass delignification and enzymatic hydrolysis for the production of reducing sugars. Ind Eng Chem Res 57:10105–10117. https://doi.org/10.1021/ACS.IECR.8B00914/ASSET/IMAGES/MEDIUM/IE-2018-009147_0007.GIF
110. Morales M, Arvesen A, Cherubini F (2021) Integrated process simulation for bioethanol production: effects of varying lignocellulosic feedstocks on technical performance. Bioresour Technol 328:124833. https://doi.org/10.1016/J.BIORTECH.2021.124833
111. Morales M, Quintero J, Conejeros R, Aroca G (2015) Life cycle assessment of lignocellulosic bioethanol: environmental impacts and energy balance. Renew Sustain Energy Rev 42:1349–1361. https://doi.org/10.1016/J.RSER.2014.10.097
112. Muratori M, Alexander M, Arent D et al (2021) The rise of electric vehicles—2020 status and future expectations. Progress in Energy 3:022002. https://doi.org/10.1088/2516-1083/ABE0AD
113. Murillo-Alvarado PE, Guillén-Gosálbez G, Ponce-Ortega JM et al (2015) Multi-objective optimization of the supply chain of biofuels from residues of the tequila industry in Mexico. J Clean Prod 108:422–441. https://doi.org/10.1016/J.JCLEPRO.2015.08.052
114. Nagy E, Mizsey P, Hancsók J et al (2015) Analysis of energy saving by combination of distillation and pervaporation for biofuel production. Chem Eng Process 98:86–94. https://doi.org/10.1016/J.CEP.2015.10.010
115. Neslen A (2012) Biofuels industry threatens to sue European Commission—renewable Carbon News. In: EURACTIV. https://renewable-carbon.eu/news/biofuels-industry-threatens-to-sue-european-commission/. Accessed 22 Jun 2022
116. Nguyen QA, Yang J, Bae HJ (2017) Bioethanol production from individual and mixed agricultural biomass residues. Ind Crops Prod 95:718–725. https://doi.org/10.1016/J.INDCROP.2016.11.040
117. Okolie JA, Nanda S, Dalai AK, Kozinski JA (2020) Hydrothermal gasification of soybean straw and flax straw for hydrogen-rich syngas production: experimental and thermodynamic modeling. Energy Convers Manag 208:112545. https://doi.org/10.1016/J.ENCONMAN.2020.112545
118. Østby H, Hansen LD, Horn SJ et al (2020) Enzymatic processing of lignocellulosic biomass: principles, recent advances and perspectives. J Ind Microbiol Biotechnol 47:623–657. https://doi.org/10.1007/s10295-020-02301-8

119. Palmqvist E, Hahn-Hägerdal B (2000) Fermentation of lignocellulosic hydrolysates. I: inhibition and detoxification. Bioresour Technol 74:17–24. https://doi.org/10.1016/S0960-852 4(99)00160-1
120. Pardo G, Moral R, del Prado A (2017) SIMS WASTE-AD—a modelling framework for the environmental assessment of agricultural waste management strategies: anaerobic digestion. Sci Total Environ 574:806–817. https://doi.org/10.1016/J.SCITOTENV.2016.09.096
121. Peng Y, Lu X, Liu B, Zhu J (2017) Separation of azeotropic mixtures (ethanol and water) enhanced by deep eutectic solvents. Fluid Phase Equilib 448:128–134. https://doi.org/10.1016/J.FLUID.2017.03.010
122. Phromphithak S, Tippayawong N, Onsree T, Lauterbach J (2022) Pretreatment of corncob with green deep eutectic solvent to enhance cellulose accessibility for energy and fuel applications. Energy Rep 8:579–585. https://doi.org/10.1016/J.EGYR.2022.07.071
123. Pienkos PT, Zhang M (2009) Role of pretreatment and conditioning processes on toxicity of lignocellulosic biomass hydrolysates. Cellulose 16:743–762. https://doi.org/10.1007/s10 570-009-9309-x
124. Putro JN, Soetaredjo FE, Lin S-Y et al (2016) Pretreatment and conversion of lignocellulose biomass into valuable chemicals. RSC Adv 6:46834–46852. https://doi.org/10.1039/C6RA09 851G
125. Qin L, Zhao X, Li W-C et al (2018) Process analysis and optimization of simultaneous saccharification and co-fermentation of ethylenediamine-pretreated corn stover for ethanol production. Biotechnol Biofuels 11:118. https://doi.org/10.1186/s13068-018-1118-8
126. Rabemanolontsoa H, Saka S (2016) Various pretreatments of lignocellulosics. Bioresour Technol 199:83–91. https://doi.org/10.1016/J.BIORTECH.2015.08.029
127. Raj T, Chandrasekhar K, Naresh Kumar A et al (2022) Recent advances in commercial biorefineries for lignocellulosic ethanol production: current status, challenges and future perspectives. Bioresour Technol 344:126292. https://doi.org/10.1016/J.BIORTECH.2021.126292
128. Rawat J (2020) Integrated approach for sustainable bio-refinery. In: 3rd EU India Advance Biofuel Conference
129. Reshmy R, Philip E, Madhavan A et al (2022) Lignocellulose in future biorefineries: Strategies for cost-effective production of biomaterials and bioenergy. Bioresour Technol 344:126241. https://doi.org/10.1016/j.biortech.2021.126241
130. Rezania S, Oryani B, Cho J et al (2020). Different pretreatment technologies of lignocellulosic biomass for bioethanol production: an overview. https://doi.org/10.1016/j.energy.2020.117457
131. Robak K, Balcerek M (2020) Current state-of-the-art in ethanol production from lignocellulosic feedstocks. Microbiol Res 240:126534. https://doi.org/10.1016/j.micres.2020.126534
132. Rulli MC, Bellomi D, Cazzoli A et al (2016) The water-land-food nexus of first-generation biofuels. Sci Rep 6:1–10. https://doi.org/10.1038/srep22521
133. Sadhukhan J, Martinez-Hernandez E, Amezcua-Allieri MA et al (2019). Economic and environmental impact evaluation of various biomass feedstock for bioethanol production and correlations to lignocellulosic composition. https://doi.org/10.1016/j.biteb.2019.100230
134. Santos SAO, Freire CSR, Domingues MRM et al (2011) Characterization of phenolic components in polar extracts of eucalyptus globulus Labill. bark by high-performance liquid chromatography-mass spectrometry. J Agric Food Chem 59:9386–9393. https://doi.org/10.1021/JF201801Q
135. Sarawan C, Suinyuy TN, Sewsynker-Sukai Y, Gueguim Kana EB (2019) Optimized activated charcoal detoxification of acid-pretreated lignocellulosic substrate and assessment for bioethanol production. Bioresour Technol 286:121403. https://doi.org/10.1016/j.biortech.2019.121403
136. Sárvári Horváth I, Sjöde A, Nilvebrant N-O et al (2004) Selection of anion exchangers for detoxification of dilute-acid hydrolysates from spruce. Appl Biochem Biotechnol 114:525–538. https://doi.org/10.1385/ABAB:114:1-3:525

137. Sasaki K, Tsuge Y, Sasaki D et al (2015) Mechanical milling and membrane separation for increased ethanol production during simultaneous saccharification and co-fermentation of rice straw by xylose-fermenting Saccharomyces cerevisiae. Bioresour Technol 185:263–268. https://doi.org/10.1016/j.biortech.2015.02.117
138. Schmitz PM, Kavallari A (2009) Crop plants versus energy plants—on the international food crisis. Bioorg Med Chem 17:4020–4021. https://doi.org/10.1016/j.bmc.2008.11.041
139. Sebastião D, Gonçalves MS, Marques S et al (2016) Life cycle assessment of advanced bioethanol production from pulp and paper sludge. Bioresour Technol 208:100–109. https://doi.org/10.1016/J.BIORTECH.2016.02.049
140. Seiler M, Jork C, Kavarnou A et al (2004) Separation of azeotropic mixtures using hyperbranched polymers or ionic liquids. AIChE J 50:2439–2454. https://doi.org/10.1002/AIC.10249
141. Sharma B, Larroche C, Dussap CG (2020) Comprehensive assessment of 2G bioethanol production. Bioresour Technol 313:123630. https://doi.org/10.1016/J.BIORTECH.2020.123630
142. Sharma S, Nair A, Sarma SJ (2021) Biorefinery concept of simultaneous saccharification and co-fermentation: challenges and improvements. Chem Eng Process Process Intensif 169:108634. https://doi.org/10.1016/j.cep.2021.108634
143. Silva JCR, Salgado JCS, Vici AC et al (2020) A novel Trichoderma reesei mutant RP698 with enhanced cellulase production. Braz J Microbiol 51:537–545. https://doi.org/10.1007/s42770-019-00167-2
144. Singh A, Bajar S, Devi A, Pant D (2021) An overview on the recent developments in fungal cellulase production and their industrial applications. Bioresour Technol Rep 14:100652. https://doi.org/10.1016/j.biteb.2021.100652
145. Singh S, Kumar A, Sivakumar N, Verma JP (2022) Deconstruction of lignocellulosic biomass for bioethanol production: recent advances and future prospects. Fuel 327:125109. https://doi.org/10.1016/J.FUEL.2022.125109
146. Singhania RR, Ruiz HA, Awasthi MK, et al (2021) Challenges in cellulase bioprocess for biofuel applications. Renew Sustain Energy Rev 151:111622. https://doi.org/10.1016/j.rser.2021.111622
147. Spaho N (2017) Distillation techniques in the fruit spirits production. Distillat Innovat Appl Model. https://doi.org/10.5772/66774
148. Sri Harjati Suhardi V, Prasai B, Samaha D, Boopathy R (2013) Combined biological and chemical pretreatment method for lignocellulosic ethanol production from energy cane. https://doi.org/10.7243/2052-6237-1-1
149. Su T, Zhao D, Khodadadi M, Len C (2020) Lignocellulosic biomass for bioethanol: recent advances, technology trends, and barriers to industrial development. Curr Opin Green Sustain Chem 24:56–60. https://doi.org/10.1016/J.COGSC.2020.04.005
150. Sun D, Lv ZW, Rao J et al (2022) Effects of hydrothermal pretreatment on the dissolution and structural evolution of hemicelluloses and lignin: a review. Carbohydr Polym 281:119050. https://doi.org/10.1016/J.CARBPOL.2021.119050
151. Sundaram V, Muthukumarappan K, Kamireddy SR (2015) Effect of ammonia fiber expansion (AFEX^{TM}) pretreatment on compression behavior of corn stover, prairie cord grass and switchgrass. Ind Crops Prod 74:45–54. https://doi.org/10.1016/j.indcrop.2015.04.027
152. Suriyachai N, Weerasaia K, Laosiripojana N et al (2013) Optimized simultaneous saccharification and co-fermentation of rice straw for ethanol production by Saccharomyces cerevisiae and Scheffersomyces stipitis co-culture using design of experiments. Bioresour Technol 142:171–178. https://doi.org/10.1016/J.BIORTECH.2013.05.003
153. Susan Van Dyk J, Li L, Leal DB et al (2016) The potential of biofuels in China
154. Takei T, Ikeda K, Ijima H, Kawakami K (2011) Fabrication of poly(vinyl alcohol) hydrogel beads crosslinked using sodium sulfate for microorganism immobilization. Process Biochem 46:566–571. https://doi.org/10.1016/j.procbio.2010.10.011
155. Taku U, Takayuki U, Akihiko N et al (2020) Convergent evolution of processivity in bacterial and fungal cellulases. Proc Natl Acad Sci 117:19896–19903. https://doi.org/10.1073/pnas.2011366117

156. Technavio (2020) Lignocellulosic feedstock-based biofuel market by conversion process and geography—forecast and analysis 2020–2024
157. Thompson DN, Campbell T, Bals B et al (2013) Chemical preconversion: application of low-severity pretreatment chemistries for commoditization of lignocellulosic feedstock. Biofuels 4:323–340. https://doi.org/10.4155/BFS.13.15
158. Tian D, Shen F, Yang G et al (2019) Liquid hot water extraction followed by mechanical extrusion as a chemical-free pretreatment approach for cellulosic ethanol production from rigid hardwood. Fuel 252:589–597. https://doi.org/10.1016/J.FUEL.2019.04.155
159. Usmani Z, Kumar V, Varjani S et al (2020) Municipal solid waste to clean energy system: a contribution toward sustainable development. Curr Dev Biotechnol Bioeng Resour Recov Wastes 217–231. https://doi.org/10.1016/B978-0-444-64321-6.00011-2
160. Verardi A, Blasi A, Marino T et al (2018) Effect of steam-pretreatment combined with hydrogen peroxide on lignocellulosic agricultural wastes for bioethanol production: analysis of derived sugars and other by-products. J Energy Chem 27:535–543. https://doi.org/10.1016/J.JECHEM.2017.11.007
161. Veza I, Djamari DW, Hamzah N et al (2022) Lessons from Brazil: opportunities of bioethanol biofuel in Indonesia. Indonesian J Comput Eng Design (IJoCED) 4:8–16. https://doi.org/10.35806/IJOCED.V4I1.239
162. Villanueva-Solís Luis A, Ruíz-Cuilty K, Camacho-Dávila A et al (2020) Lignocellulosic waste pretreatment and esterification using green solvents. Sep Purif Technol 250:117102. https://doi.org/10.1016/J.SEPPUR.2020.117102
163. Wahl D, Ness B, Wamsler C (2021) Implementing the urban food–water–energy nexus through urban laboratories: a systematic literature review. Sustain Sci 16:663–676. https://doi.org/10.1007/S11625-020-00893-9
164. Walls WD, Rusco FW (2007) Price effects of boutique motor fuels: federal environmental standards, regional fuel choices, and local gasoline prices. Energy J 28:145–163. https://doi.org/10.5547/issn0195-6574-ej-vol28-no3-8
165. Wang F, Ouyang D, Zhou Z et al (2021) Lignocellulosic biomass as sustainable feedstock and materials for power generation and energy storage. J Energy Chem 57:247–280. https://doi.org/10.1016/J.JECHEM.2020.08.060
166. Wang F, Shi D, Han J et al (2020) Comparative study on pretreatment processes for different utilization purposes of switchgrass. ACS Omega. https://doi.org/10.1021/ACSOMEGA.0C01047
167. Wang J, Wang J, Lu Z, Zhang J (2020) Adsorption and desorption of cellulase on/from enzymatic residual lignin after alkali pretreatment. Ind Crops Prod 155:112811. https://doi.org/10.1016/J.INDCROP.2020.112811
168. Wang W, Lee DJ (2021) Lignocellulosic biomass pretreatment by deep eutectic solvents on lignin extraction and saccharification enhancement: a review. Bioresour Technol 339:125587. https://doi.org/10.1016/J.BIORTECH.2021.125587
169. Wei J, Liang G, Alex J et al (2020) Research progress of energy utilization of agricultural waste in China: bibliometric analysis by citespace. Sustainability 12:812. https://doi.org/10.3390/SU12030812
170. Wibowo CS, Sugiarto B, Zikra A et al (2018) The effect of gasoline-bioethanol blends to the value of fuel's octane number. E3S Web of Conf 67:4–6. https://doi.org/10.1051/e3sconf/20186702033
171. Wirawan F, Cheng C-L, Kao W-C et al (2012) Cellulosic ethanol production performance with SSF and SHF processes using immobilized Zymomonas mobilis. Appl Energy 100:19–26. https://doi.org/10.1016/j.apenergy.2012.04.032
172. Xia J, Yang Y, Liu C-G et al (2019) Engineering Zymomonas mobilis for Robust Cellulosic Ethanol Production. Trends Biotechnol 37:960–972. https://doi.org/10.1016/j.tibtech.2019.02.002
173. Yadav VG, Yadav GD, Patankar SC (2020) The production of fuels and chemicals in the new world: critical analysis of the choice between crude oil and biomass vis-à-vis sustainability and the environment. Clean Technol Environ Policy 22:1757. https://doi.org/10.1007/S10098-020-01945-5

174. Yan Q, Liu X, Wang Y et al (2018) Cow manure as a lignocellulosic substrate for fungal cellulase expression and bioethanol production. AMB Express 8:1–12. https://doi.org/10.1186/S13568-018-0720-2/FIGURES/5
175. Yan S, Xu Y, Yu X-W (2021) From induction to secretion: a complicated route for cellulase production in Trichoderma reesei. Bioresour Bioprocess 8:107. https://doi.org/10.1186/s40643-021-00461-8
176. Yang H, Shi Z, Xu G et al (2019) Bioethanol production from bamboo with alkali-catalyzed liquid hot water pretreatment. Bioresour Technol 274:261–266. https://doi.org/10.1016/J.BIORTECH.2018.11.088
177. Yang Q, Huo D, Han X et al (2021) Improvement of fermentable sugar recovery and bioethanol production from eucalyptus wood chips with the combined pretreatment of NH4Cl impregnation and refining. Ind Crops Prod 167:113503. https://doi.org/10.1016/J.INDCROP.2021.113503
178. Yang Y, Tian Z, Lan Y et al (2021) An overview of biofuel power generation on policies and finance environment, applied biofuels, device and performance. J Traff Transp Eng (English Edition) 8:534–553. https://doi.org/10.1016/j.jtte.2021.07.002
179. Yennamalli RM, Rader AJ, Kenny AJ et al (2013) Endoglucanases: insights into thermostability for biofuel applications. Biotechnol Biofuels 6:136. https://doi.org/10.1186/1754-6834-6-136
180. Yiin CL, Yap KL, Ku AZE et al (2021) Recent advances in green solvents for lignocellulosic biomass pretreatment: potential of choline chloride (ChCl) based solvents. Bioresour Technol 333:125195. https://doi.org/10.1016/J.BIORTECH.2021.125195
181. Yılmaz E (2019) Determination of the agricultural waste capacity for bioethanol production in Turkey. Gebze Technical University
182. Yu Y, Feng Y, Xu C et al (2011) Onsite bio-detoxification of steam-exploded corn stover for cellulosic ethanol production. Bioresour Technol 102:5123–5128. https://doi.org/10.1016/J.BIORTECH.2011.01.067
183. Yu Y, Wu J, Ren X et al (2022) Steam explosion of lignocellulosic biomass for multiple advanced bioenergy processes: a review. Renew Sustain Energy Rev 154:111871. https://doi.org/10.1016/J.RSER.2021.111871
184. Zabed H, Sahu JN, Suely A et al (2017) Bioethanol production from renewable sources: current perspectives and technological progress. Renew Sustain Energy Rev 71:475–501. https://doi.org/10.1016/j.rser.2016.12.076
185. Zambrano-Prado P, Muñoz-Liesa J, Josa A et al (2021) Assessment of the food-water-energy nexus suitability of rooftops. A methodological remote sensing approach in an urban Mediterranean area. Sustain Cities Soc 75:103287. https://doi.org/10.1016/J.SCS.2021.103287
186. Zhang K, Pei Z, Wang D (2016) Organic solvent pretreatment of lignocellulosic biomass for biofuels and biochemicals: a review. Bioresour Technol 199:21–33. https://doi.org/10.1016/J.BIORTECH.2015.08.102
187. Zhang Q, Wei Y, Han H, Weng C (2018) Enhancing bioethanol production from water hyacinth by new combined pretreatment methods. Bioresour Technol 251:358–363. https://doi.org/10.1016/J.BIORTECH.2017.12.085
188. Zhang Y, Hou T, Li B et al (2014) Acetone-butanol-ethanol production from corn stover pretreated by alkaline twin-screw extrusion pretreatment. Bioprocess Biosyst Eng 37:913–921. https://doi.org/10.1007/S00449-013-1063-7/FIGURES/8
189. Zhang Y, Wang C, Wang L et al (2017) Direct bioethanol production from wheat straw using xylose/glucose co-fermentation by co-culture of two recombinant yeasts. J Ind Microbiol Biotechnol 44:453–464. https://doi.org/10.1007/S10295-016-1893-9
190. Zhao L, Sun ZF, Zhang CC et al (2022) Advances in pretreatment of lignocellulosic biomass for bioenergy production: challenges and perspectives. Bioresour Technol 343:126123. https://doi.org/10.1016/J.BIORTECH.2021.126123
191. Zhao Q, Liu Q, Wang Q et al (2021) Disruption of the Trichoderma reesei gul1 gene stimulates hyphal branching and reduces broth viscosity in cellulase production. J Ind Microbiol Biotechnol 48:12. https://doi.org/10.1093/JIMB/KUAB012

192. Zhao X, Cheng K, Liu D (2009) Organosolv pretreatment of lignocellulosic biomass for enzymatic hydrolysis. Appl Microbiol Biotechnol 82:815–827. https://doi.org/10.1007/S00253-009-1883-1
193. Zhao X, Xiong L, Zhang M, Bai F (2016) Towards efficient bioethanol production from agricultural and forestry residues: exploration of unique natural microorganisms in combination with advanced strain engineering. Bioresour Technol 215:84–91. https://doi.org/10.1016/J.BIORTECH.2016.03.158
194. Zhu J-Q, Qin L, Li B-Z, Yuan Y-J (2014) Simultaneous saccharification and co-fermentation of aqueous ammonia pretreated corn stover with an engineered Saccharomyces cerevisiae SyBE005. Bioresour Technol 169:9–18. https://doi.org/10.1016/j.biortech.2014.06.085

Bioethanol Production from Microalgae: Potentials and Challenges

Mallika Boonmee Kongkeitkajorn

Abstract Microalgae-derived ethanol is a third-generation biofuel. There are several benefits of using microalgae as feedstock, particularly in terms of land use and the use of carbon dioxide in cultivation. It is crucial to recognize and overcome the challenges to make the production of third-generation ethanol feasible. This chapter presents various aspects of using microalgae as a feedstock for fermentation-based bioethanol production. The information provided in this chapter to readers is to help them comprehend the overall picture of the process involved in producing ethanol from microalgae. As a guide for process development, the author's perspectives on the challenges of employing microalgae in ethanol production are also introduced.

Keywords Biofuels · Bioethanol · Third generation · Algae · Microalgae · Fermentation · Downstream processing · Pretreatments · Microalgae cultivation

1 Introduction

Bioethanol is a significant component of transport fuels worldwide. Currently, ethanol is produced by fermenting sugars derived from starchy biomass, such as corn and cassava, and sugar biomass, such as sugarcane and sugar beets. Although first-generation bioethanol production is effective in high fermentation yield and reducing greenhouse gas emissions compared to gasoline [63], its downside is its ecological footprint. As first-generation bioethanol production involves the agricultural sector, the emission of methane (CH_4) and nitrous oxide (N_2O) is also a result [28, 63].

First-generation bioethanol production relies on edible plants as feedstock. This has led to a discussion on food/feed or energy dilemmas. Both uses of edible plants that are bioethanol feedstocks address goals 2 (zero hunger) and 7 (affordable and

M. B. Kongkeitkajorn (✉)
Department of Biotechnology, Faculty of Technology, Khon Kaen University, Khon Kaen, Thailand
e-mail: mallikab@kku.ac.th

clean energy) of the Sustainable Development Goals (SDGs) by the United Nations [68]. Changes in land use for the cultivation of energy crops can lead to land competition and increase food security risks [102].

Lignocellulosic biomass feedstocks are used in the development of second-generation ethanol production. Biomass includes energy crops, agricultural residues, and wastes from agricultural industries. The biomass structure is composed of lignin, hemicellulose, and cellulose. Hemicellulose and cellulose are polymers consisting of sugar monomers, so they are the fractions used in ethanol fermentation. Due to the complex and high recalcitrance structure of the biomass, efficient pretreatment to obtain a high sugar yield is critical [79]. The biomass conversion process requires high energy consumption [63] and cellulase enzymes that are costlier than the amylase used in some first-generation raw materials. In addition, fermentation inhibitors such as phenol, guaiacol, vanillin, levulinic acid, furfural, and 5-hydroxymethylfurfural (5-HMF) could pose a detrimental effect on bioethanol yield, and a detoxification step would be required in the biomass conversion process [58].

The above pitfalls in producing bioethanol from first-generation and second-generation feedstocks have brought an interest in using algae. They are considered third-generation feedstocks. Algae are prokaryotic and eukaryotic organisms, ranging from unicellular to multicellular forms. Microalgae are microscopically single cells and may be prokaryotic (cyanobacteria or blue-green algae) or eukaryotic [56, 113]. They could utilize carbon dioxide (CO_2) in their growth by converting it into carbohydrates, lipids, and other useful metabolites [56]. Third-generation bioethanol production is possible as microalgae possess carbohydrates of different forms, such as glycogen, starch, alginate, and cellulose [56]. Aside from the starch that accumulates in the cells, the outer layer of cell walls may contain pectin and alginate, while the inner layer of cell walls may contain cellulose as part of their cell wall composition [24, 62].

2 Microalgae as the Feedstock for Third-Generation Bioethanol

Using microalgal biomass to produce biofuels, including ethanol, has many advantages over first- and second-generation feedstocks. Land use competition with food crops is not an issue, as microalgae cultivation could be operated on marginal land. This advantage also alleviates the food security risk from a change in land use. As the cultivation does not require actual cropping, there are no requirements for fertilizers, which reduce greenhouse gas emissions. The reduction is due to the elimination of emissions from farming, trash burning, and field emissions from the soil due to fertilizers and residues [87].

Another advantage involves water consumption. Microalgae cultivation requires much less water than crops [7]. The water footprints of the gross energy of microalgae were smaller when compared with maize and sugarcane in ethanol production across various regions in the USA, Australia, and Europe [36]. Water recycling could further reduce the water footprint of microalgae cultivation [34, 69]. In addition, fresh water is not an obligation in cultivating microalgae. Marine microalgae could utilize seawater, and some microalgae could use wastewater as nutrients for their growth [61, 81]. Controlled conditions for microalgae cultivation also make it independent of the season.

Microalgae are autotrophic organisms. Their ability to capture light and carbon dioxide is considered an advantage, especially in CO_2 mitigation. Microalgal biofixation by photosynthesis is a technique in CO_2 capture, as microalgae could convert CO_2 into organic compounds useful for their growth during cultivation [27]. Due to the large diversity of microalgae, their abilities to fix CO_2 and their tolerance to high CO_2 vary among species [17, 111]. The specific strain on the environment of the biomass production site should be considered [89].

Carbohydrate-rich microalgae are potential feedstocks for bioethanol production. The accumulated carbohydrates are the prime components serving as substrates in ethanol production from microalgae. The stored carbohydrates in prokaryote microalgae (cyanobacteria) are glycogen, while eukaryote microalgae store carbohydrates as starch [2, 7]. A study on a strain of *Chlorella sorokiniana* suggested that the molecular weight of microalgal starch was comparable with plant sources and that the amylose/amylopectin ratio was close to that of cereal starch [37]. However, only green algae accumulate starch in the same way as plants. Red algae accumulate floridean starch or semi-amylopectin [29].

Microalgae carbohydrate content, especially starch, varies depending on microalgal genera, culture conditions, and culture techniques. Examples include medium composition (e.g., nitrogen limitation, sulfur limitation), CO_2 concentration, light intensity, and cell concentration [13, 15, 64]. In suitable conditions, intracellular starch granules (Fig. 1) are visible under a transmission electron microscope [11, 15].

Chlorella spp. has been the most studied for its starch accumulation, with up to 60% of starch content in the cells reported [15]. Other microalgae that could accumulate a substantial amount of starch include *Chlamydomonas* spp. and *Scenedesmus* sp. [38, 64]. Marine microalgae were also reported to accumulate starch. They are *Klebsormidium* sp., *Tetraselmis* sp., and *Dunaliella* sp. [38, 80, 112].

3 Cultivation of Microalgae

To compete with other feedstocks and use microalgae as a feedstock for bioethanol production, a massive amount of microalgal biomass is required to meet industrial-scale fuel ethanol production. Hence, microalgal cultivation is a critical part of bioethanol production from microalgae. Microalgae can be cultivated on a large scale in a closed or open environment. Possible cultivation systems for large-scale microalgae biomass production are open ponds and photobioreactors.

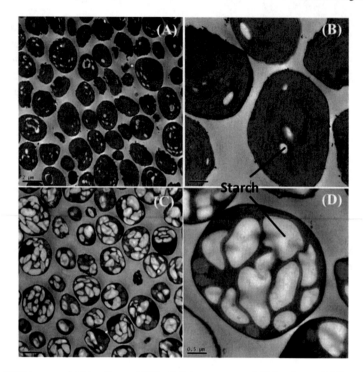

Fig. 1 TEM images of *Chlorella* sp. AE10. Cellular structure of *Chlorella* sp. AE10 at day 0 (**a, b**) and day 5 (**c, d**). This image by Cheng et al. [15] is licensed under CC BY 4.0 (http://creativec ommons.org/licenses/by/4.0/) and used without any changes

3.1 Open Ponds

Open ponds operate in an open environment. They are shallow ponds with a depth of more than 20 cm, usually between 20 and 50 cm. Deeper ponds could increase energy consumption due to the higher rotational speed of the paddle wheel to maintain sufficient circulation [23]. The systems could be tanks, big shallow ponds, circular ponds, or raceway ponds [8]. The raceway pond (Fig. 2) is the most widely used open system for industrial microalgal cultivation [72].

Raceway ponds are closed-loop recirculated structures, generally with long oval shapes. Channels are constructed to direct the flow within the structure. Recirculation of culture occurs within the pond by a paddlewheel. Many parameters are involved in designing and operating raceway ponds. Examples are pond dimensions (e.g., pond width and length, channel length, pond radius), paddlewheel dimensions (e.g., width and depth of blade), number of blades and speed of the paddlewheel, water depth, carbonation of water, and rate of water evaporation [23, 110].

Although open pond systems are simple and widely used, they are not without concerns. An obvious concern with open systems is that they are prone to contamination by other microalgae, protozoa, and bacteria [8, 56]. In addition, they are

Fig. 2 Pilot-scale raceway ponds. Modified raceway at right and standard raceway at left. This image by Cunha et al. [23] is licensed under CC BY 4.0 (http://creativecommons.org/licenses/by/4.0/) and used without any changes

dependent on the climate. Climate impacts cultivation temperature and light availability because the systems are not strict in controlling culturing conditions [8]. Although the mixing via paddlewheel requires significantly less energy than in other types of bioreactors, ineffective agitation of the culture is also to be considered [8, 72]. Open pond systems often result in low biomass density, so it is costly to harvest biomass [72].

3.2 Photobioreactors

Photobioreactors are essentially designed to enhance the sun-exposed surface area within a closed environment compared to open pond systems. They reduce the chance of contamination, which is one of the concerns in open pond systems. However, the equipment is more expensive than the open systems. The bioreactors are made from transparent materials such as glass, acrylic, polyethylene, or clear PVC to allow light penetration. Two photobioreactors generally known for microalgae cultivation are the tubular and flat plate types (Figs. 3 and 4).

3.2.1 Tubular Photobioreactors

As the name implies, tubular photobioreactors are transparent tubes where a centrifugal pump or an airlift system circulates microalgae culture. Typically, multiple tubes are used. Two types of tube layouts are normally observed: vertical and horizontal.

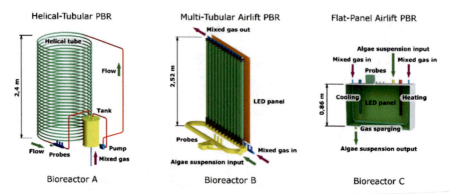

Fig. 3 Schematic representation of design of the three photo bioreactors (PBR): Helical-tubular PBR (Bioreactor A), Multi-tubular airlift PBR (Bioreactor B) and Flat panel PBR (Bioreactor C). Green parts represent sunlight illuminated zones; yellow parts represent dark zones. This image by Sukačová et al. [97] is licensed under CC BY 4.0 (http://creativecommons.org/licenses/by/4.0/) and used without any changes

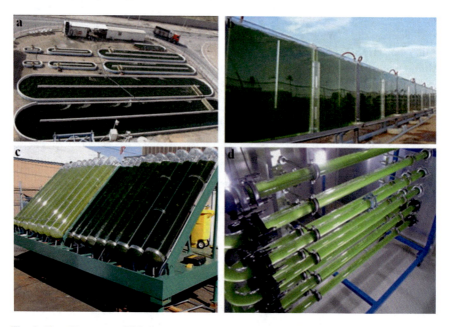

Fig. 4 Photobioreactors (PBRs) used in growing microalgae: **a** raceway pond, **b** flat-plate type, **c** inclined tubular type and **d** horizontal/continuous type. Reprinted from J. P. Bitog, I.-B. Lee, C.-G. Lee, K.-S. Kim, H.-S. Hwang, S.-W. Hong, I.-H. Seo, K.-S. Kwon, E. Mostafa (2011) Application of computational fluid dynamics for modeling and designing photobioreactors for microalgae production: a review. Comput Electron Agri 76 131–147, Copyright (2011), with permission from Elsevier

In the vertical arrangement, the transparent tubes are normally connected through interconnected tubes at both the top and bottom of the tubes (Bioreactor B in Fig. 3). An air pump is placed at the bottom section to provide air or air-CO_2 mixture to the culture. The gas leaves the vertical tubes through the headspace and the interconnected tubes on the top [97]. A variation of this arrangement is the inclined tubular type, where the tubes are placed at an angle for better light exposure (Fig. 4c).

In a horizontal arrangement, transparent tubes are connected and bent to cover the ground area, or they can be stacked in layers (Fig. 4d). A centrifugal pump circulates the culture, intending to generate turbulent flow for thoroughly mixing of the cells in the illuminated zone (tube periphery) and dark zone (tube centre). A degasser may be needed to remove oxygen that accumulates in the culture, as high oxygen concentration may decrease the productivity of the culture. Circulation by a centrifugal pump could cause shear stress to the microalgal cells. Therefore, if microalgae are shear-sensitive, airlift systems may be considered [72].

Tubular photobioreactors can be spiral-coiled (helical) tubes (Bioreactor A in Fig. 3). In this case, a transparent helical tube connects to a tank with a circulating pump. Air or an air-CO_2 mixture could be supplied through the tank [97]. It could also be seen as a degasser.

The main concern with using tubular photobioreactors is mass transfer. The flow should be turbulent to ensure that cells in the centre of the tube can be circulated to the illuminated zone. In addition, poor mixing could lead to pH gradients and concentration gradients of dissolved oxygen and CO_2, especially in a long bioreactor [8]. In addition, wall growth could present a problem in cultivation as it blocks light penetration to the cells.

3.2.2 Flat-Plate or Flat-Panel Photobioreactors

A flat-plate photobioreactor is a transparent and flat vessel (Fig. 4b). The design is to create a thin layer of microalgal culture so that the cells get maximum light exposure, hence a higher photosynthetic rate than tubular systems [8]. Regardless of the thin space in the vessel, mixing cultures is possible through air sparging [72]. In operation, many flat-plate bioreactors are placed vertically and closely to each other.

Cell production from flat-panel photobioreactors is higher than from tubular bioreactors. However, the installation cost is high. Some degree of wall growth also occurs in the flat-panel type, as in the tubular type.

3.3 Use of Wastewater and Seawater in Microalgae Cultivation

Although microalgae can utilize CO_2 as their carbon source, they also require other nutrients and minerals to support their growth. In small or mid-scale cultivation, BG-11 and Bold's basal media are largely employed as cultivation media. The medium contains sodium nitrate as a nitrogen source. Also, it has many trace elements, e.g., manganese, zinc, molybdenum, copper, and cobalt. These medium components do not significantly contribute to the production cost on experimental scales. However, medium components would contribute substantially to the production cost in large-scale production of biomass for bioethanol feedstock.

Wastewater and seawater contain trace elements such as zinc, iron, manganese, magnesium, and potassium, which could replace those in the synthetic media [14, 16, 39]. They also contain some nitrates and other nitrogen compounds. As wastewater is rich in nutrients, especially those suitable for microalgae cultivation, such as nitrates and phosphates, microalgae cultivation could be seen as a treatment option. Examples of wastewater that have been used for microalgae cultivation include brewery wastewater [3], agro-industrial wastewater [51], aquaculture wastewater [44], and urban wastewater [40].

For wastewater as a production medium, the composition of the wastewater is important as it determines whether a supply of CO_2 is necessary. Wastewaters from agricultural industries usually contain high COD and sufficient nitrogen to support microalgal growth [51]. In contrast, wastewater from aquaculture could lack a carbon source but contain enough nitrogen and micronutrients. Therefore, CO_2 must be supplied as a carbon source in cultivation [44].

The sole use of seawater in microalgae cultivation is rare. Most of the time, they are mixed with wastewater or another source of nutrients to a certain ratio to obtain desirable growth, which also depends on microalgae strains [14, 59, 81, 82]. However, growth is still observed with pure seawater, such as in *Chlorella sorokiniana* CY1 [14]. In addition, it is not a requirement to use marine microalgae when using seawater in cultivation. Freshwater algae also show an ability to grow in seawater [14, 59].

4 Microalgal Biomass Harvesting

Harvesting microalgal biomass in large-scale production for bioethanol needs careful consideration. Although numerous harvesting methods are available, low-cost harvesting methods are necessary. Harvesting generally contributes 20–30% of the production cost. As bioethanol is a low-value product, low harvesting costs and energy requirements are important to offer a chance for economic viability and a positive gain in net energy.

The nature of the microalgal cells led to difficulties in cell separation, especially when the cost is a constraint. Microalgae cultures have low sedimentation velocity

(0.1–2.6 cm/h), colloidal characteristics, and very low cell density. The harvesting process that is widely considered for microalgal biomass involves a two-step separation: thickening followed by dewatering processes. A low-cost process could be used in the thickening step before energy-consuming, and more expensive equipment is used in the dewatering step [6, 12].

4.1 Thickening of Microalgal Culture

As the cell concentration in microalgal culture broth is usually low, the thickening step could be considered a mass harvesting of the microalgal cells. The aim is to isolate the microalgal biomass from the dilute cell suspension. Methods that are often employed for this purpose include flocculation, flotation, and gravity sedimentation. The microalgae slurry would concentrate to 2–7% of the total suspended solid after the thickening step [6].

4.1.1 Flocculation

Flocculation is the aggregation of particles. It is often carried out after coagulation. The two processes are different. Coagulation is when the colloidal suspension is destabilized. The flocculation step is then followed to aggregate the particles and form flocs. The larger size of flocs results in increased microalgal cell sedimentation (settling) velocity. In high-rate algal ponds (HRAPs), microalgal flocs occur naturally due to the mixed population nature of the systems. The majority of the flocs have a settling velocity of > 1 m/h, even without flocculant [41].

Flocculation depends strongly on cells' surface properties. Microalgal species, medium composition, and growth phase could affect the flocculation process [12, 85]. Flocculation in seawater and brackish water is less effective than in freshwater due to higher salinity and requires a higher flocculant dosage [9, 98]. In addition, the excretion of algogenic organic matter (AOM) by microalgae could lower the effectiveness of flocculants. AOM consists mainly of polysaccharides and proteins. They could compete with flocculants for microalgal cell surfaces. A study on *Chlorella* sp. showed that more flocculants were needed by 1.5 to five fold when AOM was present in the culture [100].

There are many ways to induce the flocculation of microalgae. The simplest and cheapest option is 'alkali flocculation'. Alkali flocculation could be spontaneous or forced. Spontaneous alkali flocculation or autoflocculation occurs naturally in autotrophic cultures with a CO_2 supply [12, 20]. As microalgae remove dissolved CO_2 in culture media, the pH value of the media increases, resulting in an alkaline condition. Chemicals such as NaOH, KOH, $Mg(OH)_2$, and $Ca(OH)_2$ are used in forced alkali flocculation [6]. This flocculation method results in biomass at a high pH containing a high concentration of minerals. A high pH environment could cause cell damage or lysis, which could be a concern [12].

'Chemical flocculation' also involves adding chemicals referred to as flocculants. Flocculants used in harvesting microalgae could be inorganic or organic compounds. Some inorganic flocculants are $FeCl_3$, $AlCl_3$, $Fe_2(SO_4)_3$, $Al_2(SO_4)_3$, and polyaluminium chloride [6]. Note that they are all multivalent metal salts. Organic flocculants that could be used for harvesting microalgae are chitosan and cationic starch [6, 9, 99]. Although metal salts are cheaper and coagulation/flocculation occurs faster, they must be used in high doses, which increases the operational cost. In addition, certain metal salts (aluminum salts) could cause cell lysis [77]. Residual metal ions in harvested biomass could make the biomass unsuitable for further processing. Organic flocculants present some advantages over metal salts as they have less environmental impact, require a lower dosage, and have less effect on cell viability [12].

'Bioflocculation' refers to flocculation caused by secreted biopolymers [20]. The biopolymers are also known as extracellular polymeric substances (EPS). EPS is normally produced by bacteria such as *Paenibacillus* sp., *Streptomyces* sp., and *Rhizobium radiobacter* [73, 95, 106]. Flocculation of microalgal cells could be achieved by applying bioflocculants directly to the microalgal suspension or co-cultivation of microalgae and EPS-producing bacteria. Flocculation efficiency of up to 99% was reported when using polymeric bioflocculant from *Streptomyces* sp. to recover *Nannochloropsis* biomass [95]. As bacterial EPS is composed of heterogenous sugar monomers, the harvested microalgal biomass would contain a small amount of these polymers.

Apart from the above flocculation methods, physical methods for microalgae flocculation also exist. Ultrasound can be applied to capture cells. Standing waves of high frequency and low amplitude are used. Cells in the ultrasonic field move towards the area with the minimum energy (nodes) and form clumps. The cells also experience minimal shear stress as the amplitude of the sound wave is almost zero at the nodes. Although ultrasonic can be used continuously with almost no shear stress to the cell, it is unsuitable for large-scale microalgae harvesting. Compared to centrifugation, ultrasonic requires higher power consumption with inferior efficiency and lower concentration factors [10].

Another physical method for flocculating microalgae is using electrolytic cells. Briefly, the negatively charged microalgae cells attract to the anode, lose their charge, and form aggregates. The hydrogen and oxygen that are formed on the electrodes rise and carry the microalgal aggregates with them. Hence, the electrolysis process provides flocculation and flotation of the microalgae at the same time [12]. This method could be referred to as electro-flotation.

Physical methods do not involve the use of flocculants, so there would be no issue with contaminants in the harvested biomass. However, they involve high-cost equipment and require high power consumption for operation compared to traditional flocculation.

4.1.2 Gravity Sedimentation

Gravity sedimentation usually follows the coagulation/flocculation of microalgal biomass. Without prior flocculation, the low settling velocity of microalgae makes the process time-consuming in harvesting. Slow sedimentation could also lead to biomass degradation [6]. Sedimentation is a low-cost and highly energy-efficient process, making it suitable for harvesting biomass for low-value products like bioethanol.

In addition to flocculation, selection (or training) of microalgae with good settling ability could be done to aid the sedimentation process. Successful examples were the training of *Chlamydomonas reinhardtii*, *Scenedesmus rubescens*, and *Chlorella vulgaris*. The training was done by culturing the microalgae with mixing for 23 h, then stopping the mixing for 1 h. The microalgal cells that floated were discarded, the medium was exchanged for a fresh batch, and the cultivation continued. The procedure was repeated for one month. The trained microalgae immensely improved their sedimentation compared to the original cells [96].

4.1.3 Flotation

Flotation could be viewed as inverted sedimentation. Instead of settling down, the microalgal flocs float on the action of small bubbles. The process is the foundation for separating colloidal particles from an aqueous solution. Flotation is more effective in recovering microalgal biomass than sedimentation [31]. In addition, flotation gives a higher concentration factor than sedimentation due to the gravitational drainage of water from the foam [57]. As water drainage removes additional water from the microalgal cells, flotation could also be doubled as biomass dewatering [70].

For a successful flotation process, microbubbles are generated, and the particles (hence microalgal cells) must be hydrophobic and attached to the bubbles [43]. Since the microalgal cell surface is negatively charged, using coagulants is unavoidable. Coagulants destabilize the microalgal particles relative to bubbles, ensuring the attachment of microalgal particles to the bubbles [31].

Two methods that are most commonly used to generate microbubbles in flotation are dissolved air flotation and dispersed air flotation.

Dissolved air flotation

This flotation method involves pressurizing a portion of liquid (water) after the separation. The pressurized water is recycled back into the flotation tank. Depressurizing of water results in the nucleation of microbubbles with a size ranging from 10 to 100 μm [31]. Smaller bubbles are more efficient in capturing microalgal cells. Since the bubbles do not form until the pressurized water reaches the atmospheric pressure (in the tank), bubble nucleation occurs at the surface of microalgal cells, increasing the chances for particle-bubble attachment [57].

This process is energy-intensive as it requires high pressure (4.8–5.8 atm) to ensure supersaturation of air in water. In addition, the continuous pumping of clarified water to the saturator also contributes to energy usage [31, 43].

Dispersed air flotation

In this flotation method, bubbles are formed by supplying a continuous air stream to a diffuser. This method is more economical than dissolved air flotation due to simpler equipment and a lower energy requirement. Bubbles generated by these means are larger than those obtained in dissolved air flotation by order of magnitude [57]. To decrease bubble size, a finer pore size of the diffuser could be employed. However, the production of a very fine pore diffuser could be expensive and requires more expertise. Finer diffuser pore size also means that a high-pressure drop is needed to deliver the air stream [43].

Another problem in generating microbubbles through a diffuser is the detachment of bubbles from the diffuser. With a constant supply of air stream, the bubbles formed at the diffuser's pores remain attached and naturally grow until the buoyant force and the force exerted from the airstream overcomes the attachment force. The bubbles are then detached. This phenomenon hinders the generation of microbubbles [43]. Adding a surfactant may alleviate the problem by reducing the surface tension, leading to easier detachment of bubbles. However, it may not be suitable if the biomass is further used as it introduces another chemical contaminant to the harvested biomass [57].

Another approach that is effective in producing microbubbles is microflotation [43]. This method introduces a fluidic oscillator to the continuous air stream before entering the diffuser. The device transforms the air stream into an oscillatory flow. The pulse created by the flow provides extra force to the bubbles, making them easier to detach. By using the oscillator, bubble sizes generated from a 38 μm pore size diffuser decrease from an average of 1059 to 86 μm.

Selecting the thickening options is an important step. As the microalgal cells are the product of the separation, contamination of flocculants could be an issue. Flocculants could contribute to the composition and properties of final biomass products. In addition, the low-cost and energy-efficient process should be considered as biomass would be used in energy production of low value.

4.2 Dewatering of the Microalgal Cells

Dewatering is the harvesting step when most water is removed from the biomass slurry. A very concentrated biomass slurry or biomass cake is the result of dewatering. Over 90% of the total suspended solid could be achieved after dewatering. Due to the low concentration of microalgae culture, the dewatering process could be used after the thickening of culture to lower the operational load of the dewatering equipment. Filtration and centrifugation are the two most common methods for dewatering.

4.2.1 Filtration

Filtration is widely used to separate solids (insoluble) and liquids. The separation of the two phases is through filter media. Membranes (smaller pore size) could be applied in the case of microalgal separation. Filtration can also be seen as a clarification of the liquid. During the filtration process, microalgal cells are retained on the surface of a membrane, and water passes through the membrane.

Filtration modes used in harvesting microalgae cells could be dead-end or crossflow (tangential) filtration. A general problem with filtration is pore clogging or membrane fouling. This problem could be alleviated by increasing the size of the insoluble part by cell flocculation. Filtration of flocculated microalgal cells, especially in dead-end filtration, results in higher performance and harvesting efficiency [116]. As microalgal cells are small and usually secrete AOM, irreversible fouling of the membrane could result from a dense and sticky cake layer deposit on the membrane surface. Flocculation does not only create cell flocs that form a dynamic layer with voids on the membrane surface, but it also removes AOM from the cell broth. Size and floc structures affect filtration. Loose and sponge-like flocs are desirable over tight and big agglomerates as the cake layer would form more voids and channels.

Introducing flocculation before filtration reduces the energy input of the process and requires a smaller membrane area. Hence, harvesting costs are lower as compared to using filtration alone [116, 117]. Flocculation is not the only means to reduce membrane fouling. Using a wavy-patterned membrane [117], membrane vibration [116], and a tilted membrane panel with air bubbles to scrape out the foulant on the surface [33] are some other methods being tested to alleviate the problem in dead-end filtration.

Crossflow or tangential flow filtration experiences fewer fouling problems and could provide long-term continuous recovery of biomass. The perpendicular flow to the membrane surface creates some turbulence and reduces the fouling layer on the membrane surface. It should be operated at optimal transmembrane pressure (TMP) to satisfy the filtration rate and prevent membrane compaction. Although the fouling problem in crossflow filtration is not as severe as in dead-end filtration, reducing fouling is also desirable. Ultra-low-pressure membrane filtration was attempted to reduce the energy consumed by pumping [104]. Membrane fouling is riskier when using low TMP. Aeration was supplied into the filtration module just above the membrane surface to scour off the foulants.

Although there are many approaches to improving filtration efficiency and reducing operation costs, most of them are in the experimental stage. Therefore, existing industrial-scale filtration systems are the immediate choice for the large-scale process. Since a large volume of microalgal culture would be processed, continuous recovery of the cell cake would be appropriate. Dynamic crossflow filtration systems are commercially available for crossflow filtration, where continuous operation is possible as long as there is no fouling. Continuous removal of cell products in dead-end filtration is also possible. Horizontal vacuum belt and rotary drum vacuum filters are possible to apply to harvest the microalgal biomass.

4.2.2 Centrifugation

Centrifugation is the fastest cell harvesting method. It is also an energy-intensive process as the operation requires the generation of centrifugal force that is much larger than the gravitational force for the separation to occur. Centrifugation could be used as a single-step process. However, coagulation/flocculation of microalgal cells before centrifugation could help in lowering energy consumption as the volume to be processed is much reduced [6].

Large-scale centrifuges mostly come in two configurations: tubular centrifuges and disc (stack) centrifuges. A tubular centrifuge operates in batch mode. Cell slurry is introduced from the bottom of the bowl. Liquid continuously leaves the top of the centrifuge while the solid settles at the bottom of the bowl to be collected at the end of the operation.

In a disc stack centrifuge, a similar principle to that of the tubular applies where the feed is introduced from the bottom of the unit. Some designs incorporate pulse discharge of the solid or concentrated slurry of the cell.

A decanter centrifuge can be another candidate for dewatering of the microalgal biomass, especially when flocculation is done before dewatering. The decanted biomass is continuously discharged, allowing for continuous operation of the process at a large volume.

5 Pretreatment of Microalgae

Pretreatment of raw materials usually refers to the method of preparing the materials, making them easier to hydrolyze. In many cases, methods used in pretreatment also cause hydrolysis of the materials. In ethanol production from microalgae, many pretreatment methods also hydrolyze the substrate (carbohydrates) to release sugars used in fermentation. In this sense, this section would extend the meaning of the term pretreatment to cover any methods or procedures that are done before ethanol fermentation.

5.1 Characteristics of Microalgal Cell Walls

The purpose of microalgae pretreatment for ethanol production is to disrupt the microalgae cell walls and hydrolyze the stored carbohydrates to obtain fermentable sugars. As microalgae are diverse, their cell wall structures and compositions vary. Examples of cell wall composition of some common microalgae include [24, 52, 65, 66]:

- The cell walls of *Scenedesmus* consist of a pectic outer layer, an algaenan-based middle layer, and an inner cellulose layer.

- *Nannochloropsis* has a porous inner cell wall layer with a delicate fibrous composed of cellulose and glucose. Struts connect this layer to the cell membrane. The outer wall layer is a trilaminar sheath of algaenan.
- *Chlorella* has a two-layer cell wall. The inner layer is a rigid microfibrillar structure with high cellulose content and contains chitin-like glycan. These components are embedded in within a polymeric matrix consisting of uronic acids, hemicellulosic sugars, and pectin. The outer layer is the algaenan layer.
- The cell wall of *Haematococcus* also consists of an outer algaenan layer and a cellulose-based inner layer.
- *Chlamydomonas* has a regular lamellate cell wall that is composed of glycoprotein rich in hydroxyproline, which links to heterooligosaccharides composed of arabinose and galactose via glycosidic bonds. The cell wall does not have cellulose in the structure.

The cell walls of common microalgal species contain algaenans (also known as sporopollenin). They are highly resistant aliphatic polymers, indicating wall resistance [24, 30]. The presence of algaenans could obstruct the disruption of microalgae cell walls. In addition, the diverse structure of microalgal cell walls results in a different response to pretreatment in different microalgal strains. *Haematococcus* is the least susceptible to cell disruption compared to *Nannochloropsis*, while *Chlorella* is the easiest of the three to be disrupted [83, 84].

The most commonly used methods of microalgal pretreatment for ethanol production have been acid hydrolysis and enzymatic hydrolysis. A combination of the two methods is often practiced.

5.2 Acid-Thermal Pretreatment

Sulfuric acid (H_2SO_4) is the most widely used acid for microalgal pretreatment. It is used as a dilute acid with a concentration of up to 4.0% v/v [18, 35]. Most acid hydrolysis pretreatment is performed at an elevated temperature of around 100–130 °C. Both dry heat and wet heat are used [25, 78]. Pretreatment by dilute acid also hydrolyzes carbohydrates and releases sugars to be used in ethanol fermentation.

In addition to sulfuric acid, nitric acid (HNO_3), hydrochloric acid (HCl), phosphorous acid (H_3PO_3), and phosphoric acid (H_3PO_4) are reported to be used in dilute acid hydrolysis. Hydrolysis by diluted HNO_3 and HCl obtained similar results to sulfuric acid [88, 92]. With H_3PO_3 and H_3PO_4, the hydrolysis yield was much lower [74, 92]. Dilute acid hydrolysis at a relatively low temperature (90 °C) did not produce any inhibitors such as hydroxymethylfurfural (HMF) or organic acids [88].

Another variation of acid-thermal pretreatment of microalgal biomass is torrefaction. A higher temperature range is used in the process. In microwave-assisted torrefaction of wet *Chlorella* biomass using 0.2 M H_2SO_4 at 170 °C, solid biochar was formed as a co-product of the process [114]. Hydrolysate obtained from torrefaction contained high sugar concentrations. However, high temperatures in acidic conditions resulted in high hydroxymethyl furfural (HMF) concentration, which inhibited the subsequent ethanol fermentation by *Saccharomyces cerevisiae*.

Acid-thermal pretreatment delivers different results to different microalgal strains. Under the same conditions, *Picochlorum maculatum*, *Tetraselmis* sp., and *Cylindrotheca fusiformis* were more susceptible to acid hydrolysis than *Chlorella sorokiniana*, *Chlamydomonas reinhardtii*, *Nannochloropsis* sp., *Skeletonema* sp., *Nanofrustulum* sp., *Phaeodactylum tricomutum* (Necton), *Phaeodactylum tricomutum*, *Porphyridium cruentum*, and *Isochrysis galbana*. Even within the same genera, *Tetraselmis* sp. was more susceptible to acid hydrolysis than *Tetraselmis* sp. (Necton) and *Tetraselmis striata* [22].

Pretreatment by acid hydrolysis produces an acidic hydrolysate that must be neutralised before being used in ethanol fermentation. Different salts, depending on the types of acids and alkalis used, are generated due to neutralisation. Sodium hydroxide is the most common neutralising agent used. Although *Saccharomyces cerevisiae* has a moderate tolerance to salt [93], too high a salt concentration could adversely affect its fermentation performance. Using $Ca(OH)_2$ or NaOH in conjunction with electrodialysis successfully in removed Na^+, SO_4^{2-}, and NO_3^- from the acid hydrolysate of *Chlorella* sp. without removing sugars [88].

5.3 Alkali Pretreatment

Alkali pretreatment is usually followed by acid or enzymatic hydrolysis. Pretreatment by itself is uncommon. Normally, alkali pretreatment is employed to solubilize lignin from lignocellulosic materials. However, a small degree of holocellulose solubilization could occur simultaneously with delignification at a high NaOH concentration [54]. Solitary use of alkali pretreatment in microalgae has been reported in *Nannochloropsis oculate*, *Tetraselmis suecica*, *Hindakia tetrachotoma*, and *Chlorella vulgaris* [74, 81, 107].

5.4 Enzymatic Pretreatment/Hydrolysis

Enzymatic pretreatment/hydrolysis is highly energy efficient since it requires a relatively low temperature and does not involve equipment with high power consumption. However, its drawback is the high enzyme cost. Enzymatic hydrolysis of microalgal biomass could be a single-step process or follow other pretreatment methods.

The most commonly used enzymes to hydrolyze microalgal biomass for ethanol production are cellulase, amylase, and glucoamylase (amyloglucosidase). They could be used singly or in combinations. Amylases appear to be fundamental for enzymatic pretreatment for microalgae-based ethanol production as they hydrolyze the stored starch, while cellulases are for attacking the cell wall components of the microalgae.

Apart from the three major enzymes, other enzymes that could be associated with the hydrolysis of microalgal cell walls are also used, such as pectinase and xylanase [25, 35].

In several cases, a single enzyme was used to hydrolyze microalgal cells following a pretreatment step. After heat treatment at 90 °C, glucoamylase was used to hydrolyze the biomass of *Scenedesmus dimorphus* [18]. In *Chlorella* sp., hydrolysis of acid-treated biomass by glucoamylase alone resulted in the same amount of glucose as using a combination of α-amylase and glucoamylase [71].

While it seems logical to use glucoamylase alone in hydrolyzing pretreated microalgal biomass as the target substrate is starch, using just cellulase is peculiar. However, ultrasonicated cells of *Chlamydomonas Mexicana* released more carbohydrates when hydrolyzed with cellulase [32]. In addition, better saccharification results were obtained using mixed cellulase and β-glucosidase (cellobiase) than amylase, glucoamylase, and xylanase in mixed microalgal culture mainly of *Scenedesmus* sp. and *Chlorococcum* sp. [55].

When using multiple enzymes in microalgal cell hydrolysis, interest is in how to apply the enzymes in the hydrolysis process, e.g., sequential vs. simultaneous; in what order; and in what combination. As there is no definite practice due to the great diversity of microalgal strains that could differ in strain level [67], testing individual microalgal strains may be unavoidable. Examples (by case) are given below.

In the enzymatic hydrolysis of H_2SO_4-pretreated *Chlorella sorokiniana* (C1), only starch-degrading enzymes, α-amylase and glucoamylase, were used. Adding the enzymes in conventional order, i.e., α-amylase followed by glucoamylase, and adding them simultaneously resulted in the same sugar yield. Better sugar release was when the enzymes were added in reverse order, i.e., glucoamylase followed by α-amylase. After acid pretreatment, glucoamylase could provide additional disintegration of cell wall components due to the hydrolysis of glycosidic bonds in polysaccharide components of the cell wall. When α-amylase is added later, it may have a better action on intracellular starch [22].

Cellulase (including β-glucosidase) is often used along with α-amylase and glucoamylase, especially when the cell wall contains cellulose. Pretreatment to weaken the cell walls may not be applied when using cellulase to hydrolyze microalgae cells. In hydrolyzing mixed microalgae culture from a freshwater source, cellulase (CL) was introduced with α-amylase (AA) and glucoamylase (GA). Sequential addition of the enzymes (AA → GA and CL → AA → GA) suggested using all three enzymes as it resulted in a higher sugar yield [92]. A further test was on the sequence of adding the enzymes. There was no difference in sugar yields obtained in the final hydrolysis time employing different enzyme addition regimes. However, adding all three enzymes simultaneously gave a higher reaction rate [91]. Incorporating cellulase with α-amylase and glucoamylase also resulted in an additive effect in untreated *Hindakia tetrachotoma* and *Scenedesmus raciborskii* [1, 74].

The above examples imply that cell walls are obstructive to the enzymatic hydrolysis of microalgae. Pretreatments and cellulase that directly affect the integrity of cell walls have proven their importance in the enzymatic hydrolysis of many microalgae, as exampled earlier. However, in *Chlamydomonas reinhardtii*, enzyme saccharification efficiency was not obstructed by the presence of cell walls, as hydrolysis of parental and cell-wall deficient strains was similar [5].

Regarding the use of enzymes for pretreatment, enzyme mixtures used in lipid extraction that intend to target microalgal cell walls may be considered. A mixture of cellulase, protease, lysozyme, and pectinase was used to degrade the cell wall of *Nannochloropsis* sp. They showed a synergistic effect with alkaline pretreatment [108]. Another enzyme mixture used to disrupt *Nannochloropsis* sp. was cellulase and mannanase. They resulted in extensive damage to the cell wall, hence releasing intracellular components [60]. Enzymatic pretreatment of *Scenedesmus* sp. employed the mixture of pectinase, cellulase, and xylanase. They were added in a sequence where pectinase disrupted trilaminar sheaths while cellulase and xylanase targeted the inner layer composed of the two enzymes [115].

5.5 Sonication

Sonication is the method that applies sound energy (wave) to agitate particles in a medium. Application of sonication to pretreat microalgal biomass employs a frequency of 40 kHz, which is in the ultrasonic range. Extreme pressure gradients due to high-frequency waves lead to the formation of microbubbles that grow and implode, generating a large shear force that causes disruption of fibers and large particles in microalgal slurry [19]. Higher sonication intensity (amplitude) and process time promote a higher degree of cell disruption, allowing for better release of intracellular contents and better access to the hydrolyzing enzymes [19, 26].

Direct and indirect sonication could be used in pretreating microalgal biomass. An ultrasonic probe sonicator is used in direct sonication, where the probe is directly placed in a microalgal slurry. An ultrasonic bath is used in indirect sonication. Both operation modes resulted in a similar release of carbohydrates when applied to disintegrate *Scenedesmus obliquus* biomass [19]. In addition, the method has been applied to other microalgae, including *Pseudochlorella* sp., *Chlamydomonas mexicana*, and *Chl. pitschmannii* [32, 42].

For microalgal biomass pretreatment purposes, the sonication process does not involve using chemicals; therefore, contaminant problems are eliminated. However, in a large-scale process, equipment cost and energy requirements for running the process have to be considered.

5.6 Ozone Pretreatment

Ozone pretreatment for microalgal cells applies on the same basis as in water treatment or sterilization. In microalgae pretreatment, ozone is mixed with air and supplied through cell suspension. Ozone application causes physical cell disruption and hence the release of intracellular contents but does not hydrolyze carbohydrates. This was evident in mixed microalgae culture with *Scenedesmus* sp. and *Chlorococcum* sp. as the main populations. Without enzymatic hydrolysis, ozone pretreatment hardly yielded any glucose [55]. Ozone pretreatment also finds its use in disrupting microalgal cells to harvest lipids. It was claimed also to stimulate the formation of flocs in *Chlorella vulgaris* [53].

5.7 Solvent Pretreatment

Pretreatment by organic solvents is not the major or lone method used in pretreating microalgal biomass. Its use is normally intended for extracting lipid or lipid-soluble content such as carotenoids. As solvents do not hydrolyze the target substrates (starch), biomass after pretreatment is subjected to further processes such as hydrolysis or use directly in fermentation.

As the solvents dissolve lipids, the typical solvents used in lipid extraction are also used in the solvent pretreatment of microalgal biomass. The solvent mostly used is chloroform/methanol, following the Blign and Dryer method [4, 18, 45]. Other solvents are hexane/isopropanol, hexane/diethyl ether, and acetone [4, 35, 50]. The pretreated/defatted microalgae biomass is then used directly as a substrate in ethanol fermentation or further hydrolyzed by acids or enzymes.

Solvent pretreatment could be combined with other pretreatment methods to enhance performance. Acetone was used with sonication to extract carotenoids from *Chlamydomonas* sp. The process causes severe disruption of the cell wall. The cell debris could be used in ethanol fermentation by amylase-displayed *S. cerevisiae* as the carbohydrates are still intact as starch granules [50]. The pretreatment also allowed for better access to amyloglucosidase, such that simultaneous saccharification and fermentation of the lipid-extracted *Scenedesmus* sp. biomass resulted in improved ethanol production compared to untreated cells [18].

5.8 Other Potential Methods of Pretreatment

Thermal pretreatment includes methods that employ high temperatures in pretreating microalgal cells. They are normally used with acid or base pretreatment to enhance the reaction rate of the chemicals. However, it could be the pretreatment of the cells by itself. This method (at 110–150 °C wet heat) was used in *Chlorella* sp., which resulted in the release of carbohydrates that could be further hydrolyzed [71]. A similar method was employed in *Chlorella vulgaris* at 150 °C in an autoclave and an 800-W microwave. The technique increased lipid and glucose release [46]. However, *Desmodesmus* sp. did not respond well to thermal pretreatment by autoclaving at 121 °C [109].

Osmotic shock is another physical method used in microalgae pretreatment. High salt concentration has been used to create an osmotic stress environment for the cells. Once equilibrated, the cells are quickly exposed to a low osmotic pressure environment. The rapid change in osmotic pressure causes water to quickly enter cells. A sudden increase in internal cell pressure results in cell lysis. Although this method is more suitable in cells with weakened or without cell walls, it showed a positive result in *Chlorella vulgaris* [46].

A bead mill and a high-pressure homogenizer are other potential methods for the pretreatment of microalgae for ethanol production. They have successfully been used in harvesting carotenoids from *Desmodesmus* sp. and proteins from *Arthrospira platensis*, *Chlorella vulgaris*, *Haematococcus pluvialis*, *Porphyridium cruentum*, and *Nannochloropsis oculate* [84, 109]. In those applications, the bead mill and high-pressure homogenizer showed superior product recovery (implying better cell disruption) than thermal pretreatment, sonication, and alkaline pretreatment. A major concern with using these methods is their high equipment cost. However, commercial-scale units are readily available.

6 Fermentation Technology in Bioethanol Production from Microalgae

Ethanol fermentation from microalgal-derived substrates is similar to conventional ethanol production from other first- and second-generation substrates. As the substrates from microalgae are mainly starch with some cellulose from the cell wall component, glucose would be the dominant sugar for fermentation. Analysis of the sugar content of *Chlorella vulgaris* showed that glucose was the main sugar at about 71% of total carbohydrates. Xylose, arabinose, and rhamnose were also present at about 27% (combined) of total carbohydrates [25]. A slightly different hemicellulosic sugar profile was reported in *Chlorella* sp. ABC-001 where rhamnose was higher than xylose, with galactose and mannose instead of arabinose [88]. Carbohydrates in *Scenedesmus obliquus* are ~ 80% glucose and ~ 20% xylose with no other sugars present [25].

As glucose is the main sugar derived from microalgae, it is reasonable to use *Saccharomyces cerevisiae* in ethanol fermentation, as do most of the studies on ethanol fermentation using microalgae as the raw material. The bacteria, *Zymomonas mobilis*, is also another potential ethanol producer used in ethanol fermentation from microalgae [47, 48]. In addition, recombinant *S. cerevisiae* capable of producing amylase and cellulase was reported in ethanol fermentation from a slurry of acetone-treated *Chlamydomonas* sp. [50]. Using the recombinant *S. cerevisiae*, the enzymatic hydrolysis step is eliminated, along with the need for commercial cellulase and amylase.

In addition, *Pichia stipitis* could also be used as a co-fermenting yeast with *S. cerevisiae* to utilize the xylose in the microalgal hydrolysate. Although *P. stipitis* naturally ferments xylose to ethanol, the productivity is much lower than *S. cerevisiae*. For the co-fermentation of the two yeasts to succeed, the ratio between the two yeasts was important to maintaining high ethanol productivity as well as achieving high sugar consumption in the hydrolysate. In a particular case, using 75% of *S. cerevisiae*, *P. stipitis* positively contributed to xylose fermentation, where ethanol productivity increased compared to using *S. cerevisiae* alone [25].

Regarding nutrients in fermentation, several studies reported that no additional nutrients, especially nitrogen sources, were supplied to microalgal hydrolysate for ethanol production [25, 71, 88, 92]. Another work that used enzymatic hydrolysate of microalgae in ethanol fermentation reported that there was no difference in yeast performance with or without added nutrients [1]. Microalgae usually contain a high percentage of proteins that could be released during pretreatment and hydrolysis [83]. Therefore, the hydrolysate may contain enough nitrogen and other intracellular compounds that could be nutrients for yeast growth.

The fermentation modes employed in ethanol production from microalgae are the same as those used for other raw materials. Separate hydrolysis and fermentation (SHF) and simultaneous saccharification and fermentation (SSF) have been used with slight variations in each work. SHF is the fermentation of hydrolysate obtained from the hydrolysis of microalgal biomass. SSF uses pretreated biomass as a direct substrate in fermentation. In SSF, enzymes are added together with the pretreated biomass and yeast inoculum. Sugars are gradually released from enzymatic hydrolysis and are immediately used up by the yeast. Through this process, a low level of sugar is maintained throughout the fermentation. Low sugar concentration alleviates the problem of substrate inhibition that could interfere with the yeast fermenting ability to some degree.

SSF normally results in better ethanol production than SHF in ethanol fermentation from pretreated microalgal biomass [18, 32, 48]. It also requires fewer operating units as hydrolysis and fermentation occur in the same vessel [18]. However, optimal conditions for enzymatic hydrolysis and fermentation are different. Usually, enzymatic hydrolysis requires a higher temperature than yeast fermentation. Operating SSF at a temperature suitable for yeast growth would compromise the enzyme activity.

In an SHF operation, a single vessel is possible to carry out the fermentation. In this case, ethanol fermentation is carried out following the pretreatment/hydrolysis without cell debris separation [1, 32, 48]. However, a slightly lower ethanol concentration would result as cell debris could adsorb some ethanol in the fermentation broth [71].

High ethanol concentration directly results from the amount of sugar available during fermentation, which is dependent on the amount of microalgae biomass used. The high viscosity of microalgal biomass slurry limits the percentage of biomass loading. Multiple additions of microalgal biomass with enzymes could increase biomass loading and obtain a high concentration of sugars. A high ethanol concentration of 79.4 g/L was achieved by two additions of 33% w/v of *Scenedesmus raciborskii* biomass and a cellulase/amylase enzyme mixture [1].

For continual long-term ethanol production, repeated batch SSF has been tested using sonicated *Chlamydomonas mexicana*. The fermentation used *S. cerevisiae* immobilized in calcium alginate beads and could maintain high ethanol production for four cycles before the production (ethanol concentration) gradually decreased to around 50% after the 7th cycle. The rigidity of the alginate beads is a concern as cell loss due to the weakening structure of the beads and loss of immobilized cell activity after repeated use. A method of bead regeneration (reactivation) by suspending the beads in a yeast medium after each cycle could help extend its activity [32]. The method also maintains the bead integrity for a longer period compared to the non-regenerated beads (suspended in water).

Using immobilized cells in ethanol fermentation has many advantages, including increased ethanol tolerance, alleviation of substrate and product (ethanol) inhibition, and easy cell separation from fermentation broth. However, these advantages should be weighed against other procedures involved in using immobilized cells, such as the preparation of the immobilized cells and the need for regeneration to maintain the cell's activities, as these procedures also require time and resources. In addition, the number of cycles that the immobilized cells could maintain their activities should also be considered against the ease and time required for immobilized cell preparation.

7 Potentials and Challenges in Ethanol Production from Microalgae

Although ethanol production from microalgae has not been practiced on a large and commercial scale, it does show potential, as mentioned earlier in Sect. 2. The obvious potentials of using microalgae biomass as raw material can be seen in terms of environmental and social impact. Using microalgae biomass, utilizing CO_2 or wastewater as substrates in microalgae cultivation is highlighted as they involve possible CO_2 mitigation and wastewater treatment. Non-competitive land use for food and feed crops could lessen the food versus fuel dilemma, reducing the risks to food security.

To realize the full potential of using microalgae as raw material in ethanol production, there are many challenges in economics and operations to overcome to make microalgal-based bioethanol possible, even on a sizable scale of operation. Challenges in the overall process could be seen separately for microalgal cultivation and ethanol production.

In the cultivation of microalgae, the most obvious challenge is in producing and harvesting microalgae effectively and economically on a large scale. The bioethanol production scale is large and requires a huge supply of raw materials. Therefore, mass production of microalgal biomass is needed to replace or supplement the existing bioethanol production facilities. According to Vogelbusch Biocommodities GmbH, a supplier of industrial-scale bioethanol technology, the economic minimum capacity for an ethanol plant ranges from 100,000 to 300,000 L/day. Their process requires approximately 2500 kg of starchy raw materials (corn and wheat) for every ~ 1000 L of ethanol produced [103]. In total, at least 250 metric tons of starchy raw materials must be used daily if the plant is to run at full capacity. Microalgae with high starch content are desirable as they could drastically reduce the biomass required in large-scale ethanol production.

To economically produce microalgae biomass in large quantities, substrate and nutrient costs for microalgae cultivation could become an important factor. The use of waste streams such as CO_2 from flue gas and wastewater from various sources is a viable option for the large-scale production of microalgal biomass.

Another challenge with this option is that suitable strains should be used for a specific cultivating condition [89]. Using flue gas and wastewater presents wide variances in media composition as their compositions are diverse. For example, flue gas streams do not contain only CO_2; they also carry CO, O_2, NO_2, SO_x, and HCl. Their CO_2 content varies from about 9% to 14%, depending on the fuel used in boilers [76, 111]. Wastewaters from different sources have different characteristics regarding their nitrogen and phosphorus content. Most wastewaters have a higher nitrogen content than phosphorus. However, the ratio and concentration vary largely depending on the sources. Wastewaters from agro-industries and distilleries have high nitrogen concentrations and content compared to domestic wastewater [20, 51]. As the composition of nutrients using waste streams could vary widely, microalgae that are robust and specific to the cultivation environment would be required for the effective production of their biomass.

In the technical aspect, the larger-scale operation of photosynthetic microalgae would face the problem of maintaining photosynthetic efficiency [90]. This problem could be more intense in photobioreactor systems. As the growth of autotrophic microalgae and product accumulation depend largely on light availability, not enough lighting becomes a problem, especially in photobioreactors with high cell density due to the shading effect of cells. Efficient mixing of the culture in the reactor could reduce this shading problem.

Low biomass productivity and potentially high equipment costs associated with cultivation present the most challenging points in microalgal cultivation, especially when microalgal biomass would be specifically used to produce low-value biofuel like ethanol. However, an economy of scale applies to microalgae cultivation. In the production of *Nannochloropsis oceanica*, the production cost was estimated to be reduced by 18% when scaling up the production from 1 to 10 ha [101]. Another techno-economic study on microalgae production facilities in an aquaculture industry also showed that the scale of the production facility largely impacted the production cost, such that increasing the production scale from 25 to 1500 m^2 reduced the cost price by 92% [75].

For ethanol production from microalgae, challenges lie mostly in pretreatment and hydrolysis, as they are important steps to obtain fermentation substrates. Unlike second-generation raw materials, microalgae cells are relatively easier to process and obtain fermentable sugars as they lack recalcitrant lignin. Therefore, more choices of pretreatment are available to choose from, many of which avoid the use of added chemicals. Hence, the hydrolysate is less contaminated with inorganic salts, especially from acid or alkali used in chemical pretreatment.

Enzymatic hydrolysis is likely the most efficient way to hydrolyze microalgal biomass, although the cost of enzymes is a challenge. Recovery of enzymes and reusing them is one option to reduce the cost associated with enzyme hydrolysis. However, this option would involve an extra unit operation for enzyme recovery, and repeatable use of the enzyme must be assessed. Optimizing the enzyme dosage used in hydrolysis is another approach. The optimization should aim for a balance in the amount of enzyme used per substrate and the sugar obtained from the hydrolysis.

Another option is in-house enzyme production. Although it requires capital investment due to unit operation and extra operational costs, it could reduce the cost of purchasing enzymes, especially the more expensive cellulase. *Trichoderma* spp. has been used to produce mixed cellulase and the crude enzymes were used in microalgal pretreatment [5, 46].

A high ethanol concentration is desirable in fermentation as there is an economic advantage to a high ethanol concentration in the feed into the distillation system. Increasing ethanol concentration from < 1 mol.% (~ 2.6 wt% or 32 g/L) to 2 mol.% (~ 5.1 wt% or 51 g/L) could reduce the capital and operational costs by at least 50%. Further increases result in cost savings until the concentration reaches 4 mol.% (~ 10.2 wt% or 130 g/L), after which further increases do not result in as much cost savings [21].

A high sugar concentration is required to achieve a high ethanol concentration available during fermentation. This could be achieved by having a high loading of microalgal biomass during hydrolysis to produce a hydrolysate with a high sugar concentration or during SSF to allow for the release of a large amount of sugar. However, very high solid loading for hydrolysis or SSF could create problems in operation, such as mixing problems due to the high viscosity of the microalgal slurry. Hydrolysis of microalgal biomass by multiple additions of the biomass has been shown to increase sugar concentration and, hence, ethanol concentration [1].

Apart from challenges that directly relate to microalgae cultivation and processes for ethanol production from microalgae, other challenges are worth considering.

As microalgal hydrolysate may contain xylose and xylose-fermenting yeast yields poor ethanol, it is possible to use the vinasse after ethanol distillation for other purposes. Some possibilities include harvesting and purifying xylose; or using the vinasse as a substrate for producing other products from fermentation, e.g., xylitol [49], or use in the cultivation of N_2-fixing cyanobacteria, which could be used as a nitrogen source for microalgae or other microbial fermentation [86].

Microalgal cell residues remaining after hydrolysis or SSF still contain high organic matter, especially fats and proteins, depending on microalgal strains [71, 86]. They could be used as animal feed, providing that contaminants that occurred from pretreatment steps are within the regulations for a specific purpose.

In addition, as many carbohydrate-rich microalgae also accumulate a high fraction of lipids, it is possible to use the biomass to co-produce biodiesel from its lipid fraction. This application of dual biofuel products from microalgal biomass has been proven to be possible by many research studies [4, 46, 94, 105].

8 Conclusion

Using microalgae as a raw material for bioethanol production presents many advantages over sugary, starchy, and lignocellulosic raw materials. However, there are many challenges to overcome before the process becomes realizable. Major obstacles lie in the mass production of the microalgae and its preparation for use in ethanol production. Although many research works have laid the groundwork for the process development, there are still many potential areas to investigate because microalgae are diverse, and suitable strains would be required for a specific environment for biomass production.

References

1. Alam MA, Yuan T, Wenlong X, Zhang B, Lv Y, Xu J (2019) Process optimization for the production of high-concentration ethanol with *Scenedesmus raciborskii* biomass. Bioresour Technol 294:122219. https://doi.org/10.1016/j.biortech.2019.122219
2. Allen MM (1984) Cyanobacterial cell inclusions. Annu Rev Microbiol 38:1–25
3. Amenorfenyo DK, Huang X, Zhang Y, Zeng Q, Zhang N, Ren J, Huang Q, (2019) Microalgae brewery wastewater treatment: potentials, benefits and the challenges. Int J Environ Res Public Health 16(11):1910. https://doi.org/10.3390/ijerph16111910
4. Ashokkumar V, Salam Z, Tiwari ON, Chinnasamy S, Mohammed S, Ani FN (2015) An integrated approach for biodiesel and bioethanol production from *Scenedesmus bijugatus* cultivated in a vertical tubular photobioreactor. Energy Convers Manag 101:778–786. https://doi.org/10.1016/j.enconman.2015.06.006
5. Bader AN, Sanchez Rizza L, Consolo VF, Curatti L (2020) Efficient saccharification of microalgal biomass by *Trichoderma harzianum* enzymes for the production of ethanol. Algal Res 48:101926. https://doi.org/10.1016/j.algal.2020.101926
6. Barros AI, Gonçalves AL, Simões M, Pires JCM (2015) Harvesting techniques applied to microalgae: a review. Renew Sustain Energy Rev 41:1489–1500. https://doi.org/10.1016/j.rser.2014.09.037
7. Behera S, Singh R, Arora R, Sharma NK, Shukla M, Kumar S (2015) Scope of algae as third generation biofuels. Front Bioeng Biotechnol 2:90. https://doi.org/10.3389/fbioe.2014.00090
8. Bibi R, Ahmad Z, Imran M, Hussain S, Ditta A, Mahmood S, Khalid A (2017) Algal bioethanol production technology: a trend towards sustainable development. Renew Sustain Energy Rev 71:976–985. https://doi.org/10.1016/j.rser.2016.12.126
9. Blockx J, Verfaillie A, Thielemans W, Muylaert K (2018) Unravelling the mechanism of chitosan-driven flocculation of microalgae in seawater as a function of pH. ACS Sustain Chem Eng 6:11273–11279. https://doi.org/10.1021/acssuschemeng.7b04802
10. Bosma R, van Spronsen WA, Tramper J, Wijffels RH (2003) Ultrasound, a new separation technique to harvest microalgae. J Appl Phycol 15:143–153. https://doi.org/10.1023/A:1023807011027
11. Brányiková I, Maršálková B, Doucha J, Brányik T, Bišová K, Zachleder V, Vítová M (2011) Microalgae-novel highly efficient starch producers. Biotechnol Bioeng 108:766–776. https://doi.org/10.1002/bit.23016
12. Branyikova I, Prochazkova G, Potocar T, Jezkova Z, Branyik T (2018) Harvesting of microalgae by flocculation. Fermentation 4:1–12. https://doi.org/10.3390/fermentation4040093

13. Carnovale G, Rosa F, Shapaval V, Dzurendova S, Kohler A, Wicklund T, Horn SJ, Barbosa MJ, Skjånes K (2021) Starch rich *Chlorella vulgaris*: high-throughput screening and up-scale for tailored biomass production. Appl Sci 11(19):9025. https://doi.org/10.3390/app11199025
14. Chen CY, Chang JS, Chang HY, Chen TY, Wu JH, Lee WL (2013) Enhancing microalgal oil/lipid production from *Chlorella sorokiniana* CY1 using deep-sea water supplemented cultivation medium. Biochem Eng J 77:74–81. https://doi.org/10.1016/j.bej.2013.05.009
15. Cheng D, Li D, Yuan Y, Zhou L, Li X, Wu T, Wang L, Zhao Q, Wei W, Sun Y (2017) Improving carbohydrate and starch accumulation in *Chlorella* sp. AE10 by a novel two-stage process with cell dilution. Biotechnol Biofuels 10:1–14. https://doi.org/10.1186/s13068-017-0753-9
16. Cheng DL, Ngo HH, Guo WS, Chang SW, Nguyen DD, Kumar SM (2019) Microalgae biomass from swine wastewater and its conversion to bioenergy. Bioresour Technol 275:109–122. https://doi.org/10.1016/j.biortech.2018.12.019
17. Chia SR, Chew KW, Leong HY, Ho SH, Munawaroh HSH, Show PL (2021) CO_2 mitigation and phycoremediation of industrial flue gas and wastewater via microalgae-bacteria consortium: possibilities and challenges. Chem Eng J 425:131436. https://doi.org/10.1016/j.cej.2021.131436
18. Chng LM, Lee KT, Chan DJC (2017) Synergistic effect of pretreatment and fermentation process on carbohydrate-rich *Scenedesmus dimorphus* for bioethanol production. Energy Convers Manag 141:410–419. https://doi.org/10.1016/j.enconman.2016.10.026
19. Choi J-A, Hwang J-H, Dempsey BA, Abou-Shanab RAI, Min B, Song H, Lee DS, Kim JR, Cho Y, Hong S, Jeon B-H (2011) Enhancement of fermentative bioenergy (ethanol/hydrogen) production using ultrasonication of *Scenedesmus obliquus* YSW15 cultivated in swine wastewater effluent. Energy Environ Sci 4:3513. https://doi.org/10.1039/c1ee01068a
20. Christenson L, Sims R (2011) Production and harvesting of microalgae for wastewater treatment, biofuels, and bioproducts. Biotechnol Adv 29:686–702. https://doi.org/10.1016/j.biotechadv.2011.05.015
21. Collura MA, Luyben WL (1988) Energy-saving distillation designs in ethanol production. Ind Eng Chem Res 27:1686–1696. https://doi.org/10.1021/ie00081a021
22. Constantino A, Rodrigues B, Leon R, Barros R, Raposo S (2021) Alternative chemo-enzymatic hydrolysis strategy applied to different microalgae species for bioethanol production. Algal Res 56:102329. https://doi.org/10.1016/j.algal.2021.102329
23. Cunha P, Pereira H, Costa M, Pereira J, Silva JT, Fernandes N, Varela J, Silva J, Simões M (2020) *Nannochloropsis oceanica* cultivation in pilot-scale raceway ponds-from design to cultivation. Appl Sci 10(5):1725. https://doi.org/10.3390/app10051725
24. D'Hondt E, Martín-Juárez J, Bolado S, Kasperoviciene J, Koreiviene J, Sulcius S, Elst K, Bastiaens L (2017) Cell disruption technologies. In: Gonzalez-Fernandez, Muñoz (ed) Microalgae-based biofuels and bioproducts. Woodhead Publishing, Sawston, p 133–154. https://doi.org/10.1016/B978-0-08-101023-5.00006-6
25. de Farias Silva CE, Meneghello D, Bertucco A (2018) A systematic study regarding hydrolysis and ethanol fermentation from microalgal biomass. Biocatal Agric Biotechnol 14:172–182. https://doi.org/10.1016/j.bcab.2018.02.016
26. de Farias Silva CE, Meneghello D, de Souza Abud AK, Bertucco A (2020) Pretreatment of microalgal biomass to improve the enzymatic hydrolysis of carbohydrates by ultrasonication: yield vs energy consumption. J King Saud Univ Sci 32:606–613. https://doi.org/10.1016/j.jksus.2018.09.007
27. de Morais MG, de Morais EG, Duarte JH, Deamici KM, Mitchell BG, Costa JAV (2019) Biological CO_2 mitigation by microalgae: technological trends, future prospects and challenges. World J Microbiol Biotechnol 35:1–7. https://doi.org/10.1007/s11274-019-2650-9
28. Dias De Oliveira ME, Vaughan BE, Rykiel EJ (2005) Ethanol as fuel: energy, carbon dioxide balances, and ecological footprint. Bioscience 55:593–602. https://doi.org/10.1641/0006-3568(2005)055[0593:EAFECD]2.0.CO;2
29. Doan QC, Moheimani NR, Mastrangelo AJ, Lewis DM (2012) Microalgal biomass for bioethanol fermentation: implications for hypersaline systems with an industrial focus. Biomass Bioenerg 46:79–88. https://doi.org/10.1016/j.biombioe.2012.08.022

30. Dunker S, Wilhelm C (2018) Cell wall structure of coccoid green algae as an important trade-off between biotic interference mechanisms and multidimensional cell growth. Front Microbiol 9:719. https://doi.org/10.3389/fmicb.2018.00719
31. Edzwald JK (1993) Algae, bubbles, coagulants, and dissolved air flotation. Water Sci Technol 27:67–81. https://doi.org/10.2166/wst.1993.0207
32. El-Dalatony MM, Kurade MB, Abou-Shanab RAI, Kim H, Salama ES, Jeon BH (2016) Long-term production of bioethanol in repeated-batch fermentation of microalgal biomass using immobilized *Saccharomyces cerevisiae*. Bioresour Technol 219:98–105. https://doi.org/10.1016/j.biortech.2016.07.113
33. Eliseus A, Bilad MR, Nordin NAHM, Putra ZA, Wirzal MDH (2017) Tilted membrane panel: a new module concept to maximize the impact of air bubbles for membrane fouling control in microalgae harvesting. Bioresour Technol 241:661–668. https://doi.org/10.1016/j.biortech.2017.05.175
34. Farooq W, Suh WI, Park MS, Yang JW (2015) Water use and its recycling in microalgae cultivation for biofuel application. Bioresour Technol 184:73–81. https://doi.org/10.1016/j.biortech.2014.10.140
35. Fetyan NAH, El-Sayed AEKB, Ibrahim FM, Attia YA, Sadik MW (2022) Bioethanol production from defatted biomass of *Nannochloropsis oculata* microalgae grown under mixotrophic conditions. Environ Sci Pollut Res 29:2588–2597. https://doi.org/10.1007/s11356-021-15758-6
36. Gerbens-Leenes PW, Xu L, de Vries GJ, Hoekstra AY (2014) The blue water footprint and land use of biofuels from algae. Water Resour Res 50:8549–8563. https://doi.org/10.1002/2014WR015710
37. Gifuni I, Olivieri G, Krauss IR, D'Errico G, Pollio A, Marzocchella A (2017) Microalgae as new sources of starch: isolation and characterization of microalgal starch granules. Chem Eng Trans 57:1423–1428. https://doi.org/10.3303/CET1757238
38. Gifuni I, Olivieri G, Pollio A, Marzocchella A (2018) Identification of an industrial microalgal strain for starch production in biorefinery context: the effect of nitrogen and carbon concentration on starch accumulation. N Biotechnol 41:46–54. https://doi.org/10.1016/j.nbt.2017.12.003
39. Gouda VK, Banat IM, Mansour S (1993) Microbiologically induced corrosion of UNS N04400 in seawater. Corrosion 49:63–73. https://doi.org/10.5006/1.3316036
40. Gouveia L, Graça S, Sousa C, Ambrosano L, Ribeiro B, Botrel EP, Neto PC, Ferreira AF, Silva CM (2016) Microalgae biomass production using wastewater: treatment and costs. Scale-up consideration. Algal Res 16:167–176. https://doi.org/10.1016/j.algal.2016.03.010
41. Gutiérrez R, Ferrer I, Uggetti E, Arnabat C, Salvadó H, García J (2016) Settling velocity distribution of microalgal biomass from urban wastewater treatment high rate algal ponds. Algal Res 16:409–417. https://doi.org/10.1016/j.algal.2016.03.037
42. Ha GS, El-Dalatony MM, Kim DH, Salama ES, Kurade MB, Roh HS, El-Fatah Abomohra A, Jeon BH (2020) Biocomponent-based microalgal transformations into biofuels during the pretreatment and fermentation process. Bioresour Technol 302:122809. https://doi.org/10.1016/j.biortech.2020.122809
43. Hanotu J, Bandulasena HCH, Zimmerman WB (2012) Microflotation performance for algal separation. Biotechnol Bioeng 109:1663–1673. https://doi.org/10.1002/bit.24449
44. Hawrot-Paw M, Koniuszy A, Gałczynska M, Zajac G, Szyszlak-Bargłowicz J (2020) Production of microalgal biomass using aquaculture wastewater as growth medium. Water 12(1):106. https://doi.org/10.3390/w12010106
45. Hemalatha M, Sravan JS, Min B, Venkata Mohan S (2019) Microalgae-biorefinery with cascading resource recovery design associated to dairy wastewater treatment. Bioresour Technol 284:424–429. https://doi.org/10.1016/j.biortech.2019.03.106
46. Heo YM, Lee H, Lee C, Kang J, Ahn JW, Lee YM, Kang KY, Choi YE, Kim JJ (2017) An integrative process for obtaining lipids and glucose from *Chlorella vulgaris* biomass with a single treatment of cell disruption. Algal Res 27:286–294. https://doi.org/10.1016/j.algal.2017.09.022

47. Ho SH, Chen YD, Chang CY, Lai YY, Chen CY, Kondo A, Ren NQ, Chang JS (2017) Feasibility of CO_2 mitigation and carbohydrate production by microalga *Scenedesmus obliquus* CNW-N used for bioethanol fermentation under outdoor conditions: effects of seasonal changes. Biotechnol Biofuels 10:1–13. https://doi.org/10.1186/s13068-017-0712-5
48. Ho SH, Huang SW, Chen CY, Hasunuma T, Kondo A, Chang JS (2013) Bioethanol production using carbohydrate-rich microalgae biomass as feedstock. Bioresour Technol 135:191–198. https://doi.org/10.1016/j.biortech.2012.10.015
49. Hor S, Kongkeitkajorn MB, Reungsang A (2022) Sugarcane bagasse-based ethanol production and utilization of its vinasse for xylitol production as an approach in integrated biorefinery. Fermentation 8:340. https://doi.org/10.3390/fermentation8070340
50. Huang X, Bai S, Liu Z, Hasunuma T, Kondo A, Ho S-H (2020) Fermentation of pigment-extracted microalgal residue using yeast cell-surface display: direct high-density ethanol production with competitive life cycle impacts. Green Chem 22:153–162. https://doi.org/10.1039/C9GC02634G
51. Hülsen T, Hsieh K, Lu Y, Tait S, Batstone DJ (2018) Simultaneous treatment and single cell protein production from agri-industrial wastewaters using purple phototrophic bacteria or microalgae—a comparison. Bioresour Technol 254:214–223. https://doi.org/10.1016/j.biortech.2018.01.032
52. Imam SH, Buchanan MJ, Shin HC, Snell WJ (1985) The *Chlamydomonas* cell wall: characterization of the wall framework. J Cell Biol 101:1599–1607. https://doi.org/10.1083/jcb.101.4.1599
53. Kadir WNA, Lam MK, Uemura Y, Lim JW, Kiew PL, Lim S, Rosli SS, Wong CY, Show PL, Lee KT (2021) Simultaneous harvesting and cell disruption of microalgae using ozone bubbles: optimization and characterization study for biodiesel production. Front Chem Sci Eng 15:1257–1268. https://doi.org/10.1007/s11705-020-2015-9
54. Kataria R, Ghosh S (2014) NaOH pretreatment and enzymatic hydrolysis of *Saccharum spontaneum* for reducing sugars production. Energy Sources Part A Recover Util Environ Eff 36:1028–1035. https://doi.org/10.1080/15567036.2010.551268
55. Keris-Sen UD, Gurol MD (2017) Using ozone for microalgal cell disruption to improve enzymatic saccharification of cellular carbohydrates. Biomass Bioenerg 105:59–65. https://doi.org/10.1016/j.biombioe.2017.06.023
56. Khan MI, Shin JH, Kim JD (2018) The promising future of microalgae: current status, challenges, and optimization of a sustainable and renewable industry for biofuels, feed, and other products. Microb Cell Fact 17:1–21. https://doi.org/10.1186/s12934-018-0879-x
57. Laamanen CA, Ross GM, Scott JA (2016) Flotation harvesting of microalgae. Renew Sustain Energy Rev 58:75–86. https://doi.org/10.1016/j.rser.2015.12.293
58. Liu X, Xu W, Mao L, Zhang C, Yan P, Xu Z, Zhang ZC (2016) Lignocellulosic ethanol production by starch-base industrial yeast under PEG detoxification. Sci Rep 6:1–11. https://doi.org/10.1038/srep20361
59. Luangpipat T, Chisti Y (2017) Biomass and oil production by *Chlorella vulgaris* and four other microalgae—effects of salinity and other factors. J Biotechnol 257:47–57. https://doi.org/10.1016/j.jbiotec.2016.11.029
60. Maffei G, Bracciale MP, Broggi A, Zuorro A, Santarelli ML, Lavecchia R (2018) Effect of an enzymatic treatment with cellulase and mannanase on the structural properties of *Nannochloropsis* microalgae. Bioresour Technol 249:592–598. https://doi.org/10.1016/j.biortech.2017.10.062
61. Mahata C, Das P, Khan S, Thaher MIA, Quadir MA, Annamalai SN, Jabri HAl (2022) The potential of marine microalgae for the production of food, feed, and fuel (3F). Fermentation 8(7):316. https://doi.org/10.3390/fermentation8070316
62. Maia JL da, Cardoso JS, Mastrantonio DJ da S, Bierhals CK, Moreira JB, Costa JAV, Morais MG de (2020) Microalgae starch: a promising raw material for the bioethanol production. Int J Biol Macromol 165:2739–2749. https://doi.org/10.1016/j.ijbiomac.2020.10.159
63. Mat Aron NS, Khoo KS, Chew KW, Show PL, Chen WH, Nguyen THP (2020) Sustainability of the four generations of biofuels—a review. Int J Energy Res 44:9266–9282. https://doi.org/10.1002/er.5557

64. Mathiot C, Ponge P, Gallard B, Sassi JF, Delrue F, Le Moigne N (2019) Microalgae starch-based bioplastics: screening of ten strains and plasticization of unfractionated microalgae by extrusion. Carbohydr Polym 208:142–151. https://doi.org/10.1016/j.carbpol.2018.12.057
65. Miller DH, Mellman IRAS, Lamport DTA (1974) The chemical composition of the cell wall of and the concept of a plant cell wall protein. J Cell Biol 63(2):420–429. https://doi.org/10.1083/jcb.63.2.420
66. Montsant A, Zarka A, Boussiba S (2001) Presence of a nonhydrolyzable biopolymer in the cell wall of vegetative cells and astaxanthin-rich cysts of *Haematococcus pluvialis* (chlorophyceae). Mar Biotechnol 3:0515–0521. https://doi.org/10.1007/s1012601-0051-0
67. Müller J, Friedl T, Hepperle D, Lorenz M, Day JG (2005) Distinction between multiple isolates of *Chlorella vulgaris* (Chlorophyta, Trebouxiophyceae) and testing for conspecificity using amplified fragment length polymorphism and ITS rDNA sequences. J Phycol 41:1236–1247. https://doi.org/10.1111/j.1529-8817.2005.00134.x
68. Muscat A, de Olde EM, de Boer IJM, Ripoll-Bosch R (2020) The battle for biomass: a systematic review of food-feed-fuel competition. Glob Food Sec 25:100330. https://doi.org/10.1016/j.gfs.2019.100330
69. Nagappan S, Das P, AbdulQuadir M, Thaher M, Khan S, Mahata C, Al-Jabri H, Vatland AK, Kumar G (2021) Potential of microalgae as a sustainable feed ingredient for aquaculture. J Biotechnol 341:1–20. https://doi.org/10.1016/j.jbiotec.2021.09.003
70. Ndikubwimana T, Chang J, Xiao Z, Shao W, Zeng X, Ng IS, Lu Y (2016) Flotation: a promising microalgae harvesting and dewatering technology for biofuels production. Biotechnol J 11:315–326. https://doi.org/10.1002/biot.201500175
71. Ngamsirisomsakul M, Reungsang A, Liao Q, Kongkeitkajorn MB (2019) Enhanced bioethanol production from *Chlorella* sp. biomass by hydrothermal pretreatment and enzymatic hydrolysis. Renew Energy 141:482–492. https://doi.org/10.1016/j.renene.2019.04.008
72. Norsker NH, Barbosa MJ, Vermuë MH, Wijffels RH (2011) Microalgal production—a close look at the economics. Biotechnol Adv 29:24–27. https://doi.org/10.1016/j.biotechadv.2010.08.005
73. Oh HM, Lee SJ, Park MH, Kim HS, Kim HC, Yoon JH, Kwon GS, Yoon BD (2001) Harvesting of *Chlorella vulgaris* using a bioflocculant from *Paenibacillus* sp. AM49. Biotechnol Lett 23:1229–1234. https://doi.org/10.1023/A:1010577319771
74. Onay M (2019) Bioethanol production via different saccharification strategies from *H. tetrachotoma* ME03 grown at various concentrations of municipal wastewater in a flat-photobioreactor. Fuel 239:1315–1323. https://doi.org/10.1016/j.fuel.2018.11.126
75. Oostlander PC, van Houcke J, Wijffels RH, Barbosa MJ (2020) Microalgae production cost in aquaculture hatcheries. Aquaculture 525:735310. https://doi.org/10.1016/j.aquaculture.2020.735310
76. Otsuka N (2010) Fireside Corrosion. In: Cottis, Graham, Lindsay, Lyon, Richardson, Scantlebury, Stott (ed) Shreir's corrosion. Elsevier, Amsterdam, pp 457–481. https://doi.org/10.1016/B978-044452787-5.00192-X
77. Papazi A, Makridis P, Divanach P (2010) Harvesting *Chlorella minutissima* using cell coagulants. J Appl Phycol 22:349–355. https://doi.org/10.1007/s10811-009-9465-2
78. Phwan CK, Chew KW, Sebayang AH, Ong HC, Ling TC, Malek MA, Ho YC, Show PL (2019) Effects of acids pre-treatment on the microbial fermentation process for bioethanol production from microalgae. Biotechnol Biofuels 12:1–8. https://doi.org/10.1186/s13068-019-1533-5
79. Raj T, Chandrasekhar K, Naresh Kumar A, Rajesh Banu J, Yoon JJ, Kant Bhatia S, Yang YH, Varjani S, Kim SH (2022) Recent advances in commercial biorefineries for lignocellulosic ethanol production: current status, challenges and future perspectives. Bioresour Technol 344:126292. https://doi.org/10.1016/j.biortech.2021.126292
80. Ramli RN, Lee CK, Kassim MA (2020) Extraction and characterization of starch from microalgae and comparison with commercial corn starch. IOP Conf Ser Mater Sci Eng 716:012012. https://doi.org/10.1088/1757-899X/716/1/012012
81. Reyimu Z, Özçimen D (2017) Batch cultivation of marine microalgae *Nannochloropsis oculata* and *Tetraselmis suecica* in treated municipal wastewater toward bioethanol production. J Clean Prod 150:40–46. https://doi.org/10.1016/j.jclepro.2017.02.189

82. Romero Villegas GI, Fiamengo M, Acién Fernández FG, Molina Grima E (2017) Outdoor production of microalgae biomass at pilot-scale in seawater using centrate as the nutrient source. Algal Res 25:538–548. https://doi.org/10.1016/j.algal.2017.06.016
83. Safi C, Charton M, Ursu AV, Laroche C, Zebib B, Pontalier PY, Vaca-Garcia C (2014) Release of hydro-soluble microalgal proteins using mechanical and chemical treatments. Algal Res 3:55–60. https://doi.org/10.1016/j.algal.2013.11.017
84. Safi C, Ursu AV, Laroche C, Zebib B, Merah O, Pontalier PY, Vaca-Garcia C (2014) Aqueous extraction of proteins from microalgae: effect of different cell disruption methods. Algal Res 3:61–65. https://doi.org/10.1016/j.algal.2013.12.004
85. Salim S, Shi Z, Vermuë MH, Wijffels RH (2013) Effect of growth phase on harvesting characteristics, autoflocculation and lipid content of *Ettlia texensis* for microalgal biodiesel production. Bioresour Technol 138:214–221. https://doi.org/10.1016/j.biortech.2013.03.173
86. Sanchez Rizza L, Coronel CD, Sanz Smachetti ME, Do Nascimento M, Curatti L (2019) A semi-closed loop microalgal biomass production-platform for ethanol from renewable sources of nitrogen and phosphorous. J Clean Prod 219:217–224. https://doi.org/10.1016/j.jclepro.2019.01.311
87. Seabra JEA, Macedo IC, Chum HL, Faroni CE, Sarto CA (2011) Life cycle assessment of Brazilian sugarcane products: GHG emissions and energy use. Biofuels Bioprod Biorefining 5:519–532. https://doi.org/10.1002/bbb.289
88. Seon G, Kim HS, Cho JM, Kim M, Park WK, Chang YK (2020) Effect of post-treatment process of microalgal hydrolysate on bioethanol production. Sci Rep 10:1–12. https://doi.org/10.1038/s41598-020-73816-4
89. Sepulveda C, Gómez C, El Bahraoui N, Acién G (2019) Comparative evaluation of microalgae strains for CO_2 capture purposes. J CO_2 Util 30:158–167. https://doi.org/10.1016/j.jcou.2019.02.004
90. Shin YS, Choi HI, Choi JW, Lee JS, Sung YJ, Sim SJ (2018) Multilateral approach on enhancing economic viability of lipid production from microalgae: a review. Bioresour Technol 258:335–344. https://doi.org/10.1016/j.biortech.2018.03.002
91. Shokrkar H, Ebrahimi S (2018) Evaluation of different enzymatic treatment procedures on sugar extraction from microalgal biomass, experimental and kinetic study. Energy 148:258–268. https://doi.org/10.1016/j.energy.2018.01.124
92. Shokrkar H, Ebrahimi S, Zamani M (2017) Bioethanol production from acidic and enzymatic hydrolysates of mixed microalgae culture. Fuel 200:380–386. https://doi.org/10.1016/j.fuel.2017.03.090
93. Silva-Graça M, Lucas C (2003) Physiological studies on long-term adaptation to salt stress in the extremely halotolerant yeast *Candida versatilis* CBS 4019 (syn. *C. halophila*). FEMS Yeast Res 3:247–260. https://doi.org/10.1016/S1567-1356(02)00198-8
94. Sivaramakrishnan R, Incharoensakdi A (2018) Utilization of microalgae feedstock for concomitant production of bioethanol and biodiesel. Fuel 217:458–466. https://doi.org/10.1016/j.fuel.2017.12.119
95. Sivasankar P, Poongodi S, Lobo AO, Pugazhendhi A (2020) Characterization of a novel polymeric bioflocculant from marine actinobacterium *Streptomyces* sp. and its application in recovery of microalgae. Int Biodeterior Biodegrad 148:104883. https://doi.org/10.1016/j.ibiod.2020.104883
96. Su Y, Mennerich A, Urban B (2012) Comparison of nutrient removal capacity and biomass settleability of four high-potential microalgal species. Bioresour Technol 124:157–162. https://doi.org/10.1016/j.biortech.2012.08.037
97. Sukačová K, Lošák P, Brummer V, Máša V, Vícha D, Zavřel T (2021) Perspective design of algae photobioreactor for greenhouses—a comparative study. Energies 14(5):1338. https://doi.org/10.3390/en14051338
98. Sukenik A, Bilanovic D, Shelef G (1988) Flocculation of microalgae in brackish and sea waters. Biomass 15:187–199. https://doi.org/10.1016/0144-4565(88)90084-4
99. Vandamme D, Foubert I, Meesschaert B, Muylaert K (2010) Flocculation of microalgae using cationic starch. J Appl Phycol 22:525–530. https://doi.org/10.1007/s10811-009-9488-8

100. Vandamme D, Muylaert K, Fraeye I, Foubert I (2014) Floc characteristics of *Chlorella vulgaris*: Influence of flocculation mode and presence of organic matter. Bioresour Technol 151:383–387. https://doi.org/10.1016/j.biortech.2013.09.112
101. Vázquez-Romero B, Perales JA, Pereira H, Barbosa M, Ruiz J (2022) Techno-economic assessment of microalgae production, harvesting and drying for food, feed, cosmetics, and agriculture. Sci Total Environ 837:155742. https://doi.org/10.1016/j.scitotenv.2022.155742
102. Vera I, Wicke B, Lamers P, Cowie A, Repo A, Heukels B, Zumpf C, Styles D, Parish E, Cherubini F, Berndes G, Jager H, Schiesari L, Junginger M, Brandão M, Bentsen NS, Daioglou V, Harris Z, van der Hilst F (2022) Land use for bioenergy: synergies and trade-offs between sustainable development goals. Renew Sustain Energy Rev 161:112409. https://doi.org/10.1016/j.rser.2022.112409
103. Vogelbusch Biocommodities (n.d) Vogelbusch Bioethanol Technology [WWW Document]. URL https://www.vogelbusch-biocommodities.com/technology/alcohol-process-plants/bioethanol-technology/. Accessed 8.10.22
104. Wan Osman WNA, Mat Nawi NI, Samsuri S, Bilad MR, Khan AL, Hunaepi H, Jaafar J, Lam MK (2021) Ultra low-pressure filtration system for energy efficient microalgae filtration. Heliyon 7. https://doi.org/10.1016/j.heliyon.2021.e07367
105. Wang H, Ji C, Bi S, Zhou P, Chen L, Liu T (2014) Joint production of biodiesel and bioethanol from filamentous oleaginous microalgae Tribonema sp. Bioresour Technol 172:169–173. https://doi.org/10.1016/j.biortech.2014.09.032
106. Wang Y, Yang Y, Ma F, Xuan L, Xu Y, Huo H, Zhou D, Dong S (2015) Optimization of *Chlorella vulgaris* and bioflocculant-producing bacteria co-culture: enhancing microalgae harvesting and lipid content. Lett Appl Microbiol 60:497–503. https://doi.org/10.1111/lam.12403
107. Weber S, Grande PM, Blank LM, Klose H (2022) Insights into cell wall disintegration of *Chlorella vulgaris*. PLoS ONE 17:1–14. https://doi.org/10.1371/journal.pone.0262500
108. Wu C, Xiao Y, Lin W, Li J, Zhang S, Zhu J, Rong J (2017) Aqueous enzymatic process for cell wall degradation and lipid extraction from *Nannochloropsis* sp. Bioresour Technol 223:312–316. https://doi.org/10.1016/j.biortech.2016.10.063
109. Xie Y, Ho SH, Chen CNN, Chen CY, Jing K, Ng IS, Chen J, Chang JS, Lu Y (2016) Disruption of thermo-tolerant *Desmodesmus* sp. F51 in high pressure homogenization as a prelude to carotenoids extraction. Biochem Eng J 109:243–251. https://doi.org/10.1016/j.bej.2016.01.003
110. Yadala S, Cremaschi S (2016) A dynamic optimization model for designing open-channel raceway ponds for batch production of algal biomass. Processes 4(2):10. https://doi.org/10.3390/pr4020010
111. Yadav G, Dash SK, Sen R (2019) A biorefinery for valorization of industrial waste-water and flue gas by microalgae for waste mitigation, carbon-dioxide sequestration and algal biomass production. Sci Total Environ 688:129–135. https://doi.org/10.1016/j.scitotenv.2019.06.024
112. Yao C, Ai J, Cao X, Xue S, Zhang W (2012) Enhancing starch production of a marine green microalga *Tetraselmis subcordiformis* through nutrient limitation. Bioresour Technol 118:438–444. https://doi.org/10.1016/j.biortech.2012.05.030
113. Yi Z, Su Y, Brynjolfsson S, Olafsdóttir K, Fu W (2021) Bioactive polysaccharides and their derivatives from microalgae: biosynthesis, applications, and challenges. In: Rahman (ed) Studies in natural products chemistry. Elsevier, Amsterdam, pp 67–85. https://doi.org/10.1016/B978-0-323-91095-8.00007-6
114. Yu KL, Chen WH, Sheen HK, Chang JS, Lin CS, Ong HC, Show PL, Ling TC (2020) Bioethanol production from acid pretreated microalgal hydrolysate using microwave-assisted heating wet torrefaction. Fuel 279:118435. https://doi.org/10.1016/j.fuel.2020.118435
115. Zhang Y, Kang X, Zhen F, Wang Z, Kong X, Sun Y (2022) Assessment of enzyme addition strategies on the enhancement of lipid yield from microalgae. Biochem Eng J 177:108198. https://doi.org/10.1016/j.bej.2021.108198
116. Zhao Z, Li Y, Muylaert K, Vankelecom IFJ (2020) Synergy between membrane filtration and flocculation for harvesting microalgae. Sep Purif Technol 240:116603. https://doi.org/10.1016/j.seppur.2020.116603

117. Zhao Z, Muylaert K, Vankelecom IFJ (2021) Combining patterned membrane filtration and flocculation for economical microalgae harvesting. Water Res 198:117181. https://doi.org/10.1016/j.watres.2021.117181

Bioethanol Production via Fermentation: Microbes, Modeling and Optimization

Adebisi Aminat Agboola, Niyi Babatunde Ishola, and Eriola Betiku

Abstract The global demand for ethanol is rising tremendously because of fast industrialization and population expansion. The increased attention gained by ethanol is due to its application as a transport fuel because it is renewable, sustainable, eco-friendly, carbon neutral, and has a high-octane rating. The processes involved in its production include treatment and hydrolysis of substrates when starchy and lignocellulosic materials are used, microbial fermentation of substrates to bioethanol, and downstream processes such as distillation. The various steps are delineated in this chapter, but more consideration is given to the microbial fermentation of diverse substrates. Also presented in the chapter are the critical steps in adopting emerging technologies to improve the overall conversion system. Modeling and optimizing the pertinent operating variables involved in microbial fermentation to obtain bioethanol can make the process cost-effective and offer insight into understanding the behavior of the process. The benefits and downsides of some response surface methodology (RSM) screening approaches were discussed. The machine-learning techniques commonly used to model biochemical systems were also discussed extensively. It was noted that these data-mining techniques are better at handling the nonlinearity and multivariate nature of the microbial fermentation process for bioethanol production than the non-data-mining techniques, such as RSM.

Keywords Bioethanol · Fermentation · Modeling · Optimization · Machine learning technology · Microbes · Enzymes · Hydrolysis · Saccharification · Pretreatment

---

A. A. Agboola · N. B. Ishola · E. Betiku (✉)
Department of Chemical Engineering, Obafemi Awolowo University, Ile-Ife 220005, Osun State, Nigeria
e-mail: eriola.betiku@famu.edu

E. Betiku
Department of Biological Sciences, Florida Agricultural and Mechanical University, Tallahassee, FL 32307, USA

List of Abbreviations

AAD	Average absolute deviation
ACO	Ant colony optimization
ADP	Adenosine diphosphate
ANFIS	Adaptive neuro-fuzzy inference system
ANN	Artificial neural network
ATP	Adenosine triphosphate
BBD	Box Behnken design
CCD	Central composite design
EMP	Embden-Meyerhof-Parnas
GA	Genetic algorithm
MSE	Mean square error
OVAT	One variable at a time
PBD	Plackett-Burma design
PSO	Particle swarm optimization
R	Coefficient of correlation
R^2	Coefficient of determination
RMSE	Root mean square error
RSM	Response surface methodology

1 Introduction

The rate at which fossil fuels are exploited, the rise in greenhouse gas emissions brought on by utilizing them for energy, and the impact of human activities on land usage have intensified the global warming crisis. Alternative energy sources not reliant on petroleum are becoming crucial to meet demand as global energy consumption rises and world crude oil output diminishes [68]. Biofuels are clean, non-petroleum alternative energy sources that do not cause food shortages which will hamper the economy.

Bioethanol, made from biomass, is frequently used to produce sustainable biofuel. It is regarded to be a more promising sustainable renewable biofuel. As a substitute for gasoline made from sustainable sources, bioethanol offers vast potential to close a gap in the world's shared energy output since fossil fuel production rates have been steadily declining due to global warming and climate change concerns [98]. Besides having a high degree of octane number, a high level of heat of vaporization, being less poisonous, and, most importantly, reducing greenhouse gas emissions by burning to produce carbon dioxide and water, bioethanol has a variety of other advantages [68, 107]. In the transportation industry, bioethanol serves as an anti-knock agent by preventing knocking and early ignition [98]. This is realized by decreasing the properties and performance of engine oil [25].

Fundamentally, bioethanol production involves the fermentation of several feedstock [25]. The feedstock are essentially carbohydrate-rich plants that fall within the biomass high in starch, sugar, or lignocellulosic material. Sugarcane, high in sugar; corn and wheat, high in starch; and bagasse, rich in lignocellulose, have traditionally been used to produce bioethanol. Microbes are used in the fermentation process for bioethanol, and they secrete enzymes that ferment the sugar produced from the feedstock. Using inexpensive raw materials can enhance the economics of producing bioethanol through fermentation, provided the producing microbe can metabolize the specific carbohydrate.

An important technique for determining the ideal conditions for the fermentation process is modeling and optimization. As a result, both the output (desired product) and the resources required during production are maximized. A variable is altered while others are maintained constant during the process. This method may use statistically prepared trials, allowing for a more thorough knowledge of the interactions of the various process components with less testing [100]. Utilizing the experimental design with response surface methodology (RSM) will enable the bioethanol production technology to be understood easily. The use of RSM will give an insight into the interaction between the different bioethanol production operating variables and reduce the time required to achieve maximum bioethanol concentration. However, RSM cannot capture the nonlinearity associated with the bioethanol production process. The utilization of machine-learning tools, such as artificial neural networks (ANN) and adaptive neuro-fuzzy inference systems (ANFIS), has been reported to overcome this downside of RSM. In addition, the numerical optimization of RSM to obtain the input parameter combination searches for the best solution locally. It hence might not give a favorable combination of these parameters. Therefore, integrating the developed data mining technology with available global optimization techniques will help attain maximum bioethanol production with suitable process parameter combinations.

This overview offers a summary of the microbes, types of fermentation, modeling, and optimization of the bioethanol production system, which are required to determine the importance of each parameter considered during the process.

2 Processes Involved in Bioethanol Production

Bioethanol can be generated from varieties of feedstock, which undergo different pathways during production. The process involves gathering, handling, recovering, and moving the feedstock; grinding the biomass into small, homogenous particles; fractionating the polymers; separating the solid lignin component; and recovering the finished product. Bioethanol production via fermentation is a biotransformation process of biomass to bioethanol known as the first-generation process [42]. The process consists of four major unit operations: (i) pretreatment (involving milling and grinding), (ii) hydrolysis, (iii) microbial fermentation, and (iv) distillation (product purification). The complex nature of the units varies, subject to the type of feedstock used. Sugar-rich feedstock is based on the simple sugar conversion to ethanol, while starch and lignocellulosic feedstock involve a multistage conversion to ethanol [4].

2.1 Pretreatment Step

While pretreatment for starch-rich feedstock involves gelatinization, a part of the liquefaction of the starch, for accessibility of its reducing sugar, sucrose-rich feedstock merely requires shredding and milling along with the extraction of the reducing sugars. Lignocellulose biomass must be pretreated to break the cellulosic link and make cellulose easily accessible for conversion to reducing sugar. Several applications for the lignocellulosic pretreatment technology exist, including physical and mechanical pretreatment (milling and irradiation); physicochemical pretreatment (CO_2 explosion, steam explosion, steam explosion with SO_2 addition, NH_3 fiber explosion, hot-water pretreatment); chemical pretreatment (alkaline pretreatment, organosolv process, wet oxidation, ozonolysis pretreatment, acid pretreatment); biological pretreatment, i.e., using microorganisms [15, 121].

2.2 Hydrolysis Step

Hydrolysis, otherwise called saccharification, is the next stage in the bioethanol production system. This step helps break down complex sugars into monomeric sugars (hydrolysates). Starch-rich sugars and lignocellulosic materials undergo this process. Hydrolysis is characterized by chemical and enzymatic actions [1, 6, 98, 104].

In chemical hydrolysis, either concentrated or dilute acid such as H_2SO_4 and HCl is employed. For this form of hydrolysis to be carried out, the pretreatment and hydrolysis are conducted using one step. The main benefits of concentrated acid hydrolysis processes (operate in a mild environment) and mild acid hydrolysis processes (operate in high pressure and temperature environment) are that they operate at a faster rate than enzymes and can attack lignin without prior treatment. This allows for the continuous processing and high efficiency of sugar recovery [30, 75]. The main drawbacks of hydrolysis using acids are its toxicity, corrosiveness, and danger, which necessitate corrosion-resistant reactors, high initial investment costs, and inhibitor formation [1, 21]. Acid hydrolysis has become less attractive and competitive due to these significant drawbacks, which has prompted investigators to search for other environmentally benign and commercially viable approaches for obtaining sugars from lignocellulosic materials [1, 22, 56].

Contrariwise, enzymatic hydrolysis is a primary method, serving as an alternative method to chemical hydrolysis. Hemicellulose and cellulose are biochemically metamorphosed into simple sugars by an enzyme released by microorganisms like fungi and bacteria [1]. Enzymatic hydrolysis has gained more attention as a result of its significant advantages over chemical hydrolysis, which include: mild reaction conditions such as pH of 5 and temperature < 50 °C (producing better yields), low energy consumption, low environmental impact, no corrosion issues, minor inhibitory compound formation, and high glucose yield [1, 6, 46].

2.3 Fermentation Step

The third step is the fermentation process. All feedstock types for bioethanol production undergo this step. Fermentation is a natural metabolic process that occurs when microorganisms release enzymes that cause the organic substrate to alter chemically. Microbes that break down simple fermentable sugars (hydrolysates) as food during fermentation produce ethyl ethanol (bioethanol) together with other byproducts [15, 95]. Hexoses and pentoses are monomeric sugars fermented during this process to produce alcohols, lactic acid, and other desirable products [56, 66]. Various eukaryotic microbes viz *Candida tropicalis, Candida oleophila, Hanseniaspora spp., Saccharomyces uvarum, Saccharomyces cerevisiae, Saccharomyces rouxii, Kluyveromyces lactis, Kluyveromyces fragilis, Mucor rouxianus* and prokaryotic microbes *Thermoanaerobacter ethanolicus, Clostridium thermosaccharolyticum, Clostridium sporogenes, Clostridium acetobutylicum, Zymomonas mobilis,* and *Bacillus stearothermophilus* have been investigated for fermentation to produce bioethanol [110]. The performance factors for fermentation include temperature range, specificity, osmotic tolerance, genetic stability, growth rate, pH range, productivity, yield, inhibitor tolerance, and alcohol tolerance [15].

2.4 Distillation Step

Distillation technology allows for the economic recovery of dilute volatile products (such as ethanol) from streams having a mixture of impurities. This technique separates the ethanol in the liquid mixture from water; typically, a top stream high in ethanol and a bottom stream high in water are obtained [16, 68, 87]. High energy is needed for concentrating the ethanol from the azeotrope mixture of ethanol and water to 95.6%. The recovery process results in concentrated ethanol (37%) in a rectifying column to a concentration slightly lower than the azeotrope [55]. To reduce bioethanol losses, the distillation columns at the plant should have a fixed recovery rate of 99.6% [16, 68, 87].

3 Common Microbes Used for Fermentation

In the fermentation process, microorganisms metabolize carbohydrates to ethanol. These microbes are known as ethanologenic microbes. The fermentation process uses many microbes, which have their sources from bacteria, fungi, and yeast [123]. The microbes secrete fermentation enzymes which help speed up the fermentation process. *Z. mobilis, Escherichia coli, Clostridium sp.*, and *Klebsiella oxytoca* are the bacteria most frequently used, while *C. tropicalis, Candida shehatae*, and *Pichia stipites* are the most frequently used fungi. *S. cerevisiae* is the most commonly utilized yeast to produce bioethanol [13, 14, 61, 115, 135].

3.1 Saccharomyces Cerevisiae

The most popular yeast in the biotechnology sector, *S. cerevisiae*, is generally considered safe owing to its capacity to convert carbohydrates to bioethanol and its resistance to stress during fermentation. *S. cerevisiae* is the model eukaryotic cell system. It has undergone several genetic manipulations to manufacture alcohol. It is more suitable for use in industrial settings than bacteria, filamentous fungi, and other yeasts due to its many physiological properties during the manufacture of ethanol. *S. cerevisiae* has several advantages over other ethanol-producing microorganisms. Its fermentation is less sensitive to infection (other inhibitory substances) than bacteria because of its strong tolerance across a broad pH range with an acidic optimum [14, 79, 80, 100, 103, 124, 134]. *S. cerevisiae* differs from other yeasts in that it is a facultative anaerobe that can survive in aerobic and anaerobic conditions while consuming glucose [72]. It is also tolerant of high ethanol concentrations [33]. In anaerobic environments, acetaldehyde will be produced by *S. cerevisiae*, which is then converted to ethanol [126]. Up to 18% of the fermentation broth concentration can be transformed into bioethanol by this process [78, 123].

Glycolysis, commonly known as the Embden-Meyerhof-Parnas (EMP) pathway, is the main metabolic process for ethanol fermentation in *S. cerevisiae*. It culminates in the oxidation of one glucose molecule and the formation of two pyruvate molecules [12], as shown in Eq. (1).

$$C_6H_{12}O_6 + 2C_{10}H_{15}N_5O_{10}P_2 + 2Pi + 2NAD^+$$
$$\rightarrow 2\text{Pyruvate} + 2C_{10}H_{16}N_5O_{13}P_3 + 2NADH^+ + 2H^+ \qquad (1)$$

where $C_{10}H_{16}N_5O_{13}P_3$ is the Adenosine triphosphate (ATP) which is biological energy; $2C_{10}H_{15}N_5O_{10}P_2$ is adenosine diphosphate (ADP); NAD is nicotinamide adenine dinucleotide; NAD^+ and $NADH^+$ are its oxidized and reduced forms, respectively, and Pi is the inorganic phosphate. The pyruvate is reduced to bioethanol with CO_2 as a byproduct under anaerobic conditions.

3.2 Kluyveromyces Marxianus

One of the non-conventional yeast strains that have received extensive research for bioethanol production is *K. marxianus*, which is generally considered safe (GRAS). Its characteristics have some advantages for bioethanol production. The microorganism can metabolize different sugars, including xylose, lactose, sucrose, raffinose, and arabinose, due to the presence of the β-galactosidase enzyme system. It also has a higher specific growth rate than *S. cerevisiae* and is thermotolerant to temperatures between 44 and 52 °C. Its usage in bioprocesses involving cell recycling is hindered by its inability to withstand high concentrations of bioethanol, as cell viability is reduced at these concentrations [76, 84]. It is known to have the Crabtree effect; its fermentative metabolism is associated with oxygen limitation, and it cannot survive anaerobic conditions. Under simultaneous saccharification and fermentation process, *K. marxianus* has been studied as a potential microorganism for bioethanol production with various feedstock such as cheese whey [92], wheat straw and sugarcane bagasse [109], sweet sorghum bagasse [124], and sugarcane juice [77] amongst others. Research also showed that *K. marxianus* is more tolerant to inhibitory compounds when compared to *S. cerevisiae*.

K. marxianus employs the Leloir and Entner-Doudoroff pathways for bioethanol production [20]. The Leloir pathway is a metabolic pathway that catabolizes lactose and galactose to glucose involving four steps using different enzymes consecutively, such as galactose mutarotase or β-galactosidase or aldose-1-epimerase, galactokinase, galactose 1-phosphate uridylyl-transferase and UDP-galactose-4-epimerase, respectively, which releases glucose 1-phosphate. After the release of glucose 1-phosphate, the Entner-Doudoroff pathway begins, which produces pyruvate and glyceraldehydes-3-phosphate and only one molecule of ATP per glucose molecule.

3.3 Zymomonas Mobilis

Z. mobilis, an uncommon gram-negative bacterium, possesses various desirable qualities as a biocatalyst to produce ethanol. It is usually considered safe (GRAS) [44, 78]. It is a bacterium that is becoming more and more significant in global bioethanol production. It is a type of anaerobic bacterium that only thrives between the temperatures of 25 and 31 °C, with a growth drop at 15 °C [96]. Using cereals, unrefined sugar, sugarcane juice, and syrup to produce bioethanol, *Z. mobilis* has also been thoroughly explored [28, 75, 106, 135]. *Z. mobilis* solely metabolizes D-glucose, sucrose, and D-fructose as carbon and energy sources [12, 60]. *Z. mobilis* was found to have superior qualities during the manufacture of bioethanol compared to yeast by Bai et al. [12]. According to studies, *Z. mobilis* has been shown to produce 5–10% more ethanol than yeasts due to their higher tolerance to high ethanol concentrations

(> 16%) and high sugar concentrations in the fermentation solution and slow multiplication rate. As a result, it uses less energy than yeast and does not require more oxygen to complete the fermentation process [24, 97, 107].

The Entner-Doudoroff pathway is used by the anaerobic bacterium *Z. mobilis* to convert glucose into bioethanol. The process involves alcohol dehydrogenase and pyruvate decarboxylase [13, 36, 61, 135]. In contrast to the *S. cerevisiae* EMP pathway, which cleaves $C_6H_{14}O_{12}P_2$ by fructose bisphosphate aldolase to produce one molecule each of $C_3H_7O_6P$ and dihydroxyacetone phosphate, the Entner-Doudoroff pathway produces pyruvate and $C_3H_7O_6P$ by cleaving $C_6H_{11}O_9P$ to 2-keto-3-deoxy-gluconate aldolase to produce a molecule of ATP per glucose molecule [13, 61, 119]. Therefore, *Z. mobilis*, compared to *S. cerevisiae*, produces less biomass but focuses more on bioethanol fermentation carbon usage [12, 60]. The low ATP yield of *Z. mobilis* ensures that it produces ethanol at a rate typically 3–5 times more than that of *S. cerevisiae* while maintaining a high glucose flow down the Entner-Doudoroff route [13, 14, 45].

3.4 Escherichia Coli

Due to the physiology and several phenotypes that *E. coli* might adopt for synthesizing compounds with biotechnological value, researchers have extensively researched the genetic alteration of this organism [50]. Gene expression and regulation, as well as the accessibility of molecular tools for genetic engineering, are two unique advantages of *E. coli*. As a facultative anaerobic, *E. coli* can thrive heterotrophically in anaerobic and aerobic environments by utilizing various organic carbon sources [50, 71].

E. coli metabolizes glucose by using oxidative pentose phosphate and EMP pathways. A mole of glucose is converted into 2 mol of acetate, 2 mol of formate, and one mole of ethanol via the EMP route. In the presence of alcohol dehydrogenase (AdhE), acetyl-CoA is reduced into ethanol via acetaldehyde to produce ethanol. Two NADH molecules are used up in this process. Due to the usage of pyruvate formate lyase throughout the fermentation process rather than pyruvate decarboxylase, which catalyzes the reaction from pyruvate to acetaldehyde, yeast- or *Zymomonas*-like ethanol is distinct from that obtained from *E. coli*. OPPP serves as an oxidation route for NADPH synthesis. As a result, phosphotransacetylase (PTA) and acetate kinase convert acetyl-CoA to accumulated acetate [125]. The main drawback of utilizing *E. coli* as the ethanologenic microbe in fermentation is the co-production of products other than ethanol.

4 Various Fermentation Techniques

Depending on the kinetic parameters of the microbe and the kind of biomass substrate used during fermentation, several fermentation techniques and procedures can be employed to make bioethanol.

4.1 Mode of Fermentation

There are three primary fermentation methods for bioethanol production: batch, fed-batch, and continuous. The fermentation modes are based on the kinetic properties of the microorganism used and the biomass hydrolyzate [14, 81, 82, 128].

4.1.1 Batch Fermentation

The dominant and fundamental fermentation type is batch fermentation, which begins with the addition of high substrate concentrations and continues without interruption to yield a high quantity of the desired product [9, 81]. Batch fermentation can be pictured as a closed culture or closed-loop system [9, 127]. Here, microorganisms are inoculated in the fermentation media until the sugar present is depleted. The fermentation media comprises nutrients and other ingredients the microorganism utilizes [135]. It is the most modest bioreactor system with a multi-vessel, flexible, and simple control process. Complete sterilization, labor-skill-free operation, easy feedstock management, controllability, and adaptability to diverse product specifications are all benefits of batch fermentation systems [58, 62]. However, labor expenses are high and intensive, and productivity is low. A fermentation medium with excessive sugar levels could result in substrate inhibition preventing ethanol production and cell growth [9].

Repeated batch fermentation, also known as cell recycling batch fermentation, is a modification of batch fermentation that uses immobilized microbial cells in place of free cells to improve the efficiency of the system [10, 63, 82]. Cell recycling batch fermentation decreases inoculum preparation time and expense [9]. Additional benefits of repeated-batch methods include simple cell collection, dependable operation, and long-term manufacturing [9, 128].

4.1.2 Fed-Batch Fermentation

Fed-batch is an alternative fermentation method that combines batch and continuous fermentation. This process involves continuously and intermittently adding substrates to the fermentation media without removing the finished product [9, 81].

Fed-batch fermentation boosts productivity by retaining substrate at low concentrations while increasing the number of monomeric sugars that can be converted to ethanol [9, 62]. Substrate inhibition in batch fermentation technique has been addressed using this technique [9]. Given that it takes less time to ferment, produces more ethanol, employs less hazardous media ingredients, has more dissolved oxygen in the medium, and is supplied -batch fermentation costs less than non-batch fermentation [10, 30, 82]. The main disadvantage of fed-batch fermentation is that the amount of ethanol produced is limited by the feed rate and cell mass concentration [9, 88].

4.1.3 Continuous Fermentation

In this fermentation mode, nutrients, substrates, and culture medium are continuously introduced to the bioreactor, which contains active microorganisms, and products are continuously extracted at regular intervals [9, 81]. Typically, fermentation products, including bioethanol, cells, and leftover sugar, are usually removed. Continuous fermentation is mostly used to produce bioethanol at the industrial scale level [81]. Increased productivity, smaller bioreactor volumes, cheaper investment and running costs, simple process control, the elimination of tedious cleaning time, and a less labor-intensive process are all advantages of the continuous fermentation system over a batch or fed-batch fermentation system [10, 74, 82, 109].

4.2 Fermentation Classification Methods

Fermentation techniques may be categorized according to the kind of substrate employed. They are solid-state and submerged fermentation methods.

4.2.1 Solid-State Fermentation (SsF)

Solid-state fermentation (SsF) is a bioconversion technology in which a non-soluble substrate is a nutrient source and physical support for microbial cultivation and product production [36, 115]. The process occurs with little or no moisture [37, 102, 116, 120]. SsF is a 3-phase heterogeneous process with gaseous, liquid, and solids phases. The solid matrix may be used as a carbon source (together with other nutrients) or as an inert material to enable the growth of microbial cells (with impregnated growth solution) [37, 61, 116]. SsF has widely been employed to produce essential industrial enzymes, notably those from fungi and other microbes that need less moisture [60, 119]. The environment SsF offers farmed microorganisms, from which they are segregated, is as similar to their native environment as is practical. Comparing liquid fermentation carried out in a closed bioreactor to SsF allows microorganisms to work well and produce better product yields even in ideal growth and activity conditions [36]. The most important variables for microbial growth and activity in a certain substrate are moisture content and particle size [8, 104].

SsF has many benefits, such as the ability to directly process abundant, inexpensive biomaterials (starch, cellulose, lignin, hemicellulose, chitin, etc.) with little to no pretreatment, simplicity compared to submerged fermentation (SmF), and the ability to create specific microenvironments that are favorable to microbial growth and metabolic activities [131]. However, microorganisms with low water activity, such as bacteria, cannot be cultivated using this method [11, 119].

4.2.2 Submerged Fermentation (SmF)

Submerged fermentation (SmF), sometimes called liquid fermentation, involves microorganism inoculation in a liquid media to produce the desired products [115]. This method involves the release of bioactive substances into the fermentation medium [119]. Production of fermented industrial goods, such as lactic acid butanol-acetone, bioethanol, etc., can be produced by SmS [60]. SmF processes are categorized into anaerobic and aerobic processes. Enzymes and antibiotics are produced through aerobic fermentation, which occurs when oxygen is supplied into the liquid medium via SmF. The preferred fermentation operation in the industry is SmF because it incorporates well-known engineering characteristics, viz. bioreactor design, fermentation modeling, and process control [53, 60]. This fermentation method works best with bacteria and other microorganisms that require significant moisture [119]. Other advantages of SmF include easy automation and scale-up, reduced fermentation time, simpler safeguarding of ascetic conditions on an industrial scale, and simpler product purification [115, 119]. Poor productivity, medium concentration, and high production cost are the major drawbacks of SmF processes [10, 115].

4.3 Integrated Technologies

Another crucial stage in fermentation and hydrolysis is process design and integration. It involves either sequential biomass fermentation or simultaneous biomass fermentation and hydrolysis according to the conventional method [91, 109, 131]. The diverse fermentation processes can be combined to increase productivity, improve efficiency, and boost the economy while lowering production costs [90]. For the commercial manufacture of ethanol, a variety of integration methods have been modified, including separate or stepwise hydrolysis and fermentation (SHF), simultaneous or concurrent saccharification and fermentation/co-fermentation (SSF/SSCF), and consolidated bioprocessing (CBP) techniques [8, 14, 91, 109, 131].

4.3.1 Separate or Stepwise Hydrolysis and Fermentation (SHF)

The technique most frequently used to produce bioethanol from lignocellulosic material is SHF [81]. This entails individually hydrolyzing the pretreated biomass using enzymes before fermentation in different reactors [81, 108]. The primary benefit of SHF is the flexibility for ideal processing conditions at both the hydrolysis and fermentation stages to improve enzymatic hydrolysis and microbial fermentation [14, 82, 109]. The disadvantages of SHF include inhibition of the end-product of cellulolytic enzymes, which causes the hydrolysis rate to decrease gradually as cellobiose and glucose build up, high production costs, a lower yield of finished goods, and a higher risk of contamination [13, 81].

4.3.2 Simultaneous or Concurrent Saccharification and Fermentation (SSF/CSF)

Fermentation and saccharification occur concurrently in the same reactor in this integrated process [14, 82, 131]. This method maintains a low glucose concentration, preventing the accumulation of glucose inhibition on the enzyme function [81, 108]. SSF technology provides better ethanol yields while using less energy. Fermentation needs less enzyme, has a simpler process, and requires less equipment because it lowers end-product inhibition from glucose and cellobiose generated through enzymatic hydrolysis [14, 109, 131]. The constraints of SSF, however, include the difficulty of simultaneously tuning the conditions for hydrolysis and fermentation and the larger enzyme dose required, which has a favorable impact on substrate conversion but a negative impact on process costs [14, 109, 123, 131].

4.3.3 Simultaneous or Concurrent Saccharification and Co-Fermentation (SSCF)

This integrated process is an advancement in the SSF/CSF [13]. With the help of microorganisms, lignocellulosic biomass was pretreated and hydrolyzed, producing sugars. This integration approach seeks to digest all the sugars completely [130]. In SSCF, the glycolysis is accelerated by the continual discharge of hexose sugars from the enzymatic hydrolysis, resulting in a quicker and more efficient fermentation of the pentose sugars [13, 54]. The primary issue with this method is that hexose-utilizing microbes reproduce more quickly than pentose-using microorganisms, which increases the conversion of hexoses to bioethanol. Although naturally occurring microbes can employ both pentoses and hexoses, using only one that optimally assimilates both would result in high conversion and ethanol yield [17, 121, 131].

4.3.4 Consolidated Bioprocessing (CBP)

CBP is referred to as direct microbial conversion (DMC). This is another integration process; however, it only uses one microbial population to produce enzymes, hydrolyzes cellulose, and ferment carbohydrates [27, 82, 131]. This implies that a particular microorganism will conduct both fermentation and hydrolysis. As a result, CBP is preferable to alternative integration processes because it requires neither capital nor operating expenses for the system's synthesis of enzymes [82, 83, 131]. Thermophilic cellulolytic anaerobic bacteria (TCAB) viz. *Thermoanaerobacter ethanolicus, Thermoanaerobium brockii, Clostridium thermohydrosulfuricum, T. mathranii, C. thermosaccharolyticum* strain and fusarium oxysporum, a filamentous fungus, that can bioethanol from lignocellulosic materials in the CBP system [2]. The ability of TCAB to convert inexpensive biomass into ethanol and endure high temperatures makes them more favorable than yeast strains [82, 83, 131]. Due to their limited tolerance of bioethanol (2% v/v), these TCABs are not appropriate for industrial bioethanol production [15, 27, 82, 83, 130, 131].

5 Bioethanol Production System Modeling and Optimization

A mathematical model enables optimization of variables that affect the overall productivity of a process, reducing production costs and lowering energy consumption [37]. Hence, a logical understanding of different models is crucial for process optimization, modification, and scale-up. Using one variable at a time (OVAT) has some limitations, such as being time-consuming and requiring substantial resources to describe the process accurately. This challenging approach overlooks the interactions between the variables and their cumulative impact on the process [17]. The optimization method involves varying a factor at a time while retaining others. Suitable mathematical modeling techniques are employed to effectively model and optimize the bioethanol production system. These models offer greater insight into the fermentation process, experimental accuracy, and less variations from the optimal solution. The modeling of the bioethanol production process can be achieved via statistical designs embedded with statistical techniques, such as RSM, and machine learning tools, such as ANFIS and ANN. The optimization of the process to obtain optimal reaction conditions can be accomplished by employing the local optimization algorithm of RSM. In contrast, global optimization can be conducted using ant colony optimization, particle swarm optimization (PSO), Cuckoo search, and genetic algorithm (GA), amongst others. Optimization tools are applied to determine the specific input levels to maximize the output. Figure 1 shows the process modeling overview using either the experimental design method with RSM or machine learning technology to maximize bioethanol production.

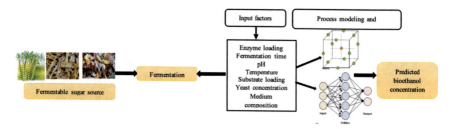

Fig. 1 Factors examined for bioethanol production using machine learning techniques

5.1 Factorial Design

Different combinations of factors and levels are frequently included in factorial designs, allowing for the screening and factor identification with the most significant influence on the response. Such techniques help do exploratory research or the first optimization stages [134]. Either actual or coded factor levels can be used to perform factorial designs. The coded factor levels are recommended because they offer a consistent framework for analyzing how a factor affects experimental results. Additionally, the coded factors are dimensionless. Consequently, they can be compared side by side. Plackett-Burma design (PBD) and Taguchi designs are commonly employed factorial designs in bioethanol production system modeling.

In some instances, bioethanol production may require screening important process factors to determine the pertinent process variables to ensure optimal bioethanol production. This approach helps filter out large chunks of undesired factors, reducing the number of factors. The PBD tool is a 2-level design developed to screen a long list of factors/variables. It was designed to improve the quality control procedure that could be utilized to research how design parameters affect system state and enable wise decisions to be made. It has been used in maximizing the amount of bioethanol produced from cassava bagasse [117], cassava peels [116], kitchen garbage [83], glycerol [34], and corn stover [85].

The Taguchi approach is an experimental design method employed for system design and process optimization studies. This method can reduce the overall number of experiments while ensuring the procedure is accurate and consistent. It is a factorial-based technique that assigns the factors chosen for an experiment using an orthogonal array (a collection of tests conducted in various contexts). The Taguchi approach has only been employed in limited research to optimize process variables in bioethanol production [38, 39]. Darvishi et al. [38] employed the Taguchi approach to boost a molasses-based industrial medium to manufacture bioethanol. The Taguchi method was used to optimize the primary factors of a medium, such as molasses, $(NH_4)_2SO_4$, urea, and pH, to maximize bioethanol production. Following medium optimization, bioethanol production in flask culture reached 10% (v/v). Das et al. [39] employed the Taguchi design to statistically optimize the different process parameters to produce bioethanol from a microwave-aided base and organosolv-pretreated wild grass mixture. The volume of mixed recombinant *C. thermocellum* hydrolytic enzymes and the volume, temperature, and pH of the inoculum of mixed fermentative bacteria were all investigated factors. The highest amount of bioethanol that was produced was 2.0 g/L under the optimum conditions of pH 4.3; 1.0, recombinant GH5 cellulase (5.7 m/g, 0.45 mg/mL); 1.5, *S. cerevisiae* (3.9×10^8 cells/mL); 0.25, *C. shehatae* (2.7×10^7 cells/mL); 35 °C of temperature; and 2.0, recombinant GH43 hemicellulase (3.7 U/mg, 0.32 mg/mL).

5.2 Response Surface Methodology (RSM)

RSM is an optimization method utilized in experiment design that combines a variety of mathematical and statistical methods frequently utilized to develop an empirical model based on experimental data gathered as part of the experimental design [46]. The primary objective is to maximize an output of interest that is induced by numerous control variables. In order to specify the provided process and attain this goal, linear or polynomial relationships are created. Defining an approximation function between the dependent and independent variables is the first step in using the RSM. This function, in some cases, can have a second or higher-order polynomial relationship, while in other cases, a linear function can also be generated between the output and the input in which the function of approximation is then referred to as the first-order model [70]. The flowchart to execute the RSM model is shown in Fig. 2.

The expression for a first-order model is given as

$$U = \alpha_0 + \sum_{i=1}^{n} \alpha_i V_i + \epsilon \quad (2)$$

While for a second-order or higher-order polynomial, a multiple regression equation is given as

$$U = \alpha_0 + \sum_{i=1}^{n} \alpha_i V_i + \sum\sum_{i<j} \alpha_{ij} V_i V_j + \sum_{i=1}^{n} \alpha_{ii} V_i^2 + \epsilon \quad (3)$$

where U is the predicted output, V_i ($i = 1, 2, ..., n$) is the first-order model coefficients for λ_i i is the linear term coefficient, j is the quadratic term coefficient, α_0 is the value of the intercept, α_{ij} is the interaction coefficients for $V_i V_j$, and α_{ii} represents the quadratic coefficients of V_i n is the number of control parameters and ϵ is random experimental error with zero mean value assumption [52].

Some experimental designs have been used for investigating the bioethanol production processes. Earlier studies have based their screening approach on the PBD, central composite design (CCD), Box–Behnken design (BBD), and Taguchi orthogonal experimental design. The commonly employed software packages for investigating different design methods include Statistica (StatSoft), MATLAB, Design Expert (Stat-Ease, Inc), and Minitab (Minitab Inc.).

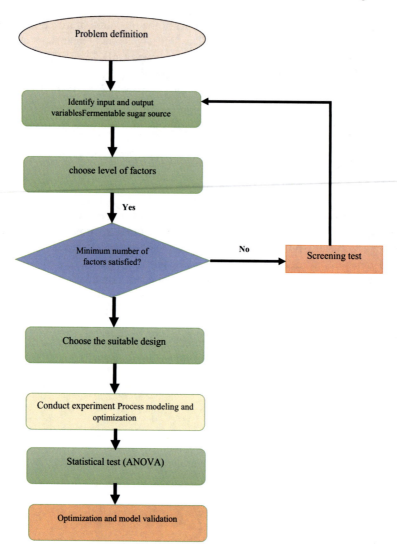

Fig.2 Flow chart for RSM execution

Box and Behnken [24] put forward the development and description of BBD. This design was employed to reduce the needed experimental runs in the fitting quadratic model. In this method, the number of factors or input variables has to be ≥ 3, and the BBD analysis should only fit the second-order polynomial equation [49]. Sivamani and Baskar [116] applied BBD to design experiments to produce bioethanol from cassava peel. The initial method for screening the design variables was the PBD. The optimal condition was attained after validation under the maximum bioethanol production of 35.010 g/L and consisted of α-amylase concentration of 24.74% (v/

v), substrate concentration of 69.82 g/L, and 5.22% (v/v) concurrent saccharification and fermentation mixture.

In the report of Kamal et al. [65], BBD was employed to optimize bioethanol production using co-culture yeast with an inoculum size ratio of 3:2 mL (*S. cerevisiae*: *C. cantarelli*) at a temperature of 35 °C and pH of 6 to obtain 92.5 g/L of bioethanol. To increase the yield of bioethanol from potato peel wastes, Hossain et al. [57] used PBD first and then BBD to improve bioethanol yield from potato peel wastes. According to the results, only KH_2PO_4, tryptone, and malt extract had a discernible impact on bioethanol production. The authors then optimized these factors using the BBD. When potato peel waste was used as the substrate, only extract from malt positively influenced ethanol synthesis. Another potato residue was investigated for *S. cerevisiae* bioethanol synthesis by Izmirlioglu and Demirci [59]. They used a previously described technique to maximize the fermentation of commercially hydrolyzed potato waste to produce bioethanol. To examine the impact of the factors, they first conducted PBD. Yeast extract, $MgSO_4 \cdot 7H_2O$, and malt extract positively impacted the bioethanol production, while $CaCl_2 \cdot 2H_2O$ and KH_2PO_4 had negative impacts. The authors chose the parameters with positive impact to maximize bioethanol yield using BBD to develop a quadratic polynomial relation for bioethanol production and data for cell population. The medium composition used to produce 24.6 g/L of ethanol contained 4.84 g/L $MgSO_4 \cdot 7H_2O$ and 50.0 g/L malt extract excluding yeast extract from the medium.

Jugwanth et al. [64] used BBD in optimizing concurrent saccharification and fermentation of sugarcane bagasse treated with alkali salt. The studied process input variables were yeast titer temperature and enzyme loading. A second-order model generated showed a p-value (0.0203) and F-value (5.21), implying the significance of the model. The prediction by the RSM model was adequate, as shown by the calculated coefficient of determination ($R^2 = 0.87$). At approximately optimal conditions of a temperature of 39 °C, yeast titer of 1, and enzyme load of 100 U/g, the highest ethanol concentration was 4.88 g/L. Dasgupta et al. [40] improved the fermentation of sugarcane bagasse with *Kluyveromyces sp.* IIPE453 to increase the production of ethanol. The authors initially used PBD to filter out the important variables that had a substantial impact on the production of ethanol. The substrate content, sodium dihydrogen phosphate concentration, pH, and fermentation time were then optimized using the BBD to maximize bioethanol production. The optimal conditions that gave maximum bioethanol concentration (17.39 g/L) are pH 4.5, Na_2HPO 0.15 g/L, fermentation time of 48 h, and substrate 40 g/L. The optimal conditions were then used to validate the model experimentally. The maximum bioethanol content was 17.44 g/L, which agrees with the model prediction. In the report of Betiku and Taiwo [18], a two-step enzymatic process was used to hydrolyze the starch from breadfruit to simple sugars. The yeast in this study could use hydrolysate from breadfruit as the carbon source to produce bioethanol. The yeast could ferment the hydrolysate both with and without dietary supplementation. However, the medium with additives showed a higher bioethanol output. The best conditions for the ethanol yield utilizing BBD were pH 5.01, hydrolysate 134.81 g/L, and time 21.33 h, with a predicted bioethanol production of 3.95% volume fraction and validated as 4.10% volume fraction in the laboratory.

CCDs are first-order (2N) designs with extra center and axial points added to them so that the tuning parameters of a second-order model can be estimated [91]. It is common practice to employ the central composite design, an effective statistical method with a hypercube geometry region, to reduce the number of experiments while preserving statistical significance [28]. It typically consists of the following components: (i) a complete or partial 2^N factorial design with factor levels coded as ($-1, +1$), known as the factorial portion; (ii) an axial portion consisting of 2N points arranged so that two points are selected on each of the axes of the control factor away from the design center (selected as the point at the origin of the coordinate system); and (iii) no center points [69]. Gawande and Patil [51] investigated bioethanol production from marred maize grain using SSF and a mixture of *S. cerevisiae* and *A. niger*. The study used CCD to examine three factors (pH, temperature, and substrate concentration). Response surface analyses determined that the best conditions to produce the highest amount of bioethanol of 4.24 (g/100 ml) were pH (5.6), temperature (31 °C), and 14% substrate concentration with 0.88 g/l/h ethanol productivity. Mazaheri and Pirouzi [89] enhanced fermenting procedure to produce bioethanol derived from potato peel (PP) waste. Enzymatic hydrolysis of the waste was initially conducted to release the fermentable sugars. CCD was utilized to examine the combination of process variables (time, initial sugar concentration, yeast extract, bacterial *Z. mobilis* dry weight, and peptone concentrations) on bioethanol production. The highest concentration of bioethanol attained was 23.3 g/L at optimum conditions of peptone weight (0.35 g), sugar concentration (61.3 g/L), fermentation time (31 h), bacterial dry cell (0.024 g), and yeast extract (0.35 g). Althuri and Banerjee [3] used RSM to achieve the optimal conditions for SHF and SSF in maximizing bioethanol production. The CCD generated 31 and 32 experimental runs for SSF and SHF, respectively. The same experimental conditions were used for the SHF and SSF, and these include inoculum age (24–72 h), temperature (30–40 °C), inoculum size (8–12% v/v), and incubation time (18–30 h). The supplementation of the substrate (15–25% w/v) was included in the SSF. The bioethanol content for SHF and SSF were 25.49 and 41.9 g/L, respectively. Table 1 summarizes the various design methods and their benefits and drawbacks. Table 2 includes previous studies that optimize various process conditions to increase ethanol output.

5.3 Machine Learning Approach to Modeling Bioethanol Production

Data-mining machine learning techniques offer a substitute for conventional modeling methodologies to deal with ambiguity, nonlinear, complex, and multiple variable behaviors of bioethanol production processes. The two techniques frequently employed in bioethanol production are ANN and ANFIS.

Table 1 Different design methods, benefits, and limitations

Design	Feature	Benefits	Drawbacks
PBD	It only examines the primary effects of factors. There are between 8 and 35 different variables that can be screened	It is chiefly employed for screening test	It cannot be used to perform optimization
Taguchi design	It uses orthogonal array. It is possible to examine the effects of many variables and levels with fewer experimental runs	Reduces cost by fewer experimental runs	Interaction between parameters cannot be accounted for
BBD	Its design is of three levels. It is used to fit second order polynomial model	Synergetic effects of parameters can be studied on the desired response. Satisfactory prediction is provided within the design space	Only limited to three levels. It excludes runs in which each factor is at its maximum or minimum level
CCD	It is a 5-level design that has 2 × factorial runs, 2 × x axial designs and n number of center points. The number of factors studied can range between 2 and 6	Interaction between the studied factors can be examined. It includes runs in which each factor is at its extreme points. It can accurately estimate first or second order polynomial models	More experimental run than BBD

ANN is the mathematical modeling of the network of interlinked nerve cells that describes how the human nervous system of the brain functions [94, 113]. Typically, a neural network comprises one input layer, one or multiple hidden layers, and one output layer. The first layer is sometimes called the input layer. It contains neurons communicating with the hidden layer neurons through synapses before sending information to the output layer neurons through other synapses. The neurons of the hidden layers help the network create complex associations between input and output parameters. The ANN can effectively handle nonlinear behaviors [32, 93]. ANN has also been used on various substrates to optimize bioethanol production. Betiku and Taiwo [18] utilized ANN to investigate the impact of pH, fermentation time, and breadfruit hydrolysate on ethanol production. The model had R^2 of 1, which shows that the model accurately predicted bioethanol volume yield.

Sebayang et al. [112] investigated the influence of reaction temperature, yeast concentration, and agitation speed as input parameters on bioethanol production with sorghum grains as biomass feedstock. The model is suitable for reliably predicting the bioethanol concentration, as evidenced by an R^2 of 0.98. Chouaibi et al. [31] examined the impact of shaking speed, yeast concentration, fermentation temperature, and pH on ethanol production employing ANN. The estimated correlation

Table 2 Experimental design approach to model and optimize bioethanol production

Design method	Fermentable source	Input variables	Response	Optimum conditions	Maximum bioethanol concentration	Reference
Taguchi	Molasses	Molasses, ammonium sulfate, urea, and pH	Bioethanol volume yield	Molasses (21% brix), ammonium sulfate (4 g/L), urea (5), and pH (5)	10% (v/v)	Darvishi and Abolhasan Moghaddami [38]
	Wild grass	Temperature, recombinant GH43 hemicellulase, recombinant GH5 cellulase, C. shehatae, pH, and S. cerevisiae	Bioethanol concentration	pH 4.3, temperature 35 °C, C. shehatae 0.25% (v/v), S. cerevisiae 1.5% (v/v), recombinant GH5 cellulase 1% (v/v), recombinant GH43, and hemicellulase 2% (v/v)	2.0 g/L	Das et al. [39]
BBD	Cassava peel	α-amylase concentration, substrate concentration, simultaneous saccharification, and fermentation mixture	Bioethanol concentration	α-amylase concentration, 24.74% (v/v), SSF, 5.22% (v/v), and substrate concentration, 69.82 g/L	35.01 g/L	Sivamani and Baskar [116]
	Corn stover	Inoculum size ratio, pH, and temperature	Bioethanol concentration	Inoculum size ratio, 3:2 mL pH, 6, and temperature, 35 °C	92.5 g/L	Kamal et al. [65]
	Potato peel waste	KH_2PO_4 concentration, malt extract, and tryptone	Bioethanol concentration	Tryptone 10 g/L, KH_2PO_4, 2.5 g/L, and malt extract, 25 g/L	21.3 g/L	Hossain et al. [57]
	Potato waste	$MgSO_4 \cdot 7H_2O$ concentration yeast extract, and malt extract	Bioethanol concentration	$MgSO_4 \cdot 7H_2O$ concentration, 4.84 g/L Yeast extract, 0.0 g/L, and malt extract, 50.0 g/L	26.6 g/L	Izmirlioglu and Demirci [59]
	Sugarcane bagasse	Yeast titer, temperature, and enzyme loading	Bioethanol concentration	Yeast titer, 1 time, temperature, 39 °C, and enzyme loading, 100 U/g	4.88 g/L	Jugwanth et al. [64]

(continued)

Table 2 (continued)

Design method	Fermentable source	Input variables	Response	Optimum conditions	Maximum bioethanol concentration	Reference
	Sugarcane bagasse	Fermentation time, Na_2HPO_4 concentration, pH, and substrate concentration	Bioethanol concentration	Fermentation time, 48 h, pH, 4.5, substrate concentration, 40 g/L, and Na_2HPO_4 concentration, 0.15 g/L	14.4 g/L	Dasgupta et al. [40]
	Breadfruit starch	Breadfruit starch hydrolysate concentration, time, and pH	Bioethanol yield	pH, 4.5, time, 24 h, and breadfruit starch hydrolysate concentration, 120 g/L	4.10%	Betiku and Taiwo [18]
CCD	Corn grains	Substrate concentration, pH, and temperature	Bioethanol productivity	Substrate concentration, 12%, pH, 5.6, and temperature, 31 °C	0.88 g/L/h	Gawande and Patil [51]
	Potato peel wastes	Meat peptone, bacterial dry cell, initial sugar concentration, time, and yeast extract	Bioethanol concentration	Time, 31 h, sugar concentration 61.3 g, bacterial dry cell, 0.024 g, meat peptone, 0.35 g, and yeast extract, 0.35 g	23.3 g/L	Mazaheri and Pirouzi [89]
	Lignocellulosic and biomass mixture	Inoculum volume and inoculum age, temperature, incubation time, and substrate load variable in the case of SSF	Bioethanol concentration	SHF: inoculum age, 24 h, incubation time, 27.33 h, temperature, 38.18 °C, and inoculum volume, 8% v/v SSF: inoculum age, 44.84 h, incubation time, 30 h, temperature, 38.18 °C, substrate concentration, 25%, and inoculum volume, 8% v/v	SHF: 25.49 g/L SSF: 41.9 g/L	Althuri and Banerjee [3]

coefficient (R) was 0.9999, while the value of root means square error (RMSE) was determined for testing, training, and validation as 0.989, 0.7968, and 0.05924, respectively, showing that the implemented ANN model was reliable. Sathendra et al. [111] used an ANN model to predict the bioethanol yield produced using palm wood hydrolysate. The studied input parameters are inoculum size, pH, temperature, substrate concentration, and agitation rate. The R-value obtained was 0.96213, which shows that the model could adequately predict the bioethanol yield. In the work of Dave et al. [41], ANN was designed to investigate the effects of the following factors on the bioethanol yield produced from locally available marine macroalgal biomass: agitation rate, substrate concentration, fermentation time, inoculum size, temperature, and pH. The estimated R was 0.96015, suggesting that the model was reliable. ANN was employed to predict bioethanol productivity from cashew apple juice in the report of da Silva et al. [37]. The process input parameters studied were substrate concentration, cell concentration, temperature, and stirring speed. The apparent error estimated for the process was 8.3%, indicating that the model is efficient and capable of predicting the examined bioethanol productivity. Jahanbakhshi and Salehi [61] utilized the ANN to predict bioethanol concentration from watermelon waste. The studied input parameters were agitation speed and different levels of yeast. The R^2 for the model was 0.9895 showing that the model has a prediction ability for bioethanol production. The flowchart for modeling with ANN is shown in Fig. 3. Fuzzy systems paired with neural networks are used to create ANFIS. It combines the ideas from both methodologies into a single concept, which could allow it to benefit from the complementary strengths [132].

These advantages include neural network processing power, easy learning, and the ability of fuzzy systems to illustrate uncertainty [87]. Contrary to many other study areas, studies of the bioethanol production process have few reports on utilizing ANFIS. The detailed development of the ANFIS model has been described elsewhere [6]. Ezzatzadegan et al. [47] used ANFIS to examine the impact of input parameters (glucose content, fermentation time, and temperature) on bioethanol concentration from corn stover. The developed ANFIS model had R^2 and RMSE values of 0.99 and 0.637, respectively, which shows that the developed model accurately predicted the bioethanol concentration. In another recent report by the same authors [48], the model was developed to predict bioethanol concentration made from oil palm trunk sap adequately. The studied input parameters were sugar content, pH, fermentation time, and temperature. The estimated R and RMSE were 0.997 and 0.3447, respectively, showing that the developed ANFIS model is precise and reliable. Jahanbakhshi and Salehi [61] employed the ANFIS model to predict bioethanol concentration from watermelon waste using agitation speed and different yeast amounts as the process input parameters. The estimated R^2 for the developed ANFIS model was 0.9993, indicating a good predictive capability of the model for the bioethanol production system. The flow chart to implement ANFIS modeling is described in Fig. 4.

Fig. 3 Flow chart for typical neural network execution

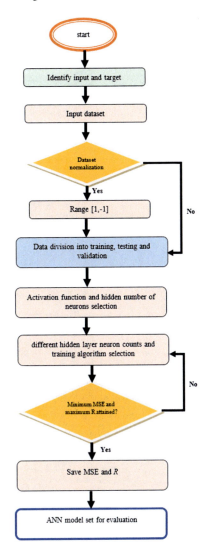

5.4 Process Optimization of Bioethanol Production Process Using Global Approaches

When the process is modeled using ANN or ANFIS, global optimization approaches, viz. ant colony optimization (ACO), PSO, and GA, must be employed to get the optimal conditions for bioethanol production. Unlike RSM modeling, which has an inbuilt optimization algorithm and searches the solution space within the given range (local optimization), ANN or ANFIS needs to be connected to either GA, ACO, or PSO, etc., which then searches the solution space globally (i.e., within and outside

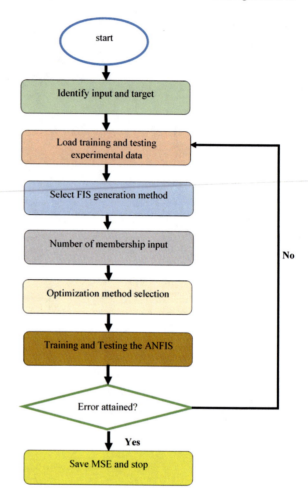

Fig. 4 Flow chart for typical ANFIS execution

the input variable ranges). PSO is a population-based evolutionary algorithm optimization method that produces globally optimal solutions for complicated non-linear bioprocesses more rapidly than previous statistical optimization methods [43]. Due to its non-deterministic bioinspired population optimization problem, PSO can be used for optimizing non-linear problems in a multivariate system. It possesses characteristics similar to how real things move to achieve the best results, including birds flocking and fish schooling [19, 37]. In PSO, those interacting parameters will promote helpful communication, information sharing, low-cost computing, and simple implementation. PSO has been used mainly in the optimization of biofuel. In bioethanol production, PSO has been integrated with ANN in ethanol production from apple cashew juice [37]. The optimum conditions estimated when ANN was coupled with PSO in this study were initial substrate concentration (127 g/L), initial cell concentration (5.8 g/L), temperature (35 °C), agitation speed (111 rpm), and fermentation time of 7 h. The maximum productivity and efficiency obtained were 8 g/L/h and 91.5%, respectively. Ezzatzadegan et al. [47] integrated PSO with an ANFIS model to obtain the highest ethanol concentration of 30.01 g/L under the favorable conditions of glucose content (0.23 g/L), temperature (34.5 °C) and fermentation time (69.39 h). Similarly, in another report by these authors [48], the maximum bioethanol concentration (44.1484 g/L) was obtained under the optimal conditions of sugar content (1.3588 g/L), pH (4.54), fermentation time (16.18 h) and temperature (27.3 °C).

GA is another evolutionary optimization approach. A stochastic optimization tool called a genetic algorithm can be used to maximize or minimize a particular function to search for the best solution to a particular computational problem. This optimization technique is based on the natural selection of Darwin's theory and the survival of the fittest by imitating the biological process of reproduction [63, 117]. By using genetic operators like population size, mutation, selection, and crossover that imitate the principle of natural evolution, it evolves through generations of subsequent iterations starting with a population of initial sets of solutions, with each population represented by a set of chromosomes or an individual [63, 113]. In order to get a close match, the two best individual chromosomes are chosen as the parents and then arbitrarily united via "crossing over," which mimics the biological process of natural selection. The ability of the algorithm to explore is increased because, during the crossover, parents are encouraged to mate to create a single new offspring. A prospective global optimal solution is created repeatedly until an optimum threshold is reached [113]. In contrast to previous search techniques, the GA acts over the population, producing multiple solutions simultaneously. It can be coupled with various statistical models to achieve optimal solutions such as ANN and ANFIS. In the work of [18], ANN was coupled with GA to maximize bioethanol yield. The maximum bioethanol yield obtained was 4.21% under optimum conditions of pH (4.5), breadfruit hydrolysate concentration (120 g/L), and hydraulic retention time (24 h) and [111] coupled the developed ANN model in their study in order to maximize bioethanol concentration. The best combination of input parameters of temperature (45 °C), substrate concentration (8% (v/v)), inoculum size (3.2% (v/v)), agitation rate (156 rpm), pH (5), and gave maximum bioethanol concentration of 22.90 g/L. ANN-GA was utilized to optimize ethanol production from *Ulva prolifera* biomass [41]. Utilizing 10% (v/v) yeast inoculum, pH 6.0, substrate concentration 30 g/L, and no extra nutrients under the optimal process parameters of 50 rpm, 30 °C, for 48 h, the maximum experimental bioethanol yield was 0.242 g/g.

The foraging concept of ant colonies is employed in ant colony algorithms. Ants can determine the quickest route from food to their nest. The technique is used to solve optimization issues, and this is accomplished by ants communicating with one another inadvertently by leaving chemical pheromone trails [22]. ANN has been integrated with ACO in the study of Sebayang et al. [112]. The best combination of the process input factors attained was agitation speed (181 rpm), yeast concentration (1.3 g/L), and reaction temperature (35.6 °C), with the highest bioethanol concentration of 82.11 g/L.

5.5 Ranking of ANFIS, ANN, and RSM for Bioethanol Production

The predictive capabilities of RSM, ANN, and ANFIS can be assessed using a variety of statistical measures, including R, R^2, mean square error (MSE), root MSE (RMSE), and average absolute deviation (AAD), along with others. These indicators have proved efficient in assessing the predictive capability of the studied models.

Betiku and Taiwo [18] assessed the performance capacity of RSM and ANN modeling and optimizing bioethanol production from breadfruit. For the RSM model, R^2 (0.9882) and AAD (1.67367) were obtained, while for the ANN model, R^2 (0.9930) and AAD (0.190823) were estimated. The authors concluded that the ANN was better than the RSM. Similarly, Chouaibi et al. [31] employed ANN and RSM in modeling the bioethanol production system with pumpkin peel wastes. These authors reported that the ANN model with R (0.9849) was superior to the RSM model with R^2 (0.9762). Furthermore, Sathendra et al. [111] examined the RSM and ANN models in bioethanol production. The estimated values of R^2 for both RSM and ANN models were 0.8300 and 0.9257. These results revealed that the ANN model was more precise than the RSM model. The performance ability of the ANN and ANFIS model for bioethanol production was also conducted in the report of Jahanbakhshi and Salehi [61]. It was concluded that the ANFIS model with an estimated R^2 value of 0.9993 was superior to the ANN model with an R^2 value of 0.9895. Nevertheless, both models can accurately prognosticate the concentration of bioethanol. The previously reported works on the processes to model and optimize ethanol production are described in Table 3.

6 Conclusion and Future Investigations

Bioethanol has been shown to be a potential biofuel that can substitute for fossil fuels. This review provided a complete analysis of the modeling and optimization of the fermentation process used to produce bioethanol based on investigations utilizing various substrates. This work also provided information on the steps involved during bioethanol production, including pretreatment, hydrolysis, fermentation, and distillation. In addition, common microorganisms used for bioethanol production were reviewed. These microorganisms comprise *S. cerevisiae*, due to its high ethanol tolerance across a broad pH range, *K. marxianus*, due to its higher specific growth rate and its thermotolerant nature at temperatures between 44 and 52 °C, *Z. mobilis*, a bacterium that produces ethanol that is 5–10% more than yeast and uses less energy, and *E. coli*, a facultative gram-negative bacterium that thrives heterotrophically in an aerobic and anaerobic environment. Besides, the various modeling techniques widely utilized in bioethanol production were highlighted in this review, along with how these techniques may be used to describe the impacts of the fermentation conditions considered on the substrates used and the bioethanol produced.

The efficacy of various modeling and optimization tools often used for bioethanol production systems was discussed. In conclusion, more studies need to be conducted using machine learning or data-driven techniques due to their ability to capture the nonlinear nature of the bioethanol production system.

Table 3 Summary of ANN and ANFIS-based modeling and optimization

Model	Fermentation source	Input parameter	Best prediction accuracy	Optimum condition	Maximum bioethanol concentration/ yield/ productivity	Reference
ANN	Breadfruit hydrolysate	pH, hydraulic retention time, and breadfruit hydrolysate concentration	$R^2 = 1.000$	Breadfruit hydrolysate concentration (120 g/L), hydraulic retention time (24 h), and pH (4.5) with GA	4.21% yield	Betiku and Taiwo [18]
	Sorghum grains	Agitation speed, yeast concentration, and temperature	$R^2 = 0.9800$	Agitation speed (181 rpm), yeast concentration (1.3 g/L), and temperature (35.6 °C) with ACO	82.11 g/L concentration	Sebayang et al. [112]
	Pumpkin peel waste	Yeast concentration, temperature, shaking speed, and pH	$R = 0.999$	pH (5.06), yeast concentration (1.95 g/L), temperature (45 °C), and shaking speed (188.5 rpm) The optimization technique used was not reported	84.36 g/L concentration	Chouaibi et al. [31]
	Palm wood hydrolysate	pH, inoculum size, temperature, substrate concentration, and agitation rate	$R = 0.9621$	pH (5), inoculum size (3.2% (v/v)), substrate concentration (8% (v/v)), temperature (45 °C), and agitation rate (156 rpm) with GA	22.90 g/L concentration	Sathendra et al. [111]

(continued)

Table 3 (continued)

Model	Fermentation source	Input parameter	Best prediction accuracy	Optimum condition	Maximum bioethanol concentration/ yield/ productivity	Reference
	Marine macroalgal biomass	Agitation speed, fermentation time, substrate concentration, pH, inoculum size, and temperature	$R = 0.96015$	Fermentation time (48 h), substrate concentration (30 g/L), temperature (30 °C), agitation speed (50 rpm), and inoculum size (10% (v/v)) with GA	0.242 g/g yield	Dave et al. [41]
	Cashew apple juice	Initial substrate concentration, temperature, initial cell concentration, and stirring speed	Apparent error = 8.3%	Initial substrate concentration (127 g/L), initial cell concentration (5.8 g/L), temperature (35 °C), agitation speed (111 rpm), and fermentation time (7 h) with PSO	8 g/L/h productivity	da Silva Pereira et al. [37]
ANN and ANFIS	Watermelon waste	Different level of yeast and agitation speed	For ANN $R^2 = 0.9895$ For ANFIS, $R^2 = 0.9993$	Agitation speed (120 rpm) and different level of yeast (5 g) for both models No optimization technique was mentioned	35.5 g yield	Jahanbakhshi and Salehi [61]
ANFIS	Corn stover	Glucose content, temperature, and fermentation time	$R^2 = 0.9900$	Glucose content (0.23 g/L), temperature (34.5 °C), and fermentation time (69.39 h) with PSO	30.01 g/L yield	Ezzatzadegan et al. [47]

(continued)

Table 3 (continued)

Model	Fermentation source	Input parameter	Best prediction accuracy	Optimum condition	Maximum bioethanol concentration/ yield/ productivity	Reference
	Palm trunk sap	Sugar content, pH, fermentation time, and temperature	$R = 0.9970$	Sugar content (1.3588 g/L), pH (4.54), time (16.18 h), and temperature (27.3 °C) with PSO	44.1485 g/L concentration	Ezzatzadegan et al. [48]

References

1. Abo BO, Gao M, Wang Y, Wu C, Ma H, Wang Q (2019) Lignocellulosic biomass for bioethanol: an overview on pretreatment, hydrolysis and fermentation processes. Rev Environ Health 34(1):57–68
2. Ali SS, Nugent B, Mullins E, Doohan FM (2016) Fungal-mediated consolidated bioprocessing: the potential of *Fusarium oxysporum* for the lignocellulosic ethanol industry. AMB Express 6(1):1–13
3. Althuri A, Banerjee R (2019) Separate and simultaneous saccharification and fermentation of a pretreated mixture of lignocellulosic biomass for ethanol production. Biofuels 10(1):61–72
4. Amelio A, Van Der Bruggen A, Lopresto BC, Verardi A, Calabro V, Luis P (2016) Pervaporation membrane reactors: biomass conversion into alcohols place. Woodhead publishing, pp 331–381
5. Amit K, Nakachew M, Yilkal B, Mukesh Y (2018) A review of factors affecting enzymatic hydrolysis of pretreated lignocellulosic biomass. Res J Chem Environ 22:62–67
6. Arafeh L, Singh H, Putatunda SK (1999) A neuro fuzzy logic approach to material processing. IEEE Trans Syst Man Cybern Part C (Appl Rev) 29 (3):362–370
7. Arora R, Behera S, Kumar S (2015) Bioprospecting thermophilic/thermotolerant microbes for production of lignocellulosic ethanol: a future perspective. Renew Sustain Energy Rev 51:699–717
8. Auria R, Palacios J, Revah S (1992) Determination of the inter-particular effective diffusion coefficient for CO_2 and O_2 in solid state fermentation. Biotechnol Bioeng 39:898–902
9. Azhar SHM, Abdulla R, Jambo SA, Marbawi H, Gansau JA, Faik AAM, Rodrigues KF (2017) Yeasts in sustainable bioethanol production: a review. Biochem Biophys Rep 10:52–61
10. Babbar N, Oberoi HS (2014) Enzymes in value-addition of agricultural and agro-industrial residues. In: Brar SK, Verma M (eds) Enzymes in value-addition of wastes. Nova Publishers, pp 29–50
11. Babu KR, Satyanarayana T (1996) Production of bacterial enzymes by solid state fermentation. J Sci Ind Res 55:464–467
12. Bai FW, Anderson WA, Moo-Young M (2008) Ethanol fermentation technologies from sugar and starch feedstocks. Biotechnol Adv 26:89–105
13. Balat M (2008) Global trends on the processing of biofuels. Int J Green Energy 5(3):212–238
14. Balat M (2011) Production of bioethanol from lignocellulosic materials via the biochemical pathway: a review. Energy Convers Manage 52:858–875
15. Balat M, Balat H, Öz C (2008) Progress in bioethanol processing. Prog Energy Combust Sci 34:551–573
16. Banerjee S, Mudliar S, Sen R, 1 Giri, D Satpute, T Chakrabarti, RA Pandey (2010) Commercializing lignocellulosic bioethanol: technology bottlenecks. Biofuels, Bioproducts Biorefining 4:77–93

17. Baş D, Boyaci HI (2007) Modelling and optimization II: comparison of estimation capabilities of response methodology with artificial neural networks in a chemical reaction. J Food Eng 78:846–854
18. Betiku E, Taiwo AE (2015) Modeling and optimization of bioethanol production from breadfruit starch hydrolyzate vis-a-vis response surface methodology and artificial neural network. Renewable Energy 74:87–94
19. Bhattacharya S, Dineshkumar R, Dhanarajan G, Sen R, Mishra S (2017) Improvement of ε-polylysine production by marine bacterium *Bacillus licheniformis* using artificial neural network modeling and particle swarm optimization technique. Biochem Eng J 126:8–15
20. Bilal M, Ji L, Xu Y, Xu S, Lin Y, Iqbal HM, Cheng H (2022) Bioprospecting *Kluyveromyces marxianus* as a robust host for industrial biotechnology. Front Bioeng Biotechnol 10:1–18
21. Binod P, Janu KU, Sindhu R, Pandey A (2011) Hydrolysis of Lignocellulosic biomass for bioethanol production. alternative feedstocks and conversion processes. Academic Press, Waltham
22. Blum C (2005) Ant colony optimization: Introduction and recent trends. Phys Life Rev 2(4):353–373
23. Bochner B, Gomez V, Ziman M, Yang S, Brown SD (2010) Phenotype microarray profiling of *Zymomonas mobilis* ZM4. Appl Biochem Biotechnol 161(1–8):116–123
24. Box GE, Behnken DW (1960) Some new three level designs for the study of quantitative variables. Technometrics 2(4):455–475
25. Bušić A, Marđetko N, Kundas S, Morzak G, Belskaya H, Šantek MI, Komes D, Novak S, Šantek B (2018) Bioethanol production from renewable raw materials and its separation and purification: a review. Food Technol Biotechnol 56(3):289–311
26. Carere CR, Sparling R, Cicek N, Levin DB (2008) Third generation biofuels via direct cellulose fermentation. Int J Mol Sci 9:1342–1360
27. Cazetta ML, Celligoi MAPC, Buzato JB, Scarmino IS (2007) Fermentation of molasses by *Zymomonas mobilis*: effects of temperature and sugar concentration on ethanol production. Biores Technol 98(15):2824–2828
28. Chaibaksh N, Abdul Rahman MB, Basri M, Salleh AB, Rahman RNZRA (2009) Effect of alcohol chain length on the optimum conditions for lipase-catalyzed synthesis of adipate esters. Biocatal Biotransform 27(5–6):303–308
29. Cheng NG, Hasan M, Kumoro AC, Ling CF, Tham M (2009) Production of ethanol by fed-batch fermentation. Pertanika J Sci Technol 17(2):399–408
30. Cheung S, Anderson B (1996) Ethanol production from wastewater solids. Water Environ Technol 8(5):55–60
31. Chouaibi M, Daoued KB, Riguane K, Rouissi T, Ferrari G (2020) Production of bioethanol from pumpkin peel wastes: comparison between response surface methodology (RSM) and artificial neural networks (ANN). Ind Crops Prod 155:112822
32. Ciric A, Krajnc B, Heath D, Ogrinc N (2020) Response surface methodology and artificial neural network approach for the optimization of ultrasound-assisted extraction of polyphenols from garlic. Food Chem Toxicol 135:110976
33. Claassen PAM, van Lier JB, Lopez Contreras AM, van Niel EWJ, Sijtsma L, Stams AJM, de Vries SS, Weusthuis RA (1999) Utilisation of biomass for the supply of energy carriers. Appl Microbiol Biotechnol 52:741–755
34. Cofré O, Ramírez M, Gómez JM, Cantero D (2012) Optimization of culture media for ethanol production from glycerol by *Escherichia coli*. Biomass Bioenerg 37:275–281
35. Conway T (1992) The Entner-Doudoroff pathway: history, physiology and molecular biology. FEMS Microbiol Rev 103(1):1–27
36. Costa JA, Treichel H, Kumar V, Pandey A (2018) Advances in solid-state fermentation. In: Current developments in biotechnology and bioengineering. Elsevier, pp 1–17
37. da Silva PA, Pinheiro ÁDT, Rocha MVP, Gonçalves LRB, Cartaxo SJM (2021) Hybrid neural network modeling and particle swarm optimization for improved ethanol production from cashew apple juice. Bioprocess Biosyst Eng 44(2):329–342

38. Darvishi F, Abolhasan Moghaddami N (2019) Optimization of an industrial medium from molasses for bioethanol production using the Taguchi statistical experimental-design method. Fermentation 5(1):14
39. Das SP, Das D, Goyal A (2014) Statistical optimization of fermentation process parameters by Taguchi orthogonal array design for improved bioethanol production. J Fuels 2014
40. Dasgupta D, Suman SK, Pandey D, Ghosh D, Khan R, Agrawal D, Jain RK, Vadde VT, Adhikari DK (2013) Design and optimization of ethanol production from bagasse pith hydrolysate by a thermotolerant yeast Kluyveromyces sp. IIPE453 using response surface methodology. SpringerPlus 2(1):1–10
41. Dave N, Varadavenkatesan T, Selvaraj R, Vinayagam R (2021) Modelling of fermentative bioethanol production from indigenous Ulva prolifera biomass by Saccharomyces cerevisiae NFCCI1248 using an integrated ANN-GA approach. Sci Total Environ 791:148429
42. Devi S, Suhag M, Dhaka A, Singh J (2011) Biochemical conversion process of producing bioethanol from lignocellulosic biomass. Int J Microb Resour Technol 1:28–32
43. Dhanarajan G, Mandal M, Sen R (2014) A combined artificial neural network modeling–particle swarm optimization strategy for improved production of marine bacterial lipopeptide from food waste. Biochem Eng J 84:59–65
44. Dien BS, Cotta MA, Jeffries TW (2003) Bacteria engineered for fuel ethanol production: current status. Appl Microbiol Biotechnol 63(3):258–266
45. El Naggar NE, Deraz S, Khalil A (2014) Bioethanol production from lignocellulosic feedstocks based on enzymatic hydrolysis: current status and recent developments. Biotechnology 13(1):1–21
46. Ethaib S, Omar R, Mazlina M, Radiah A, Syafiie S (2016) Development of a hybrid PSO–ANN model for estimating glucose and xylose yields for microwave-assisted pretreatment and the enzymatic hydrolysis of lignocellulosic biomass. Neur Comput Appl 1–11
47. Ezzatzadegan L, Morad NA, Yusof R (2016) Prediction and optimization of ethanol concentration in biofuel production using fuzzy neural network. Jurnal Teknologi 78(10)
48. Ezzatzadegan L, Yusof R, Morad NA, Shabanzadeh P, Muda NS, Borhani TN (2021) Experimental and artificial intelligence modelling study of oil palm trunk sap fermentation. Energies 14(8):2137
49. Ferreira SC, Bruns R, Ferreira HS, Matos GD, David J, Brandão G, da Silva EP, Portugal L, Dos Reis P, Souza A (2007) Box-Behnken design: an alternative for the optimization of analytical methods. Anal Chim Acta 597(2):179–186
50. Förster AH, Gescher J (2014) Metabolic engineering of *Escherichia coli* for production of mixed-acid fermentation end products. Front Bioeng Biotechnol 2(16):1–12
51. Gawande SB, Patil ID (2018) Experimental investigation and optimization for production of bioethanol from damaged corn grains. Mater Today: Proc 5(1):1509–1517
52. Ghafari S, Aziz HA, Isa MH, Zinatizadeh AA (2009) Application of response surface methodology (RSM) to optimize coagulation–flocculation treatment of leachate using poly-aluminum chloride (PAC) and alum. J Hazard Mater 163(2–3):650–656
53. Gutierrez-Correa M, Villena GK (2003) Surface adhesion fermentation: a new fermentation category. Revista Peruana Biologia 10:113–124
54. Hahn-Hagerdal B, Galbe M, Gorwa-Grauslund MF, Liden G, Zacchi G (2006) Bio-ethanol—the fuel of tomorrow from the residues of today. Trends Biotechnol 24:549–556
55. Hamelinck CN, van Hooijdonk G, Faaij APC (2005) Ethanol from lignocellulosic biomass: techno-economic performance in short-, middle- and long-term. Biomass Bioenerg 28(4):384–410
56. Hossain N, Zaini JH, Mahlia TMI (2017) A review of bioethanol production from plant-based waste biomass by yeast fermentation. Int J Technol 1:5–18
57. Hossain T, Miah AB, Mahmud SA, Mahin A-A (2018) Enhanced bioethanol production from potato peel waste via consolidated bioprocessing with statistically optimized medium. Appl Biochem Biotechnol 186(2):425–442
58. Ivanova V, Petrova P, Hristov J (2011) Application in the ethanol fermentation of immobilized yeast cells in matrix of alginate/magnetic nanoparticles, on chitosanmagnetite microparticles and cellulose coated magnetic nanoparticles. Int Rev Chem Eng 3(2):289–299

59. Izmirlioglu G, Demirci A (2015) Enhanced bio-ethanol production from industrial potato waste by statistical medium optimization. Int J Mol Sci 16(10):24490–24505
60. Jagatee S, Behera S, Dash PK, Sahoo S, Mohanty RC (2015) Bioprospecting Starchy feedstocks for bioethanol production: a future perspective. J Microbiol Res Rev 3(3):24–42
61. Jahanbakhshi A, Salehi R (2019) Processing watermelon waste using Saccharomyces cerevisiae yeast and the fermentation method for bioethanol production. J Food Process Eng 42(7):e13283
62. Jain A, Chaurasia SP (2014) Bioethanol production in membrane bioreactor (MBR) system: a review. Int J Environ Res Dev 4(4):387–394
63. Josiah A, Abimbola E (2011) Optimization of fermentation processes using evolutionary algorithms—a review. Sci Res Essays 6(7):1464–1472
64. Jugwanth Y, Sewsynker-Sukai Y, Kana EG (2020) Valorization of sugarcane bagasse for bioethanol production through simultaneous saccharification and fermentation: optimization and kinetic studies. Fuel 262:116552
65. Kamal S, Rehman S, Rehman K, Ghaffar A, Bibi I, Ahmed T, Maqsood S, Nazish N, Iqbal HM (2022) Sustainable and optimized bioethanol production using mix microbial consortium of Saccharomyces cerevisiae and Candida cantarelli. Fuel 314:122763
66. Kang Q, Appels L, Tan T, Dewil R (2014) Bioethanol from lignocellulosic biomass: current findings determine research priorities. Scie World J 1–13
67. Karuppiah R, Peschel A, Martín M, Grossmann IE, Martinson W, Zullo L (2007) Energy optimization for the design of corn-based ethanol plants. Paper presented at the Special Symposium-EPIC-1: European Process Intensification Conference—1, Copenhagen
68. Khalil SRA, Abdelhafez AA, Amer EAM (2015) Evaluation of bioethanol production from juice and bagasse of some sweet sorghum varieties. Ann Agric Sci 60(2):317–324
69. Khuri AI, Mukhopadhyay S (2010) Response surface methodology. Wiley Interdisc Rev Comput Stat 2(2):128–149
70. Kiran B, Pathak K, Kumar R, Deshmukh D (2016) Statistical optimization using central composite design for biomass and lipid productivity of microalga: a step towards enhanced biodiesel production. Ecol Eng 92:73–81
71. Koppolu V, Vasigala VKR (2016) Role of *Escherichia coli* in biofuel production. Microbiol Insights 9:29–35
72. Krantz M, Nordlander B, Valadi H, Johansson M, Gustafsson L, Hohmann S (2004) Anaerobicity prepares *Saccharomyces cerevisiae* cells for faster adaptation to osmotic shock. Eukaryot Cell 3(6):1381–1390
73. Kumar S, Dheeran P, Singh SP, Mishra IM, Adhikari DK (2015) Continuous ethanol production from sugarcane bagasse hydrolysate at high temperature with cell recycle and in-situ recovery of ethanol. Chem Eng Sci 138:524–530
74. Lee W-C, Huang C-T (2000) Modeling of ethanol fermentation using *Zymomonas mobilis* ATCC 10988 grown on the media containing glucose and fructose. Biochem Eng J 4(3):217–227
75. Lenihan P, Orozco A, O'Neill E, Ahmad MNM, Rooney DW, Walker GM (2010) Dilute acid hydrolysis of lignocellulosic biomass. Chem Eng J 156(2):395–403
76. Leonel LV, Arruda PV, Chandel AK, Felipe MGA, Sene L (2021) *Kluyveromyces marxianus*: a potential biocatalyst of renewable chemicals and lignocellulosic ethanol production. Crit Rev Biotechnol 41(8):1131–1152
77. Limtong S, Sringiew C, Yongmanitchai W (2007) Production of fuel ethanol at high temperature from sugar cane juice by a newly isolated *Kluyveromyces marxianus*. Biores Technol 98:3367–3374
78. Lin Y, Tanaka S (2006) Ethanol fermentation from biomass resources: current state and prospects. Appl Microbiol Biotechnol 69:627–642
79. Lin Y, Zhang W, Li C, Sakakibara K, Tanaka S, Kong H (2012) Factors affecting ethanol fermentation using *Saccharomyces cerevisiae* BY4742. Biomass Bioenerg 47:395–401
80. Liu C-G, Li K, Wen Y, Gena B-Y, Liu Q, Lin Y-H (2018) Bioethanol: new opportunities for an ancient product. Adv Bioenergy 4:1–34

81. Lugani Y, Rai R, Prabhu AA, Maan P, Hans M, Kuma V, Kumar S, Chandel AK, Sengar RS (2020) Recent advances in bioethanol production from lignocelluloses: a comprehensive review with a focus on enzyme engineering and designer biocatalysts. Biofuel Res J 28:1267–1295
82. Lynd LR, Van Zyl WH, Mcbride JE, Laser M (2005) Consolidated bioprocessing of cellulosic biomass: an update. Curr Opin Biotechnol 16:577–583
83. Ma H, Wang Q, Zhang W, Xu W, Zou D (2008) Optimization of the medium and process parameters for ethanol production from kitchen garbage by *Zymomonas mobilis*. Int J Green Energy 5(6):480–490
84. Madeira-Jr JV, Gombert AK (2018) Towards high-temperature fuel ethanol production using *Kluyveromyces marxianus*: On the search for plug-in strains for the Brazilian sugarcane-based biorefinery. Biomass Bioenerg 119:217–228
85. Madhuvanthi S, Jayanthi S, Suresh S, Pugazhendhi A (2022) Optimization of consolidated bioprocessing by response surface methodology in the conversion of corn stover to bioethanol by thermophilic *Geobacillus thermoglucosidasius*. Chemosphere 135242
86. Madson PW, Lococo DB (2000) Recovery of volatile products from dilute high-fouling process streams. Appl Biochem Biotechnol 84–86:1049–1061
87. Malik Z, Rashid K (2000) Comparison of optimization by response surface methodology with neurofuzzy methods. IEEE Trans Magn 36(1):241–257
88. Margaritis A, Merchant FJA (1987) The technology of anaerobic yeast growth. Yeast Biotechnol 84
89. Mazaheri D, Pirouzi A (2020) Valorization of Zymomonas mobilis for bioethanol production from potato peel: fermentation process optimization. Biomass Convers Biorefinery 1–10
90. Mojović L, Pejin D, Grujić O, Markov S, Pejin J, Rakin M, Vukašinović M, Nikolić S, Savić D (2009) Progress in the production of bioethanol on starch-based feedstocks. Chem Ind Chem Eng Q/CICEQ 15(4):211–226
91. Montgomery DC (2005) Design and analysis of experiments: response surface method and designs. John Wiley and Sons, Inc., New Jersey
92. Murari CS, Machado WRC, Schuina GL, Del Bianchi VL (2019) Optimization of bioethanol production from cheese whey using *Kluyveromyces marxianus* URM 7404. Biocatal Agric Biotechnol 20:1–8
93. Musa KH, Abdullah A, Al-Haiqi A (2016) Determination of DPPH free radical scavenging activity: application of artificial neural networks. Food Chem 194:705–711
94. Nagata U, Chu KH (2003) Optimization of a fermentation medium using neural networks and genetic algorithms. Biotech Lett 25:1837–1842
95. Naik SN, Goud VV, Rout PK, Dalai AK (2010) Production of first and second generation biofuels: a comprehensive review. Renew Sustain Energy Rev 14:578–597
96. Nassir TH, Al-Sahlany STG (2021) Bioethanol production from agricultural wastes by *Zymomonas mobilis* and used in vinegar production. J Microbiol Biotechnol Food Sci 11(3):e3709
97. Nazhad MM, Ramos LP, Paszner L, Saddler JN (1995) Structural constraints affecting the initial enzymatic hydrolysis of recycled paper. Enzyme Microb Technol 17(1):68–74
98. Nurfahmi MM, Ong HC, Jan BM, Kusumo F, Sebayang AH, Husin H, Silitonga AS, Mahlia TMI, Rahman SMA (2019) Production process and optimization of solid bioethanol from empty fruit bunches of palm oil using response surface methodology. Processes 7(10):1–16
99. Ortiz-Muñiz B, Carvajal-Zarrabal O, Torrestiana-Sanchez B, Aguilar-Uscanga MG (2010) Kinetic study on ethanol production using *Saccharomyces cerevisiae* ITV-01 yeast isolated from sugar canemolasses. J Chem Technol Biotechnol Adv 85(10):1361–1367
100. Osunkanmibi OB, Owolabi TO, Betiku E (2015) Comparison of artificial neural network and response surface methodology performance on fermentation parameters optimization of bioconversion of cashew apple juice to gluconic acid. Int J Food Eng 11(3):393–403
101. Pandey A, Soccol CR, Mitchell D (2000) New developments in solid state fermentation: I-bioprocesses and products. Process Biochem 35(10):1153–1169

102. Prasertwasu S, Khumsupan D, Komolwanich T, Chaisuwan T, Luengnaruemitchai A, Wongkasemjit S (2014) Efficient process for ethanol production from Thai Mission grass (*Pennisetum polystachion*). Biores Technol 163:152–159
103. Quintero-Ramirez R (ed) (2014) Hydrolysis of lignocellulosic biomass. In: Sugarcane bioethanol—R&D for productivity and sustainability. Editora Edgard Blücher, São Paulo
104. Renge VC, Khedkar SV, Nandurkar NR (2012) Enzyme synthesis by fermentation method. Sci Rev Chem Commun 2(4):585–590
105. Rogers PL, Joachimsthal EL, Haggett KD (1997) Ethanol from lignocellulosics: potential for a Zymomonas-based process. Australas Biotechnol 7(5):304–309
106. Rutkis R, Kalnenieks U, Stalidzans E, Fell DA (2013) Kinetic modelling of the *Zymomonas mobilis* Entner–Doudoroff pathway: insights into control and functionality. Microbiology 159(Pt_12):2674–2689
107. Sambo S, Faruk UZ, Shahida AA (2015) Ethanol production from fresh and dry water hyacinth using ruminant microorganisms and ethanol producers. Glob Adv Res J Biotechnol 4(1):023–029
108. Sánchez ÓJ, Cardona CA (2008) A review: trends in biotechnological production of fuel ethanol from different feedstocks. Biores Technol 99:5270–5295
109. Sandoval-Nuñez D, Arellano-Plaza M, Gschaedler A, Arrizon J, Amaya-Delgado L (2018) A comparative study of lignocellulosic ethanol productivities by *Kluyveromyces marxianus* and *Saccharomyces cerevisiae*. Clean Technol Environ Policy 20(7):1491–1499
110. Sarris D, Papanikolaou S (2016) Biotechnological production of ethanol: biochemistry, processes and technologies. Eng Life Sci 16:307–329
111. Sathendra ER, Baskar G, Praveenkumar R, Gnansounou E (2019) Bioethanol production from palm wood using *Trichoderma reesei* and *Kluveromyces marxianus*. Biores Technol 271:345–352
112. Sebayang AH, Masjuki HH, Ong HC, Dharma S, Silitonga AS, Kusumo F, Milano J (2017) Optimization of bioethanol production from sorghum grains using artificial neural networks integrated with ant colony. Ind Crops Prod 97:146–155
113. Sewsynker-Sukai Y, Faloye F, Kana EBG (2017) Artificial neural networks: an efficient tool for modelling and optimization of biofuel production (a mini review). Biotechnol Biotechnol Equip 31(2):221–235
114. Sharma N, Sharma N (2018) Second generation bioethanol production from lignocellulosic waste and its future perspectives: a review. Int J Curr Microbiol App Sci 7(5):1285–1290
115. Sharma R, Oberoi HS, Dhillon GS (2016) Fruit and vegetable processing waste: renewable feedstocks for enzyme production. Agro-industrial wastes as feedstock for enzyme production. Academic Press
116. Sivamani S, Baskar R (2015) Optimization of bioethanol production from cassava peel using statistical experimental design. Environ Prog Sustain Energy 34(2):567–574
117. Sivamani S, Baskar R (2018) Process design and optimization of bioethanol production from cassava bagasse using statistical design and genetic algorithm. Prep Biochem Biotechnol 48(9):834–841
118. Stewart GG, Panchal CJ, Russel I, Sills AM (1983) Biology of ethanol-producing microorganisms. Crit Rev Biotechnol 1:161–188
119. Subramaniyam R, Vimala R (2012) Solid state and submerged fermentation for the production of bioactive substances: a comparative study. Int J Sci Nat 3(3):480–486
120. Sun Y, Cheng J (2002) Hydrolysis of lignocellulosic materials for ethanol production: a review. Biores Technol 83(1):1–11
121. Taherzadeh MJ, Karimi K (2008) Pretreatment of lignocellulosic waste to improve ethanol and biogas production. a review. Int J Mol Sci 9:1621–1651
122. Taylor MP, Eley KL, Martin S, Tuffin MI, Burton SG, Cowan DA (2009) Thermophilic ethanologenesis: future prospects for second-generation bioethanol production. Trends Biotechnol 27:398–405
123. Tesfaw A, Assefa F (2014) Current trends in bioethanol production by saccharomyces cerevisiae: substrate, inhibitor reduction, growth variables, coculture, and immobilization. A review article. Int Schol Res Notices 1–11

124. Tinôco D, Genier HLA, da Silveira WB (2021) Technology valuation of cellulosic ethanol production by *Kluyveromyces marxianus* CCT 7735 from sweet sorghum bagasse at elevated temperatures. Renewable Energy 173:188–196
125. Trotter EW, Rolfe MD, Hounslow AM, Craven CJ, Williamson MP, Sanguinetti G, Poole RK, Green J (2011) Reprogramming of *Escherichia coli* K-12 metabolism during the initial phase of transition from an anaerobic to a micro-aerobic environment. PLoSONE 6(7):e25501
126. Tse TJ, Wiens DJ, Reaney MJT (2021) Production of bioethanol—a review of factors affecting ethanol yield. Fermentation 7(4):268
127. Vasíc K, Knez Ž, Leitgeb M (2021) Bioethanol production by enzymatic hydrolysis from different lignocellulosic sources. Molecules 26(3):1–23
128. Vazirzadeh M, Karbalaei-Heidari HR, Mohsenzadeh M (2012) Bioethanol production from white onion by yeast in repeated batch. Iranian J Sci Tech 36(4):477–480
129. Vazirzadeh M, Robati R (2013) Investigation of bio-ethanol production from waste potatoes. Ann Biol Res 4(1):104–110
130. Vohra M, Manwar J, Manmode R, Padgilwar S, Patil S (2014) Bioethanol production: feedstock and current technologies. a review. J Environ Chem Eng 2:573–584
131. Wang L, Yang ST (2007) Solid state fermentation and its applications. Bioprocessing for value-added products from renewable resources 465–489
132. Yaghoobi A, Bakhshi-Jooybari M, Gorji A, Baseri H (2016) Application of adaptive neuro fuzzy inference system and genetic algorithm for pressure path optimization in sheet hydroforming process. Int J Adv Manuf Tech 86(9):2667–2677
133. Yang ST, Liu X, Zhang Y (2007) Metabolic engineering—applications, methods, and challenges. Bioprocess Value-Added Prod Renew Resour 73–118
134. Yoshida T, Oshima Y, Matsumura Y (2004) Gasification of biomass model compounds and real biomass in supercritical water. Biomass Bioenerg 26(1):71–78
135. Zabed H, Faruq G, Sahu JN, Azirun MS, Hashim R, Nasrulhaq Boyce A (2014) Bioethanol production from fermentable sugar juice. A review. Sci World J 1–11

Bioethanol Recovery and Dehydration Techniques

Babatunde Oladipo, Abiola E. Taiwo, and Tunde V. Ojumu

Abstract Within the framework of the circular economy, bioethanol made from biological resources is regarded as a viable renewable energy option. However, the dehydration process is critical to bioethanol production suitable for use as fuel. Due to the ever-increasing demands placed on the yield and purity of anhydrous bioethanol, numerous methods have been developed and improved over time to achieve the desired product. In this chapter, we first briefly reviewed the conventional distillation method for bioethanol recovery, which results in azeotropic ethanol. Second, for industrial applications, we assessed the mode of operation, benefits and drawbacks of past and current purification technologies used in bioethanol dehydration, such as the adsorption process, extractive distillation, azeotropic distillation, membrane process, supercritical fluid extraction and hybrid process. The review showed that the membrane technique might significantly increase the efficiency of bioethanol purification because it offers the benefits of high selectivity, energy-efficient, eco-friendly (low waste generation) and cost-effective continuous operation to handle large volumes of feedstock. The economic impacts and future directions of bioethanol recovery and dehydration processes were also appraised.

Keywords Bioethanol · Distillation · Dehydration · Azeotropic distillation · Extractive distillation · Membrane process · Adsorption process · Supercritical fluid extraction · Hybrid process

B. Oladipo · T. V. Ojumu (✉)
Department of Chemical Engineering, Cape Peninsula University of Technology, Bellville Campus, Bellville, Cape Town 7535, South Africa
e-mail: ojumut@cput.ac.za

A. E. Taiwo
Faculty of Engineering, Mangosuthu University of Technology, Durban 4000, South Africa

© The Author(s), under exclusive license to Springer Nature Switzerland AG 2023
E. Betiku and M. M. Ishola (eds.), *Bioethanol: A Green Energy Substitute for Fossil Fuels*, Green Energy and Technology,
https://doi.org/10.1007/978-3-031-36542-3_9

1 Introduction

There are significant advances in bioethanol production from various feedstock [3, 47, 52]. However, the applied technology, process conditions and origin of bioethanol feedstock influence the purity and quality of anhydrous (fuel-grade) bioethanol produced. In practice, producing anhydrous bioethanol entails three crucial stages: fermentation, distillation/recovery (pre-concentration) and dehydration (purification). Following alcoholic fermentation, the broth is primarily made up of ethanol and total water. In a typical fermentation process, the effluent stream contains 5–12 wt% bioethanol, which must be processed to obtain pure bioethanol [23, 55]. As a result, efficient separation from the fermentation broth is of great essence.

Distillation has many advantages for recovering bioethanol from fermentation broths, making it the preferred choice for industrial applications. These advantages include high alcohol recovery (> 99%), simplicity of operation, adequate energy efficiency and capacity to simulate the process with computer applications–making easy the integration of mass and energy. Unfortunately, conventional distillation has some drawbacks, such as the formation of an azeotropic mixture (which requires additional separating equipment to produce anhydrous ethanol), high energy consumption at low alcohol feed concentration and, most notably, high working temperature levels (which deactivate the microbes, enzymes and proteins) [53]. Thus, finding effective recovery and dehydration methods is a big concern, as they often account for ~ 40% (and sometimes ~ 80%) of the costs associated with bioethanol production [32].

In the recovery step, which is the first stage of separation and purification, the conventional distillation produces ethanol of ~ 37 wt% along with residues containing a lot of water. After centrifugation, approximately 25% of the solid residues are returned to the fermentation process. Before entering the dehydration stage, the pre-concentrated ethanol is refined in a rectifying column to create azeotropic ethanol [19, 38].

In the dehydration step, traditional distillation cannot separate the ethanol–water mixture to produce a satisfactory purity level of anhydrous bioethanol (99.6 wt%). This is because ethanol–water systems tend to form azeotropes, which boil at a lower temperature (78.2 °C) than their individual components (ethanol boils at 78.4 °C while water boils at 100 °C). In comparison to pure ethanol, the ethanol–water mixture is more volatile. Since boiling does not alter the properties of the azeotrope, no further separation occurs because the vapour phase retains the same proportion as the liquid phase. Thus, methods such as the adsorption process, extractive distillation, azeotropic distillation, membrane process, catalytic dehydration, membrane pervaporation, supercritical fluid extraction and hybrid techniques have been employed to separate ethanol–water systems beyond the azeotropic threshold [23]. These techniques involve incorporating an entrainer into the system which changes the volatility of the constituents. It also involves the use of some special membranes and molecular sieves to facilitate the transport of only specific components. Due to the advantages and disadvantages peculiar to each of these dehydration methods, their economic viability is determined by the amount of anhydrous bioethanol generated, the affordability and ease of access to entrainers and the desired level of product purity [46]. To compete with traditional energy (fossil fuel), it is necessary to lower bioethanol production costs related to fermentation and product separation (pre-concentration and purification). Figure 9.1 illustrates a simplified schematic diagram of the stages involved in anhydrous bioethanol production [29].

To implement the techniques mentioned for industrial applications and achieve certain objectives, there is a pressing need to evaluate the techniques. In this context, this chapter will address the technologies in two sections: (i) conventional distillation applicable to recovering bioethanol from diluted solutions and (ii) technologies that are effective in dehydrating relatively concentrated bioethanol stream.

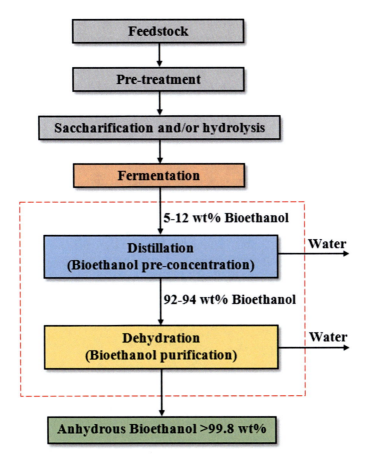

Fig. 9.1 Schematic diagram of stages involved in anhydrous bioethanol production. *Source* modified from [30]

2 Bioethanol Recovery (Pre-concentration): Distillation Process

The overall cost and complexity of ethanol recovery are driven mainly by the necessity of pre-conditioning procedures like degassing, heating and particle removal. Distillation is a method for separating and purifying liquids based on their varying boiling points. In fermentation, distillation is used to separate ethanol from water and other impurities from the rest of the liquid mixture produced during fermentation. Ethanol, which is a target component in the fermented broth, has a lower boiling point (78.4 °C) than water (100 °C). Therefore, when the liquid mixture is heated, the ethanol will vaporize before the other components. The vapour is then cooled, causing the ethanol to condense back into a liquid, which can be collected. This is the basic principle behind the distillation process of fermentation products. Modern distillation systems for ethanol separation consists of multiple columns, typically including a pre-column, a rectification column and a stripping column [4, 25].

3 Bioethanol Dehydration (Purification) Technologies

The production of an azeotropic mixture presents the most significant challenge in ethanol dehydration. When the ethanol–water mixture is partially boiled, a vapour is formed with the same constituents as the original mixture. An azeotrope is also known as a constant boiling mixture since partial boiling does not change their composition, rendering subsequent separation via ordinary distillation impossible. Azeotropic mixture separation is hindered by the presence of hydrogen bonds that result in stronger intermolecular interactions between the components [23, 36]. As a result, specialized separation techniques are required to eliminate leftover water to produce anhydrous bioethanol. Some selected processes involved in bioethanol dehydration are elaborated on in this section.

3.1 Azeotropic Distillation

Azeotropic distillation allows azeotropic mixtures formed through the conventional distillation method to be separated into their individual constituents. For this approach to be effective, a third volatile chemical component called an entrainer must be introduced; this entrainer then produces a ternary azeotropic mixture with the target component(s) due to polarity differences [31, 46]. Incorporating an entrainer thus modifies the relative volatilities, and hence the separation factor in the distillation unit. Azeotropic distillation systems typically consist of a modern distillation and an azeotropic column. Modern distillation columns concentrate ethanol to 92.4 wt% to achieve its binary azeotropic composition. After that, an azeotropic column is fed with the concentrated ethanol for additional purification. In the azeotropic column, a supplementary feed with a high entrainer content is introduced on the top tray and the entrainer combines with the mixture to form a solvent. The bottom stream from the column consists of highly purified ethanol (> 99 wt%), whereas the top stream comprises solvent, water vapour and a small amount of ethanol; which is very similar to the composition of the ternary azeotrope. As the top stream passes through a decanter, it separates into organic (rich in ethanol-entrainer) and aqueous streams (rich in water-entrainer). The organic phase is then returned to the azeotropic column [31]. The aqueous phase is treated in an entrainer recovery column such that the entrainer can be recycled for further use [21]. However, in a way to achieve the desired ethanol purity, a considerable measure of entrainer is necessary, which drives up the capital and energy costs. To produce anhydrous ethanol, several solvents like benzene, acetone, cyclohexane, isooctane, n-pentane, n-heptane, diethyl ether, hexane and polymers have been employed as an entrainer in the azeotropic distillation. However, benzene and cyclohexane have been the most often used of these. The flow diagram of a typical azeotropic distillation for the separation of ethanol–water mixture using an entrainer is shown in Fig. 9.2.

As a result of the challenges associated with high capital cost and energy usage, susceptibility to feedstock contaminants and dependency on solvents that are both carcinogenic (benzene) and flammable (cyclohexane), azeotropic distillation is a less commonly used technology. To address these limitations, various research groups have shifted their focus to using entrainers with less toxicity and a lower boiling point, resulting in alternative technology, such as extractive distillation.

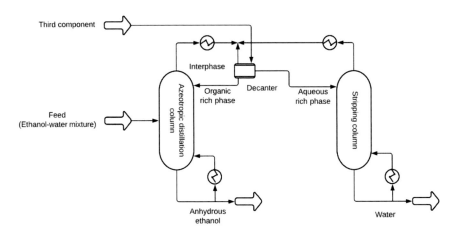

Fig. 9.2 Process flow diagram of azeotropic distillation process for ethanol–water mixture. *Source* modified from [31]

3.2 *Extractive Distillation*

Extractive distillation for bioethanol dehydration can be considered both a liquid–liquid and liquid–vapour process. In this technique, a liquid solvent is used to selectively extract component(s) from a liquid mixture. The extracted component will have a higher solubility in the solvent than in the original mixture. Distillation, a liquid–vapour process, separates the solvent and the extract, and the desired component is recovered. Using a selective high-boiling solvent in this method increases the separation factor, which raises the relative volatility by modifying the activity coefficients [12, 50]. Since just a small quantity of entrainer is evaporated during extractive distillation, it uses less energy and requires fewer resources than azeotropic distillation. However, the cost of the process is determined by the kind of third component used as an extractive/separating agent. Previous investigations have identified various classifications of extractive distillation depending on the type and nature of the third component [21, 33], which are further described in this section.

3.2.1 Extractive Distillation Using Liquid Solvent

In this process, a liquid solvent that is relatively non-volatile is supplied into a distillation column. The boiling temperatures of typical liquid solvents employed in extractive distillation are generally high. To achieve feed component separation in the column, the solvent modifies the volatility of one feed component relative to the other. For instance, the solvent used for the process may make ethanol more volatile than water. This scenario would allow ethanol to emerge in the overhead product. Also, water can be made more volatile than ethanol. Here, water would emerge in the overhead product. Both scenarios would contain solvent in the bottom product that needs to be removed before being recycled back to the column for a continuous run. Solvents with high selectivity for the desired component can enhance the efficiency of the process. In extractive distillation, anhydrous ethanol has been produced using a variety of liquid solvents as entrainers. Examples include diethyl ether, ethylene glycol, furfural and toluene. Using ethylene glycol offers the benefits of a good product attribute, applicability to mass production, low volatilization and reduced solvent usage [13, 45]. However, the energy consumption is enormous due to the large quantity of recycled solvent. To circumvent the drawbacks of using liquid solvents, salts are often used as entrainers.

3.2.2 Extractive Distillation Using Dissolved Salt

The process of extractive distillation using soluble salts and that of extractive distillation using liquid solvents are comparable. As salt is added, it alters the vapour–liquid equilibrium, which disrupts azeotropic formation. The non-volatile salt is supplied to the distillation column near the top tray, flows downwards and is collected with the bottom product [34]. When dissolved in liquids containing two volatile components, salts may form liquid phase complexes, changing the relative volatility of the component(s) and allowing for separation. As the dissolved salt binds more strongly to the molecules of one liquid component than the other, the solubility relationship between the two changes, and one component becomes "*salted out*". As a result, the relationship between the activities of the two volatile components of the liquid solution is altered so that the equilibrium vapour phase has a different composition despite the lack of salt in it. When compared with extractive distillation using liquid solvents to produce anhydrous bioethanol, soluble salts offer the following advantages: reduced toxicity of some salts, substantial energy savings and generation of a distillate devoid of the salt employed as an extractive agent. Salts frequently used in the extractive distillation of ethanol–water systems include $CaCl_2$, $NiCl_2$, $SrBr_2$, $CoCl_2$, CH_3COONa, CH_3COOK, NaI, KI, $Ca(NO_3)_2$, $HgCl_2$, $LiCl$ and $CuCl_2$ [31]. Some of the disadvantages associated with using dissolved salt as an entrainer include; corrosion of the distillation column due to high salt concentrations, salt solubility limitations, increased energy costs, reduced ethanol purity and waste disposal problems.

3.2.3 Extractive Distillation Using Mixture of Liquid Solvent and Dissolved Salt

In this process of extractive distillation, the salt effect and liquid solvent are combined. The process flowchart in this method is identical to that of extractive distillation with either a liquid solvent or dissolved salt. In most cases, a minimal quantity of salt was all that was necessary for the liquid solvent and dissolved salt combination. The dissolved salt is used to modulate the solubility of the components in the solvent, and it does that by changing the activity coefficient of the components in the solvent. This technique is useful in situations where the use of a pure solvent is not feasible or when the use of pure solvent results in poor selectivity or separation efficiency. Examples of dissolved salts used in this type of extractive distillation are $SrCl_2$, KNO_3, $CaCl_2$, K_2CO_3, $NaCl$, $Cu(NO_3)_2$, $Al(CO_3)_3$ and CH_3COOK [21]. A few factors to consider when selecting an industrial salt are its minimal toxicity, chemical stability, high solubility, little equipment corrosion and cheap cost. The salt-in-solvent method of extractive distillation uses less energy usage, fewer theoretical plates, less solvent recycled and lower equipment costs over conventional extractive distillation. The disadvantages of this technique include; corrosion, increased cost, difficulty of solvent and salt recovery and complexity of the process.

3.2.4 Extractive Distillation Using Ionic Liquid

The separation technique of extractive distillation using ionic liquids employs the selective solubility features of ionic liquids to separate a mixture of liquid components. The ionic liquid functions as an extractive agent, selectively removing one or more constituents from a mixture. By distillation, the extracted components are subsequently separated from the ionic liquid. Utilizing ionic liquids in extractive distillation has various advantages, including good selectivity, low volatility and recyclability, which makes it an appealing alternative to conventional distillation techniques. Studies have been conducted using ionic liquids (ILs) in extractive distillation to separate an ethanol–water mixture [11, 20]. In comparison to extractive distillation (which uses a mixture of liquid solvent and solid salt), this method facilitates good separation capacity, relatively low melting point, ease of operation, negligible vapour pressure and no concerns with solvent entrainment [44, 56]. ILs can significantly increase the relative volatility of ethanol over water when used as a separating agent. ILs are attractive separation agents owing to their beneficial features, such as high thermal stability, good solubility and lesser corrosiveness [15]. However, ionic liquids may exhibit a low separation efficiency due to their high viscosity.

Among the ILs employed as extractive agents in ethanol dehydration are $[EMIM]^+[BF_4]^-$ [56], $[EMIM]^+[Cl]^-$ [11] and $[EMIM]^+[OAc]^-$ [11]. The anions influenced selectivity significantly more than the cations [17]. ILs with $[OAc]^-$ and $[Cl]^-$ are the most promising, as their relative volatility is not noticeably higher than that of $[BF_4]^-$ [43]. Other features to consider when selecting ILs for industrial processes are toxicity, viscosity, recyclability, chemical stability, simplicity of recovery, flammability, and cost. A low-viscosity IL is often preferred.

3.2.5 Extractive Distillation Using Hyperbranched Polymers

Hyperbranched polymers are highly branched, high molecular weight three-dimensional structure polymers, which makes them highly effective at binding to and dissolving specific molecules. Their unique chemical structure makes them suitable as solvents in various applications. In extractive distillation practice, hyperbranched polymers are unique separating agents employed as a liquid solvent to dehydrate ethanol from aqueous solutions. In comparison to linear polymers, hyperbranched polymers have improved selectivity, capacity, and thermal stability with lower viscosity [21]. As a result, they are used as entrainers for separating azeotropes. Examples of polymeric entrainers employed for dehydrating ethanol are poly(ethylene glycol), polyglycerol, and poly(acrylic acid). Nevertheless, hyperbranched polymers are rarely used in industry because of their high cost and difficulties in sourcing raw materials.

3.3 Membrane Processes

The membrane process is a mass transfer unit-based operation that separates liquids and gases. Essentially, a membrane is a semipermeable barrier that separates two fluids by allowing some species to flow through. Whenever a membrane is used to process feeds, the feed allowed to pass through it is called the permeate, and the feed that cannot pass is known as the retentate [6].

Membrane separation by pervaporation is a technique for separating azeotropes that is energy-saving and efficient [1, 32]. Since membrane separation does not rely on liquid–vapour equilibrium, it can effectively be used to separate azeotropes and close boiling liquids [9]. As pervaporation membranes are nonporous, molecular separation depends on the affinity of molecules for the membranes [2, 41]. Low-affinity molecules are upheld, while high-affinity molecules are taken in and dispersed across the membrane.

In terms of their water permeability, membranes are classified into hydrophilic (water permselective) and hydrophobic (ethanol permselective). Since water molecules are smaller, the majority of membranes are hydrophilic. Also, there are three types of membranes distinguished by the materials employed in their fabrication: inorganic, polymeric and composite membranes.

In general, anhydrous ethanol is produced by vacuum and sweep gas pervaporation. In the vacuum pervaporation process, hydrophilic laminated membranes are used to separate ethanol and water based on their differences in diffusion resistances and partial pressures. This process involves contacting the membrane on the feed side (upstream or retentate) with the solution to be separated. Also, in the permeate (downstream) area, a vacuum pump is employed to reduce the partial pressure of the pervaporated permeate. Dehydrated ethanol is retained as the retentate, while water permeates the membrane. On the other hand, sweep gas pervaporation uses a hydrophilic hollow fiber membrane to increase the separation factor and an inert sweep gas such as nitrogen to reduce the partial pressure of the permeate during pervaporation [21, 23].

Despite drawbacks such as the high cost of membrane, technical complexity and limited membrane lifespan, pervaporation is a potential technique that offers the benefits of high selectivity, energy-efficient, eco-friendly (low waste generation) and continuous operation to handle large volumes of feedstock. This has made the process a preferred choice for industrial application in anhydrous bioethanol production. However, future research must focus on designing a pervaporation membrane that is both more cost-effective and durable than existing membranes while still being highly selective toward bioethanol.

3.4 Adsorption Processes

The ethanol–water separation involves two types of adsorption: (i) liquid-phase adsorption from the fermentation broth and (ii) vapour-phase adsorption from the distillation stream [5]. In most cases, vapour-phase adsorption is used, where overhead condensate from rectification columns is evaporated and superheated, or just the overhead vapour is superheated, before being exposed to the molecular sieve bed [53]. In adsorption operations, molecular sieves are utilized to absorb water preferentially since water molecules are significantly smaller than those of ethanol.

Molecular sieves of diameter 3 Å are employed to dehydrate ethanol, as they can adsorb water molecules of 2.5 Å diameter. However, ethanol molecules which are typically of 4 Å diameter cannot pass through and hence flow around the material.

Adsorbents made of inorganic materials, such as molecular sieves [7], silica gel [37] and calcium chloride [8], have all shown great promise for removing water molecules from ethanol–water mixtures through the vapour phase. Additionally, bio-based adsorbents have also been studied. This type of adsorbent can be further subdivided into lignocellulosic (such as rice straws and bagasse) and starch-based adsorbents (such as cornmeal). For high-purity applications such as in anhydrous bioethanol production, the adsorption process has several disadvantages: capacity limitations, poor selectivity, low efficiency, and high cost in adsorbent regeneration.

3.5 Supercritical Fluid Extraction

The solvent recovery columns and high-pressure extractors are the essential components of a basic bioethanol dehydration process using supercritical light hydrocarbons [16, 42]. In operation, the supercritical fluid solvent stream is fed from below the extraction column, while the ethanol–water solution enters from above. During operation, the column is subjected to pressures just above critical and temperatures near the critical point of the solvents. A valve lowers the pressure of the extract and delivers it to a distillation column to recover the solvent. A near-critical solvent entrains water in this column, which recovers the solvent and completely dehydrates the bioethanol. The bottom product, practically pure bioethanol, is produced in the distillation unit and the distillate is recycled as a supercritical solvent back into the extractor. Numerical outcomes in this process are highly dependent on how accurately the thermodynamic model predicts important phase equilibrium properties [42].

By using supercritical CO_2, ethanol can be removed or extracted in ethanol fermentation [18]. Some of the advantages of this technique are; reduced processing time, reduced use of organic solvents, improved efficiency and enhanced product purity. However, because it inhibits the metabolism of some yeasts and bacteria, supercritical CO_2 has limited use for in situ extractive fermentation. An increase in CO_2 solubility at high-pressure results in an acidic pH, which is responsible for the inhibitory impact [27]. Challenges with the complexity of the process, high capital and operational costs and environmental concerns are the drawbacks of using supercritical fluid extraction for anhydrous bioethanol production. The merits and demerits of bioethanol dehydration techniques are presented in Table 9.1.

Table 9.1 Comprehensive summary of dehydration techniques in terms of their merits and demerits

Dehydration technology	Merits	Demerits
Azeotropic distillation	**High purity**: Azeotropic distillation can produce high-purity ethanol (>99.5%) by forming an azeotrope with an entrainer that allows for the separation of ethanol and water **Simple process**: It is a relatively simple process that can be easily integrated into existing distillation processes for bioethanol production **High yield**: Azeotropic distillation can achieve high ethanol yields, resulting in more efficient use of raw materials and lower costs **Versatility**: It can be used with a variety of feedstocks and ethanol concentrations, making it a flexible option for bioethanol production **Waste**: Azeotropic distillation can result in reduced waste compared to other distillation processes, as it requires fewer separation stages	**Additional equipment requirements**: Azeotropic distillation requires additional equipment, such as a reflux drum and a decanter, which can increase the capital and operating costs of the distillation process **Energy consumption**: The system typically requires more energy, which can lead to higher operating costs and carbon emissions **Environmental concerns**: Azeotropic distillation often requires additional solvents, such as benzene or cyclohexane, which can be hazardous and require special handling and disposal procedures **Limited scalability**: The process is not easily scalable, and the process may need to be modified or redesigned for larger production volumes **Product purity**: Azeotropic distillation may not always produce high-purity bioethanol, which can impact the quality of the final product and potentially lead to additional processing steps or re-distillation

(continued)

Table 9.1 (continued)

Dehydration technology	Merits	Demerits
Extractive distillation using liquid solvent	**Higher purity**: A liquid solvent can help remove impurities from the bioethanol stream, resulting in higher purity of the final product **Energy savings**: Extractive distillation can require less energy, which can result in lower operating costs **Increased efficiency**: The use of a solvent can improve the efficiency of the distillation process by reducing the number of separation stages required **Reduced water usage**: Using a solvent can help reduce the amount of water required in the distillation process, which can be beneficial in areas with water scarcity **Environmental benefits**: The use of a liquid solvent can reduce the emission of volatile organic compounds (VOCs), which can have environmental and health impacts. Additionally, some solvents can be recycled and reused, reducing waste and environmental impact	**Cost**: The use of a liquid solvent can add to the cost of the distillation process, especially if the solvent is expensive or needs to be disposed of as hazardous waste **Compatibility**: The choice of solvent is critical, and some solvents may not be compatible with the bioethanol stream or other process conditions, which can lead to reduced efficiency or quality of the final product **Solvent recovery**: The recovery of the solvent can be challenging and may require additional equipment, which can increase capital costs and operating expenses **Safety concerns**: Some solvents used in extractive distillation can be flammable or toxic, which can present safety risks if not handled properly **Regulatory issues**: The use of some solvents may be subject to regulations and environmental requirements, which can increase compliance costs and time

(continued)

Table 9.1 (continued)

Dehydration technology	Merits	Demerits
Extractive distillation using dissolved salt	**Availability and affordability**: Salt is a widely available and affordable entrainer, making it an attractive option for bioethanol production **Improved separation**: The addition of salt can improve the separation of the components in the mixture, leading to higher purity and yield of ethanol **Lower operating temperatures**: Using salt as an entrainer can allow the distillation process to occur at lower temperatures, reducing energy consumption and operational costs **Lower pressure requirements**: Salt can also reduce the pressure required for distillation, which can reduce capital costs for equipment **Reduced water consumption**: Using salt as an entrainer can reduce water usage during the distillation process, which can be important in areas with water scarcity	**Corrosion**: High salt concentrations can cause corrosion in the distillation column and related equipment, leading to maintenance and repair costs **Solubility limitations**: The solubility of the salt in the mixture to be distilled may limit its ability to act as an entrainer **Increased energy costs**: Dissolving salt into the mixture increases the boiling point, requiring more heat to maintain the distillation process, which can lead to increased energy costs **Reduced ethanol purity**: The use of salt as an entrainer can result in the reduction of ethanol purity, as the salt and other components of the mixture can be carried over into the distilled product **Waste disposal**: The salt-rich waste generated from the distillation process can be difficult to dispose of and may cause environmental problems

(continued)

Table 9.1 (continued)

Dehydration technology	Merits	Demerits
Extractive distillation using mixture of liquid solvent and dissolved salt	**Increased selectivity**: The presence of a solvent and salt mixture can increase the selectivity of the distillation process, allowing for the separation of bioethanol from other components more effectively **Higher bioethanol concentration**: The use of a solvent and salt mixture can result in higher bioethanol concentrations in the distillate, which can reduce downstream processing costs **Improved energy efficiency**: The use of a solvent and salt mixture can also improve the energy efficiency of the distillation process by reducing the number of distillation stages required **Flexibility**: The use of a solvent and salt mixture allows for flexibility in the choice of solvent and salt combination, which can be tailored to specific feedstocks and operating conditions	**Complexity**: The process becomes more complex with the addition of a second component, and the selection of the correct solvent and salt combination can be challenging **Corrosion**: The presence of salt in the mixture can lead to corrosion of the distillation equipment, leading to maintenance and repair costs **Solvent and salt recovery**: After the distillation, the recovery of the solvent and salt mixture can be difficult and time-consuming, adding to the overall cost of the process **Environmental impact**: The disposal of the used solvent and salt mixture can have a negative impact on the environment if not handled properly

(continued)

Table 9.1 (continued)

Dehydration technology	Merits	Demerits
Extractive distillation using ionic liquid	**Reduced energy consumption**: Extractive distillation using ionic liquids can also require less energy, which can result in lower operating costs **Improved efficiency**: The use of an ionic liquid can improve the efficiency of the distillation process by reducing the number of separation stages required **Reduced environmental impact**: Ionic liquids have low vapour pressure and can be recycled, reducing waste and environmental impact **Versatility**: Ionic liquids can be tailored to meet specific process conditions and requirements, making them a versatile option for bioethanol production	**Cost**: Ionic liquids can be expensive to produce or purchase, which can increase the cost of the distillation process **Solvent recovery**: The recovery of ionic liquids can be challenging, and some ionic liquids may require specialized equipment or processes for effective recovery **Environmental concerns**: Some ionic liquids can be toxic in the environment, which can lead to environmental concerns if not handled and disposed of properly **Compatibility**: The choice of ionic liquid is critical, and some ionic liquids may not be compatible with the bioethanol stream or other process conditions **Limited knowledge**: The use of ionic liquids in extractive distillation is still taking the attention of researchers, and there is limited knowledge of their long-term effects on equipment and processes

(continued)

Table 9.1 (continued)

Dehydration technology	Merits	Demerits
Membrane processes	**High purity**: Membrane separation technology can achieve high levels of ethanol purity (up to 99.5%), which is suitable for fuel-grade ethanol production **Energy efficiency**: Membrane separation technology can be more energy-efficient because it operates at lower temperatures and pressures, reducing the energy required for separation **Lower capital and operating costs**: Membrane separation technology can require lower capital and operating costs, especially for smaller-scale operations **Continuous process**: Membrane separation technology can operate continuously, which can reduce the downtime associated with traditional batch distillation processes **Scalability**: Membrane separation technology can be easily scaled up or down, making it suitable for both small and large-scale production **Smaller footprint**: Membrane separation technology requires less space and equipment, which can reduce the overall footprint of the production facility	**Membrane fouling**: The membranes used in the separation process can become fouled by impurities in the feedstock, which can reduce efficiency **Limited ethanol concentration**: Membrane separation technology is typically limited to lower ethanol concentrations, and the process becomes less efficient as the ethanol concentration increases **Sensitivity to operating conditions**: Membrane separation technology is sensitive to changes in operating conditions, such as temperature and pressure, which can impact the efficiency of the process **Higher operating costs**: Membrane separation technology can have higher operating costs compared to traditional distillation processes, particularly if additional purification steps are required to remove impurities

(continued)

Table 9.1 (continued)

Dehydration technology	Merits	Demerits
Adsorption processes	**High purity**: Adsorption technology can produce high-purity ethanol, typically up to 99.9%, which is suitable for a wide range of industrial applications **Energy efficiency**: The process is more energy-efficient and has lower operational costs because it requires less energy for separation and does not require high temperatures or pressure **Continuous operation**: Adsorption technology can operate continuously, reducing downtime and increasing efficiency **Scalability**: The process is easily scalable and can be used in both small and large-scale bioethanol production facilities **Reduced waste**: Adsorption technology can reduce the amount of waste produced during the dehydration process, as it does not require the use of chemicals or generate any harmful byproducts	**Adsorbent regeneration**: The adsorbent material used in the dehydration process must be periodically regenerated, which can be a complex and energy-intensive process **Adsorbent stability**: The adsorbent material can degrade over time, reducing its effectiveness and requiring replacement **Adsorbent disposal**: The used adsorbent material must be disposed of properly, which can create additional waste management challenges **High capital cost**: Adsorption technology can require a significant initial investment in equipment and infrastructure, which can be a barrier to entry for some bioethanol producers **Limited capacity**: Adsorption technology has a limited capacity for ethanol dehydration, which can make it unsuitable for large-scale production **Sensitivity to impurities**: Adsorption technology can be sensitive to impurities, which can reduce the effectiveness of the separation process and require additional purification steps

Sources [8, 14, 24, 26, 28, 44, 50]

3.6 Hybrid/integrated Dehydration

Recovery of bioproducts entails a sequence of unit operations that are carried out to purify and concentrate the target product [51]. Anhydrous bioethanol production benefits from the integration of separation methods due to the complementary nature of the processes involved [26]. Research has been done into the use of hybrid methods for dehydrating ethanol, such as distillation in combination with pervaporation and distillation integrated with membrane-based processes [40, 48]. It has been found that the energy needed to purify bioethanol utilizing a highly integrated process for its recovery or dehydration is significantly lower when compared with using conventional distillation alone [31]. The use of hybrid techniques that combine membrane separation and distillation is a cost-effective and energy-efficient approach to purifying bioethanol. According to Novita et al. [40], a hybrid extractive distillation process was suggested that uses a cellophane membrane and high-selectivity pervaporation.

It was reported that the hybrid arrangement could save up to 41% of energy costs compared to the typical extractive distillation of ethanol dehydration. However, some limitations prevent the commercialization of hybrid technology, such as the high cost of the membrane, shorter lifetime, poor separation factor and membrane fouling. Thus, membranes that will eliminate these drawbacks should be developed to make hybrid distillation-membrane processes commercially feasible and competitive with other hybrid technologies. The merits and demerits of the hybrid technologies in bioethanol dehydration are presented in Table 9.2.

4 Economic Implications of Bioethanol Recovery and Dehydration Techniques

The selection of bioethanol dehydration is influenced by the capacity of the fermentation unit. Batch extractive distillation, pervaporation, or adsorption methods are suitable methods for producing low volumes of bioethanol. Based on an economic comparison of hybrid technologies, Khalid et al. [26] reported that hybrid techniques that combine membrane separation and distillation provide a cost-effective solution for bioethanol dehydration. In order to dehydrate ethanol to a concentration of 99.5 wt%, Kaminski et al. [22] investigated the cost efficacies of distillation and pervaporation. Adsorption and azeotropic distillation with cyclohexane cost more than pervaporation for a production rate of 100 L/day by factors of 2 and 1.5, respectively. The pervaporation cost increases linearly with the rate of production for rates greater than 100 L/h. At water concentrations under 10 wt%, pervaporation is economically advantageous. In this instance, there is 1–100 ppm of residual water in the ethanol. At production rates greater than 5000 L/h, distillation was recommended [10].

Additionally, hybrid methods that combine steam stripping with vapour permeation and pervaporation with distillation have also been shown to be cost-effective

Table 9.2 Detailed summary of the common hybrid dehydration techniques in terms of their merits and demerits

Hybrid dehydration technology	Merits	Demerits
Adsorption and distillation	**High efficiency**: The hybrid technology has high efficiency in dehydrating bioethanol due to the synergy between adsorption and distillation, leading to higher productivity and lower energy consumption **Reduced energy consumption**: Adsorption and distillation is more energy-efficient than traditional distillation processes, leading to lower operating costs and reduced greenhouse gas emissions **Reduced capital costs**: The capital cost of setting up an adsorption and distillation system can be lower than other hybrid technologies such as membrane distillation and adsorption **Reduced waste**: The system produces less waste, making it a more environmentally friendly option **Scalability**: The technology is highly scalable, making it suitable for both small and large-scale bioethanol production **Flexibility**: The system can handle a wide range of feedstock compositions and can be easily adjusted to meet specific product requirements	**Complexity and cost**: The system require complex equipment and specialized materials, which can increase the cost of production. The need for multiple stages further adds to the complexity, making it harder to optimize **Energy consumption**: The hybrid system require a significant amount of energy, particularly during the regeneration phase of the adsorbent material. The energy requirements can increase the production cost and may reduce the sustainability of the process **Adsorbent material performance**: The performance of the adsorbent material can be affected by various factors such as impurities, temperature, and humidity. This can affect the efficiency of the dehydration process and may require additional steps to maintain the quality of the adsorbent material in the hybrid system **Capacity limitations**: The capacity of the adsorbent material is limited by its saturation point, which means that the hybrid process needs to be frequently paused for regeneration of the material. This can result in production interruptions and increased costs **Safety concerns**: The system involve the use of flammable and potentially hazardous materials, which can pose a safety risk if not handled properly

(continued)

Table 9.2 (continued)

Hybrid dehydration technology	Merits	Demerits
Membrane distillation and adsorption	**Energy efficiency**: Membrane distillation and adsorption hybrid technology requires less energy than traditional distillation processes because it operates at lower temperatures and pressures. The use of a hydrophobic membrane also allows for the reuse of the waste heat generated during the process **High purity**: The hybrid system allows for high-purity ethanol production. The hydrophobic membrane removes water, while the adsorbent material removes any remaining water vapour, resulting in high-purity ethanol **Reduced waste**: The hybrid technology can reduce the amount of waste produced during the dehydration process, as it does not require the use of chemicals or generate any harmful byproducts **Scalability**: Membrane distillation and adsorption hybrid technology can be used in both small and large-scale bioethanol production facilities, making it easily scalable **Continuous operation**: The system can operate continuously, making it more efficient than batch processes	**High capital costs**: The initial capital cost of installing a membrane distillation and adsorption system can be high **Limited scale-up**: The scale-up of membrane distillation and adsorption systems can be limited due to the potential for fouling of the membrane and adsorbent material at larger scales **Membrane degradation**: The membrane used in the hybrid system can degrade over time due to exposure to heat and chemicals, leading to reduced separation efficiency and higher operating costs **Specific feed composition**: The membrane distillation and adsorption system requires specific feed composition to operate efficiently, which can be a limitation for certain feedstocks **Complexity**: The hybrid nature of membrane distillation and adsorption systems can lead to greater complexity in operation and maintenance **Maintenance**: Membrane distillation and adsorption systems require regular maintenance to prevent fouling and degradation of the membrane and adsorbent material

(continued)

Table 9.2 (continued)

Hybrid dehydration technology	Merits	Demerits
Membrane pervaporation and distillation	**High purity**: Membrane pervaporation and distillation can achieve high purity levels of bioethanol, up to 99.9%, which is important for meeting industry and regulatory standards **Reduced waste**: The system can reduce the amount of waste produced during the dehydration process, making it a more environmentally friendly option **Energy efficiency**: The system is more energy-efficient, as it requires lower temperatures and pressures **Flexibility**: The system can operate with a range of feedstocks, including bioethanol solutions with varying concentrations, and can be easily adjusted to meet specific product requirements **Continuous operation**: The hybrid technology can operate continuously, reducing downtime and increasing efficiency **Reduced capital costs**: The capital cost of setting up a membrane pervaporation and distillation system can be lower than other hybrid technologies	**Membrane fouling**: Similar to membrane distillation and adsorption, membrane pervaporation and distillation can experience fouling of the membrane due to impurities in the bioethanol feedstock, leading to reduced separation efficiency and increased operating costs **Energy consumption**: While membrane pervaporation and distillation can be more energy-efficient, it still requires significant energy input, particularly for the distillation step **Limited scalability**: The size of the membrane module used in the system can limit scalability in large-scale bioethanol production **Sensitivity to operating conditions**: The system is sensitive to changes in operating conditions such as temperature, pressure, and flow rate, which can impact system performance

Sources [10, 26, 35, 46, 48]

ways to produce bioethanol [48]. The overall cost of a membrane is lower when it has a higher flux and separation factor. Membrane pervaporation properties affect the performance of the whole system and, consequently, its economics. Advancements in membrane materials and properties have improved the economics of pervaporation/distillation hybrid systems [48]. This method for breaking the azeotrope can save up to 52.4% of the energy used during ethanol production [55].

Efficient ethanol recovery typically involves making trade-offs between recovery rate, process cost and separation system sustainability [49]. An alternate method different from the commonly used distillation for recovering ethanol must be economically viable to be used on an industrial scale. Vane and Alvarez [54] conducted a technical economic analysis based on energy consumption for several ethanol recovery processes. The cost of feedstocks, enzymes and pretreatment methods all affect operational costs during the production of ethanol [55]. Consequently, comparing separation techniques based solely on energy demand is not adequate. Thus, additional factors like capital and operating expenditures must be considered for valid comparison. Comparing economic costs per product unit between the same technology and the alternatives is difficult due to the wide variety of feedstocks and operational conditions. Therefore, robust appraisal of different separation processes using methods such as techno-economic analysis, conducting research under similar operating conditions and factoring in all costs is recommended.

5 Current Challenges and Future Directions

There is conflicting and varied information on the thermodynamic model, feed composition, process synergistic data, cost correlations, and total energy requirement in literature, so it is difficult to assess technological alternatives for bioethanol dehydration. Therefore, it is critical to evaluate bioethanol separation alternatives under the same working parameters, needed energy values and cost estimates to enable industrial adoption.

Researchers have significantly advanced hybrid distillation-membrane separation technologies to become more energy efficient than distillation. Nevertheless, many barriers like membrane fouling, shorter lifespan, poor separation factor and expensive membrane cost still need to be overcome before commercializing the method. Significant development has occurred in recent years in the pervaporation process as many obstacles have been resolved. While pervaporation saves energy, its use in industry is still constrained. This is primarily caused by the unavailability of large-scale, economically viable and sustainable membrane tools for use in industrial operations. Despite this, membrane technology is progressing at an ever-increasing rate, and much effort is going into addressing the technical challenges related to its design, scaling up and economics.

Most of the published research on the simulation of the bioethanol production process has concentrated on its design parts. However, there is comparatively little information on the issue of process control. Another key hindrance to mass bioethanol production is the limited design and control data availability. As a result, expanding bioethanol production on a commercial scale requires further research into the control and dynamic modelling of novel separation methods.

Therefore, it is necessary to work towards developing a common model (*if possible*) that can accurately describe the separation dynamics (equilibrium and kinetic data) while also being capable of portraying operating conditions. Improved system performance can also be attained through process optimization. Conventional separation models can benefit greatly from using modern optimization algorithms (especially on heat integration), resulting in significant energy savings. Developing and implementing advanced optimization techniques to reduce energy consumption is necessary. If all these factors are further developed, it will increase the wide acceptability of membrane technology on a commercial scale.

6 Conclusion

Anhydrous bioethanol is an effective alternative to gasoline for its low pollution levels during combustion. This chapter gave an overview of the distillation process for bioethanol recovery with the numerous conventional and non-conventional technologies employed in dehydrating bioethanol. Conventional distillation is still the method of choice for recovering and concentrating ethanol from fermented broths due to its proven principles and operational simplicity. Nonetheless, the technique has several drawbacks at lower ethanol concentrations. Extractive and azeotropic distillation processes are not commonly employed due to their high energy consumption, operational costs, and inadequate environmental sustainability. As a result, adsorption and membrane techniques, which have the benefits of requiring a small amount of energy to operate and eco-benign attributes, are implemented. However, poor selectivity and low efficiency limit the use of the adsorption process. High selectivity, cost-effective and energy-efficient benefits of the pervaporation-membrane technique made the process a preferred choice for commercial application. For industrial bioethanol production to be both technically and economically viable, a low-cost hybrid system that achieves high productivity in an eco-sustainable way must be developed. Membrane pervaporation combined with distillation shows great potential as an energy-efficient and cost-effective method for bioethanol purification. Improved material options and fabrication methods will open up new possibilities for membrane hybrid models to be used effectively in bioethanol purification at both the pilot scale and full-industrial levels.

References

1. Aptel P, Challard N, Cuny J, Neel J (1976) Application of the pervaporation process to separate azeotropic mixtures. J Membr Sci 1:271–287
2. Baker RW, Cussler E, Eykamp W, Koros W, Riley R (1991) Membrane separation system: recent developments and future directions. Noyes Data Corporation, Park Ridge, NJ, p 451
3. Balat M (2011) Production of bioethanol from lignocellulosic materials via the biochemical pathway: a review. Energy Convers Manage 52(2):858–875
4. Balat M, Balat H, Öz C (2008) Progress in bioethanol processing. Prog Energy Combust Sci 34(5):551–573
5. Beery KE, Ladisch MR (2001) Adsorption of water from liquid-phase ethanol–water mixtures at room temperature using starch-based adsorbents. Ind Eng Chem Res 40(9):2112–2115
6. Chapman PD, Oliveira T, Livingston AG, Li K (2008) Membranes for the dehydration of solvents by pervaporation. J Membr Sci 318(1–2):5–37
7. Chen W-C, Sheng C-T, Liu Y-C, Chen W-J, Huang W-L, Chang S-H, Chang W-C (2014) Optimizing the efficiency of anhydrous ethanol purification via regenerable molecular sieve. Appl Energy 135:483–489
8. de Jesús Hernández-Hernández E, Cabrera-Ruiz J, Hernández-Escoto H, Gutiérrez-Antonio C, Hernández S (2022) Simulation study of the production of high purity ethanol using extractive distillation: Revisiting the use of inorganic salts. Chem Eng Process Process Intensification 170:108670
9. Farhan NM, Ibrahim SS, Leva L, Yave W, Alsalhy QF (2022) The combination of a new PERVAPTM membrane and molecular sieves enhances the ethanol drying process. Chem Eng Process Process Intensification 174:108863
10. Frolkova A, Raeva V (2010) Bioethanol dehydration: state of the art. Theor Found Chem Eng 44:545–556
11. Ge Y, Zhang L, Yuan X, Geng W, Ji J (2008) Selection of ionic liquids as entrainers for separation of (water + ethanol). J Chem Thermodyn 40(8):1248–1252
12. Gerbaud V, Rodriguez-Donis I, Hegely L, Lang P, Denes F, You X (2019) Review of extractive distillation. Process design, operation, optimization and control. Chem Eng Res Des 141:229–271
13. Gil I, García L, Rodríguez G (2014) Simulation of ethanol extractive distillation with mixed glycols as separating agent. Braz J Chem Eng 31:259–270
14. Gil I, Uyazán A, Aguilar J, Rodríguez G, Caicedo L (2008) Separation of ethanol and water by extractive distillation with salt and solvent as entrainer: process simulation. Braz J Chem Eng 25:207–215
15. Gjineci N, Boli E, Tzani A, Detsi A, Voutsas E (2016) Separation of the ethanol/water azeotropic mixture using ionic liquids and deep eutectic solvents. Fluid Phase Equilib 424:1–7
16. Gros H, Díaz S, Brignole E (1998) Near-critical separation of aqueous azeotropic mixtures: process synthesis and optimization. J Supercrit Fluids 12(1):69–84
17. Gutiérrez JP, Meindersma GW, de Haan AB (2012) COSMO-RS-based ionic-liquid selection for extractive distillation processes. Ind Eng Chem Res 51(35):11518–11529
18. Güvenç A, Mehmetoglu Ü, Calimli A (1998) Supercritical CO_2 extraction of ethanol from fermentation broth in a semicontinuous system. J Supercrit Fluids 13(1–3):325–329
19. Hamelinck CN, Van Hooijdonk G, Faaij AP (2005) Ethanol from lignocellulosic biomass: techno-economic performance in short-, middle-and long-term. Biomass Bioenergy 28(4):384–410
20. Han J, Fan Y, Yu G, Yang X, Zhang Y, Tian J, Li G (2021) Ethanol dehydration with ionic liquids from molecular insights to process intensification. ACS Sustain Chem Eng 10(1):441–455
21. Huang H-J, Ramaswamy S, Tschirner U, Ramarao B (2008) A review of separation technologies in current and future biorefineries. Sep Purif Technol 62(1):1–21
22. Kaminski W, Marszalek J, Ciolkowska A (2008) Renewable energy source—Dehydrated ethanol. Chem Eng J 135(1–2):95–102

23. Karimi S, Karri RR, Tavakkoli Yaraki M, Koduru JR (2021) Processes and separation technologies for the production of fuel-grade bioethanol: a review. Environ Chem Lett 19(4):2873–2890
24. Karimi S, Yaraki MT, Karri RR (2019) A comprehensive review of the adsorption mechanisms and factors influencing the adsorption process from the perspective of bioethanol dehydration. Renew Sustain Energy Rev 107:535–553
25. Karuppiah R, Peschel A, Grossmann IE, Martín M, Martinson W, Zullo L (2008) Energy optimization for the design of corn-based ethanol plants. AIChE J 54(6):1499–1525
26. Khalid A, Aslam M, Qyyum MA, Faisal A, Khan AL, Ahmed F, Lee M, Kim J, Jang N, Chang IS (2019) Membrane separation processes for dehydration of bioethanol from fermentation broths: Recent developments, challenges, and prospects. Renew Sustain Energy Rev 105:427–443
27. Khosravi-Darani K, Vasheghani-Farahani E (2005) Application of supercritical fluid extraction in biotechnology. Crit Rev Biotechnol 25(4):231–242
28. Kiss AA, David J, Suszwalak P (2012) Enhanced bioethanol dehydration by extractive and azeotropic distillation in dividing-wall columns. Sep Purif Technol 86:70–78
29. Kiss AA, Ignat RM (2012) Innovative single step bioethanol dehydration in an extractive dividing-wall column. Sep Purif Technol 98:290–297
30. Klemes JJ, Varbanov PS, Liew PY (2014) 24th European Symposium on Computer Aided Process Engineering: Part A and B. Elsevier
31. Kumar S, Singh N, Prasad R (2010) Anhydrous ethanol: a renewable source of energy. Renew Sustain Energy Rev 14(7):1830–1844
32. Le NL, Wang Y, Chung T-S (2011) Pebax/POSS mixed matrix membranes for ethanol recovery from aqueous solutions via pervaporation. J Membr Sci 379(1–2):174–183
33. Lei Z, Li C, Chen B (2003) Extractive distillation: a review. Sep Purif Rev 32(2):121–213
34. Ligero E, Ravagnani T (2003) Dehydration of ethanol with salt extractive distillation—a comparative analysis between processes with salt recovery. Chem Eng Process 42(7):543–552
35. Loy Y, Lee X, Rangaiah G (2015) Bioethanol recovery and purification using extractive dividing-wall column and pressure swing adsorption: an economic comparison after heat integration and optimization. Sep Purif Technol 149:413–427
36. Mahdi T, Ahmad A, Nasef MM, Ripin A (2015) State-of-the-art technologies for separation of azeotropic mixtures. Sep Purif Rev 44(4):308–330
37. Mekala M, Neerudi B, Are PR, Surakasi R, Manikandan G, Kakara VR, Dhumal AA (2022) Water removal from an ethanol-water mixture at azeotropic condition by adsorption technique. Adsorpt Sci Technol
38. Michaels W, Zhang H, Luyben WL, Baltrusaitis J (2018) Design of a separation section in an ethanol-to-butanol process. Biomass Bioenergy 109:231–238
39. Mulder M, Smolders C (1984) On the mechanism of separation of ethanol/water mixtures by pervaporation I. Calculations of concentration profiles. J Membr Sci 17(3):289–307
40. Novita FJ, Lee H-Y, Lee M (2018) Energy-efficient and ecologically friendly hybrid extractive distillation using a pervaporation system for azeotropic feed compositions in alcohol dehydration process. J Taiwan Inst Chem Eng 91:251–265
41. Ong YK, Shi GM, Le NL, Tang YP, Zuo J, Nunes SP, Chung T-S (2016) Recent membrane development for pervaporation processes. Prog Polym Sci 57:1–31
42. Paulo CI, Diaz MS, Brignole EA (2009) Energy consumption minimization in bioethanol dehydration with supercritical fluids. In: Computer aided chemical engineering, vol 27. Elsevier, pp 1833–1838
43. Pereiro A, Araújo J, Esperança J, Marrucho I, Rebelo L (2012) Ionic liquids in separations of azeotropic systems—a review. J Chem Thermodyn 46:2–28
44. Quijada-Maldonado E, Aelmans T, Meindersma G, de Haan A (2013) Pilot plant validation of a rate-based extractive distillation model for water–ethanol separation with the ionic liquid [emim][DCA] as solvent. Chem Eng J 223:287–297
45. Ravagnani M, Reis M, Maciel Filho R, Wolf-Maciel M (2010) Anhydrous ethanol production by extractive distillation: a solvent case study. Process Saf Environ Prot 88(1):67–73

46. Saini S, Chandel AK, Sharma KK (2020) Past practices and current trends in the recovery and purification of first generation ethanol: A learning curve for lignocellulosic ethanol. J Clean Prod 268:122357
47. Sims R, Taylor M, Saddler J, Mabee W (2008) From 1st-to 2nd-generation biofuel technologies. International Energy Agency (IEA) and Organisation for Economic Co-Operation and Development, Paris, pp 16–20
48. Singh A, Rangaiah GP (2017) Review of technological advances in bioethanol recovery and dehydration. Ind Eng Chem Res 56(18):5147–5163
49. Strods M, Mezule L (2017) Alcohol recovery from fermentation broth with gas stripping: system experimental and optimisation. Agron Res 15:897–904
50. Sun S, Lü L, Yang A, Shen W (2019) Extractive distillation: advances in conceptual design, solvent selection, and separation strategies. Chin J Chem Eng 27(6):1247–1256
51. Taiwo A, Madzimbamuto T, Ojumu T (2020) Development of an integrated process for the production and recovery of some selected bioproducts from lignocellulosic materials. Valorization of Biomass to Value-Added Commodities, pp 439–467
52. Taiwo AE, Madzimbamuto TN, Ojumu TV (2018) Optimization of corn steep liquor dosage and other fermentation parameters for ethanol production by Saccharomyces cerevisiae type 1 and anchor instant yeast. Energies 11(7):1740
53. Vane LM (2008) Separation technologies for the recovery and dehydration of alcohols from fermentation broths. Biofuels Bioprod Biorefin 2(6):553–588
54. Vane LM, Alvarez FR (2008) Membrane-assisted vapor stripping: energy efficient hybrid distillation–vapor permeation process for alcohol–water separation. J Chem Technol Biotechnol Int Res Process Environ Clean Technol 83(9):1275–1287
55. Zentou H, Abidin ZZ, Yunus R, Awang Biak DR, Korelskiy D (2019) Overview of alternative ethanol removal techniques for enhancing bioethanol recovery from fermentation broth. Processes 7(7):458
56. Zhu Z, Ri Y, Li M, Jia H, Wang Y, Wang Y (2016) Extractive distillation for ethanol dehydration using imidazolium-based ionic liquids as solvents. Chem Eng Process Process Intensification 109:190–198

Ethanol Utilization in Spark-Ignition Engines and Emission Characteristics

Roland Allmägi, Marcis Jansons, Kaie Ritslaid, and Risto Ilves

Abstract The chapter gives an overview of ethanol as a fuel and the effects of ethanol and ethanol blends on spark ignition engines. Subjects such as the effect of ethanol fuel properties on engine-out emissions, the advantages and disadvantages of ethanol fuel, and the change of these properties as ethanol is blended with gasoline are highlighted. The octane number, the heat of evaporation, the air-to-fuel (A/F) ratio, the heating value of ethanol, and engine characteristics such as volumetric efficiency, knock, construction, etc., are discussed. Emission components, including nitrous oxides (NOx), unburned hydrocarbons (HC), carbon monoxide (CO), and particle matter (PM), are considered. In addition, the cold start characteristics of ethanol fuel.

Keywords Ethanol · Ethanol blends · Heat of vaporisation · Calorific value · Octane number · Volumetric efficiency · Spark ignition engine · NOx · HC · Emissions

1 Introduction

Ethanol as a motor fuel has been around for a long time, and several ethanol blends have been introduced. Ethanol is seen as an alternative fuel for polluting fossil fuels. However, the sourcing of first-generation ethanol is by no means environmentally friendly and depends on foodstuff to be produced. At the same time, second and third-generation bioethanol, which is produced from hay or biowaste, gives great promise. Thus, the research in the fields of ethanol production and usage as fuel is relevant

---

R. Allmägi · K. Ritslaid · R. Ilves (✉)
Institute of Forestry and Engineering, Estonian University of Life Sciences, Kreutzwaldi 1, 51006 Tartu, Estonia
e-mail: risto.ilves@emu.ee

M. Jansons
Mechanical Engineering Department, College of Engineering, Wayne State University, Detroit, MI 48202, USA

© The Author(s), under exclusive license to Springer Nature Switzerland AG 2023
E. Betiku and M. M. Ishola (eds.), *Bioethanol: A Green Energy Substitute for Fossil Fuels*, Green Energy and Technology,
https://doi.org/10.1007/978-3-031-36542-3_10

even today. As the physicochemical properties of ethanol differ from gasoline, spark ignition engines must be modified to run effectively on ethanol and ethanol blends. Various ethanol blends are available worldwide, and engines with corresponding adaptations must be used for individual blends to ensure smooth operation and low emissions. Flex-fuel engines can detect ethanol ratios in the fuel tank and adjust engine parameters accordingly to ensure optimum operating conditions [7, 29, 47].

The chapter provides an overview of ethanol and ethanol blends in spark ignition engines. The chapter discusses ethanol properties and their effects on the engine and emissions.

2 Globally Used Ethanol Fuels for SI Engines

The main ethanol fuel blends globally available are E5, E10 and E85 [9, 25]. Whereas several countries use E15, E20, E25, E70, E75 and E100 blends. E15 is a fuel with up to 15% ethanol content primarily used in the USA. This fuel blend is used without restrictions in cars introduced after 2001, as the effects of ethanol fuel on older engines are unknown. E20 and E25 are used in Brazil, where most vehicles are adapted to run on various blends of ethanol, but neat gasoline cannot be used as engine knock may occur. In other parts of the world, cars adapted for ethanol fuels are less common than in Brazil, and thus the usage of E15-E25 fuels is also lower. In these regions, E5 and E10 are used since these fuels can be used in conventional gasoline engines without any issues and no restrictions. E85 is the most prevalent high ethanol-content fuel on the market, but only specialised flex-fuel vehicles can use it. E70 and E75 are a variation of E85 with higher gasoline content to support cold starts during colder seasons and, as a result, are not independently labelled. E100 is pure ethanol fuel with a water content of up to 4.4%. Due to the high latent heat of hydrous ethanol, issues with cold starting at ambient temperatures below 15 °C occur. As a result, the fuel is only available in the warm climate of Brazil [29, 47].

The properties of ethanol fuels are standardised, and Table 10.1 gives an overview of the standards of different countries.

The climate of individual countries governs the requirements set for fuels. Table 10.2 presents the Canadian standards CAN/CGSB-3.511–2021 and CAN/CGSB3.5122018. The purpose of the table is to present the standardised properties of ethanol fuels and their parameters.

As seen in Table 10.2, the octane number for higher ethanol-content fuels is not measured. The same practice is followed for other standards in Table10.1.

Many of the parameters in Table 10.2 are homologized for a range of different ethanol fuels. The difference in high and low ethanol content fuels lies in the water and manganese content and octane number.

Table 10.1 Overview of the standards of ethanol fuel

Standard	Fuel	Remark
EN 228	Gasoline (E0) and ethanol E5, E10	Sets regulation for gasoline and gasoline-ethanol blends
EN 15,376	Ethanol E100	Sets regulation for ethanol
ASTM D 4806	Ethanol E100	Standard specification for ethanol fuel use as SI automotive fuel
ASTM D 5798	ethanol and blends E85, E95	Standard specification for ethanol fuel blends for flexible automotive engines
CAN/CGSB-3.511–2021	Oxygenated automotive gasoline containing ethanol (E1-E10 and E11-E15)	Fuels which are intended for use in spark-ignition engines under a wide range of climatic conditions
CAN/CGSB3.5122018	Automotive ethanol fuel (E50-E85 and E20-E25)	Fuels are intended for use in flexible fuel vehicles over a wide range of climatic condition
GB 18,351–2017	Ethanol gasoline for motor vehicles (E10)	Fuels which are intended for use in gasoline engines
GB 35,793–2018	Ethanol gasoline for motor vehicles (E85)	Fuels are intended for use in flexible fuel vehicles over a wide range of climatic condition

Source [69] (GB 35,793–2018; GB 18,351–2017; CAN/CGSB3.5122018; CAN/CGSB-3.511–2021)

3 Effect of Ethanol Fraction on Key Fuel Parameters

When blended into gasoline-range motor fuels, the physicochemical properties of ethanol impact various aspects of combustion engine performance, emissions and efficiency. Characteristics including oxygen content, heating value, liquid density, the heat of vaporisation, surface tension, viscosity, vapour pressure and flame speed are well known to significantly affect particular engine behaviours such as soot formation, mileage, cold-starting ability, emissions and knock propensity [6, 16]. The charge cooling effect resulting from ethanol's high heat of vaporisation and its high-Octane numbers, in particular, extend knock limits and allow operation at higher, more efficient compression ratios [10, 36, 37]. However, ethanol's high flame speed has been correlated to increased occurrence of low-speed pre-ignition [15, 25, 34], which limits boost levels in down-sized turbocharged engines. By elevating reaction front propagation speed and promoting transient detonation development, ethanol addition may also lead to stronger knock intensity once detonation is initiated [51]. Effects depend on the type of combustion, and ethanol fuel blends can be optimised for particular combustion modes, including Homogenous Charge Compression Ignition (HCCI) [64, 72], Reactivity Controlled Compression Ignition (RCCI) [19, 26], Spark Assisted Compression Ignition (SACI) [59], Turbulent Jet Ignition (TJI) [10] or downsized, high boost, direct injection (DI), spark ignition (SI) [25].

Table 10.2 Requirements of ethanol fuel blends

Parameter	Requirement			
	E10	E11-E15	E20-E25	E50-E85
Denatured fuel ethanol, at point of blending, % by volume	1–10	10–15	20–25	50–85
Ethanol, % by volume	–	–	18–25	46–84
Sulphur, mg/kg	Max 80	Max 80	Max 80	Max 80
Methanol, % by volume	Max 0.5	Max 0.5	Max 0.5	Max 0.5
Water, % by mass	–	–	Max 1	Max 1
Copper strip corrosion, 3 h at 50 °C	Class 1	Class 1	Class 1	Class 1
Lead, mg/L	Max 5	Max 5	Max 5	Max 5
Manganese content, mg/L	Max 18	Max 18	–	–
Benzene content, % by volume	Max 1.5	Max 1.5	Max 1.5	Max 1.5
Oxidation stability, (Induction period)	Min 240	Min 240	Min 240	–
Methanol content % by volume	Max 0.30	Max 0.30	Max 0.50	Max 0.50
Antiknock index (RON + MON)/2 Min	87–93	87–93	91	–
Octane number, MON/RON (min)	-/82	-/82	–	–
Vapour pressure kPa	35–110	35–110	35–110	35–110
Final boiling point, °C	Max 225	Max 225	–	–

Source (CAN/CGSB-3.511–2021; CAN/CGSB3.5122018)

3.1 Engine Fuel Requirements—Octane Number

Combustion in SI systems is initiated by a spark plug or other ignition source that produces a flame kernel. The resulting flame propagates outward into the unburned mixture in a deflagration process, sustained by the exothermic reactions in the flame front and consuming the unburned mixture as it advances. As the burned gas fraction rises to flame temperatures, the accompanying pressure increase propagates ahead of the flame front throughout the cylinder, compressing the unburned mixture in front of it. The temperature rise due to the compression can initiate autoignition in the unburned end-gas, resulting in uncontrolled high rates of heat release and the undesirable condition known as knock, which is characterised by high-frequency pressure fluctuations of elevated magnitude that can lead to engine failure. The autoignition process occurs more rapidly at higher pressures and particularly higher temperatures. The maximum allowable engine cylinder pressure and temperature are then determined by the autoignition chemistry of the fuel [48], and engine operating conditions for a given fuel are selected such that knock is avoided. Since typical means of increasing engine efficiency, such as raising compression ratios, increasing boost levels, and advancing spark timing, are typically knock-limited, a primary concern for SI combustion systems is the anti-knock quality of the fuel.

Octane ratings are conventionally used to quantify a fuel's knock resistance relative to primary reference fuels (PRF) defined as binary mixtures of n-heptane and 2,2,4-trimethyl pentane (iso-octane). For historical reasons, the anti-knock index (AKI), or the average of the research and motor octane numbers (RON and MON) [42, 57] is used in the United States as a fuel specification, with the RON used in most E.U. countries. RON and MON are both determined using the Cooperative Fuel Research carburetted, variable compression ratio, single-cylinder, four-stroke engine. The research engine is fuelled with a test fuel and the compression ratio is raised until a standard level of knock is attained with the engine running at the prescribed conditions. Two PRF blends are then selected to bracket the knock intensity at the previously determined compression ratio [6]. The Octane Number of a fuel is thus defined as the volumetric percentage of iso-octane in the PRF blend that exhibits similar knocking characteristics at the same engine conditions and indicates in-cylinder engine conditions.

On this scale, pure iso-octane has RON and MON of 100, and pure n-heptane has an Octane Number of zero. A test fuel having an Octane Number of 50 knocks with the same intensity at the same compression ratio as a 50:50 blend of iso-octane and n-heptane, by volume, at the RON or MON test conditions. The main difference between the standards is that the RON test prescribes an inlet air temperature of 52 °C and speed of 600 RPM, while the MON procedure is performed at 149 °C and 900 RPM. The two test conditions result in different temperature–pressure trajectories during the air–fuel charge compression process [46, 75] and account for changes in the relative ranking of knock resistance among a set of fuels as the engine conditions change [67]. With its higher temperature, during the compression stroke of the MON test, the charge avoids the temperature–pressure conditions of the low-temperature heat release (LTHR) regime prior to entering the negative temperature coefficient region (NTC), where chemical kinetic rates slow with an increase in temperature. The lower inlet air temperature test conditions of the RON result in a charge temperature–pressure trajectory with much stronger LTHR. Consequently, fuels having two-stage ignition characteristics, such as the paraffinic n-heptane and iso-octane used in PRFs, have similar RON and MON. In contrast, fuels such as aromatics, olefins [41] and ethanol [75] are single-stage ignition fuels with minimal low-temperature reactivity and little LTHR and thus have larger differences between RON and MON. Differences between RON and MON are quantified by the Sensitivity (S) factor.

$$S \equiv RON - MON \qquad (10.1)$$

Although in-cylinder temperature and pressure conditions achieved by the RON and MON tests bracketed the range of those encountered in the automotive fleet when the standards were devised in 1932, they increasingly deviated from the operating conditions of modern engines [48]. The octane response of modern engines correlates better with RON than MON [54].

To account for varying knock intensity observed in fuels of different compositions but the same Octane Numbers, an Octane Index (OI) has been found to define best the anti-knock fuel quality [31, 32]. The performance parameter K is introduced to scale cylinder charge temperature and pressure conditions between the RON and MON tests and must be empirically determined. The value of K, then, identifies a particular type of combustion that varies with the combustion system and operating mode. Since K can take on negative values, a fuel of lower MON (but higher sensitivity) for a given RON can have a higher OI and more knock resistance. Thus, for $0 \leq K \leq 1$, the temperature–pressure trajectory lies between those of the RON and MON tests, with $K > 1$ representing higher temperature and lower pressure trajectories than in the MON test and $K < 0$ representing engines with in-cylinder compression processes typically occur at lower temperatures but higher pressures than the trajectory prescribed by the RON test.

$$OI \equiv RON - KS = (1 - K)RON + K \cdot MON \qquad (10.2)$$

Carburetted engine systems are described by values of K between zero and unity [75]. The values of K vary in HCCI engines, from mildly positive at near-stoichiometric, heavier loads, to negative values as low as -1.75 [33, 58], at the leanest operating conditions. Importantly, values of K are distinctly negative at operating conditions of boosted and fuel-injected engines with lower inlet temperature but higher-pressure conditions of the RON test conditions [30, 32, 48, 67, 75]. Data over the previous 80 years show engine operating conditions described by a continuous decrease in the value of K [48]. This long-term trend is expected to continue [53] as the penetration of turbo-charging technologies climbs over 35% of the US vehicle market [22] for the model year 2020. The implication is that for the future vehicle fleet, anti-knock properties of fuel can be enhanced not only by higher RON but also by raising sensitivity. Electron delocalisation has been identified as the mechanism determining low-temperature reactivity and sensitivity in hydrocarbons, including alcohols [73], where C–O bond fission in the early fuel decomposition process produces relatively stable aldehydes, delaying autoignition until higher temperatures are reached and quenching low-temperature reactivity. Provided the management system of the engine is calibrated to take advantage of the fuel's OI, increased fuel sensitivity can improve engine efficiency [55]. Using two gasoline-ethanol blends, when mixed in different proportions and resulting in different OI, has been suggested to dynamically fuel an engine with just the required anti-knock quality for the instantaneous operating condition, thus reducing the fuel consumption of the higher OI fuel [50].

Higher Octane Numbers generally indicate greater anti-knock characteristics of fuel and allow for higher engine efficiencies. Simulations suggest that introducing standard grades of 98 and 100 RON gasoline to the US market would reduce fleet fuel consumption by 4.5–6.0% if new vehicles were manufactured to take advantage of such fuel [12]. An accompanying increase in fuel sensitivity can enhance the impact of raising fuel RON on the economy, efficiency gains of 1.3–3.1% have been documented by increasing RON from 92 to 98 with a fuel Sensitivity of 5, which increase to 3.6–5.3% when the sensitivity increased to 10 [55]. Thus, ethanol, with

RON and sensitivity values of 18.4 and 108, respectively, considerably higher than that of gasoline, is an effective octane and sensitivity booster and has characteristics that, when blended into fuel, offer a means to reduce fleet fuel consumption, provided the engine is calibrated to take advantage of the additional anti-knock tolerance provided.

3.2 Blending Rules

Using a volumetric basis is standard practice for producing motor fuels of desired properties such as heating values, composition and density using different refinery streams. However, chemical properties that involve gas-phase phenomena, such as vapour pressure, are better described using molar concentration [4]. Although, in general, the RON of gasoline-ethanol blends increases with ethanol volume fraction [28], the relationship between these parameters is complex. Gasoline, being a distillate, comprises a range of aliphatic and aromatic compounds commonly grouped into five classes, linear paraffins, branched paraffins, naphthene, olefins and aromatics. When blended, synergistic or antagonistic chemical interactions can occur between ethanol and the various gasoline components depending on the composition of the base fuel [2, 14], resulting in RON values that deviate from the linear relationship between blend RON and ethanol volume or molar fraction. While hydrocarbons of the same class interact linearly with one another, synergistic or antagonistic interactions between compounds of different classes produce RON values above or below, respectively, RON or MON values expected based on a linear function of volume or molar fraction [24]. Although earlier measurements suggested the RON of gasoline-ethanol blends increases linearly with ethanol volume [28], or molar fraction [4, 6], an API survey of 13 blend stocks sourced across the US market showed more nuanced relationships, with a linear increase in RON as the ethanol fraction is raised from zero to 20% by volume, with varying behaviours at higher blend fractions [20]. Although the results uniformly showed ethanol to be a more effective octane booster for lower RON base gasoline, its effects on blend RON clearly vary depending on the particular composition of the blend. Experiments of ternary ethanol-PRF blends show this interaction is synergistic, with RON increasing steeply with initial ethanol fraction and RON count per ethanol fraction increasing as the RON of the base PRF decreases [24]. RON gains fall off at higher ethanol fractions, the initial 30% (vol/vol ethanol content raises the RON of the ethanol/n-heptane blend from zero to 53, but the blend RON increases only nine points from 100 to 109 when 70% ethanol content is increased to neat ethanol. As a point of interest and illustrating the strength of the synergy, the RON values of iso-octane/ethanol blends over the ethanol range 20–80% (vol/vol exceed the RON value of neat ethanol, peaking at a RON value of 110.2 at 40% ethanol fraction by volume. In contrast, toluene blends antagonistically with ethanol, moderating the synergistic behaviour of iso-octane and n-heptane. As the toluene fraction is increased of base mixtures of RON = 91, the synergistic effects of ethanol interaction with iso-octane and n-heptane are inhibited and the rate of RON increase with ethanol fraction becomes more linear. Tri-methyl benzene has also been shown to have an antagonistic interaction with ethanol [3], suggesting the antagonistic behaviour may hold for the general aromatic class. Ethanol also serves as a Sensitivity booster, with the steepest increases achieved at lower ethanol fractions [24].

Examining the composition and fuel properties of nearly three hundred ethanol-gasoline blends and identifying base gasoline octane sensitivity S_g or saturate and aromatic fractions Sat_g, $Arom_g$ [5] developed a blending model predicting the RON of the blend, RON_{blend} as a function of the base gasoline RON, RON_g and ethanol mole fraction, x_e using a coefficient Z_{RON} to quantify the degree of synergistic/antagonistic blending.

$$RON_{blend} = (1 - x_e)RON_g + x_e(108.6 + Z_{RON}(1 - x_e)) \qquad (10.3)$$

$$Z_{RON} = 15.0 - 1.76 S_g + 11.3 x_e \qquad (10.4)$$

$$Z_{RON} = -47.8 + 64.3 Sat_g + 24.6 Arom_g + 12.0 x_e. \qquad (10.5)$$

The ethanol molar fraction x_e can be found from the more common volume fraction C_e using Eq. (10.4), where M_e, M_g and ρ_e, ρ_g are the molecular weights and densities of ethanol and base gasoline, respectively.

$$x_e = C_e/\left[C_e + (1 - C_e)(M_e/\rho_e)(M_g/\rho_g)\right] \qquad (10.6)$$

A mathematically similar description has been developed by AlRamadan for ethanol blending with PRF and toluene blends [2] and a gasoline surrogate whose palette comprises eight compounds representing all five of the hydrocarbon classes [3]. Blending rules have also been developed for ethanol in HCCI combustion modes [72].

The synergetic interaction between ethanol and paraffins can be effectively utilized for fuel design. The interaction allows a particular RON fuel specification to be met with both lower ethanol and aromatic content and suggests greater fleet fuel economy gains can be achieved by distributing available ethanol widely rather than concentrating it in E85 fuels [24].

3.3 Heat of Vaporization and Charge Cooling

The heat of vaporization of ethanol is 2.6 times that of gasoline on a unit mass of fuel basis but 4.2 times that of gasoline per unit mass of stoichiometric mixture due to its lower stoichiometric air-to-fuel ratio [66]. Using ethanol in gasoline blends can provide significant charge-cooling effects that increase monotonically with ethanol content. Since knock-inducing preflame reactions and their subsequent autoignition are strong temperature functions, reducing charge temperature by evaporative fuel cooling is an effective means to suppress engine knock. Knock-limited spark timing can then be advanced to optimise combustion phasing or increase compression ratio better, increasing engine efficiency [39, 40, 65]. Although the mass-based heat of vaporization of ethanol has been found to increase with ethanol volume fraction in gasoline blends non-monotonically [35], more recent work suggests mass-based heat of vaporization increases linearly with ethanol volume fraction over a range of temperatures and for different hydrocarbon blend stocks [13].

Since in an engine, the fuel spray will partially impinge on the piston, cylinder wall or inlet port surfaces, extracting heat from metal and not air, not all the heat of vaporization will be extracted from the charge. The maximum thermodynamic charge cooling effect then varies not only with the ethanol content of the fuel but also with the fuel injection system and engine operating parameters such as intake air temperature and load. Theoretical maximum charge cooling effects have been calculated to be in the range of 20–80 °C as the ethanol content is swept from 0 to 100% for gasoline-ethanol blends, with experimental cooling being 70% of the theoretical values in a direct-injection engine across the range of ethanol fraction [36, 37].

The knock resistance provided by ethanol is the consequence of both charge cooling and chemical effects. Using a knock-limit predictive model together with a turbocharged spark-ignited engine operated with both port and direct fuel injection systems, Kasseris and Heywood separated the two effects and showed that the chemical effect's contribution to an effective octane number, or knock resistance, is non-linear with increasing ethanol content [36, 37] when blended with gasoline. Knock resistance is initially significantly improved as ethanol fraction increases up to 30–40% by volume. Increased ethanol blending beyond this fraction provides a little anti-knock benefit based on chemistry alone. On the other hand, the anti-knock effect of charge cooling increases nearly linearly over the whole range of ethanol fraction. Therefore, anti-knock benefits become minimal when ethanol blend fraction goes above 30–40% by volume for port fuel injected systems, where fuel impinges on port surfaces providing the heat of evaporation. However, higher ethanol content will provide additional anti-knock benefits in direct injection systems with minimal fuel surface wetting. Evaporative cooling effects are roughly equivalent to 5 octane number points for gasoline, increasing to 18 octane number points for an E85 blend of ethanol [36, 37].

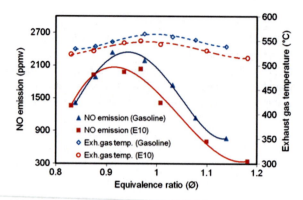

Fig. 10.1 NOx emissions and exhaust gas temperatures dependent of equivalents ratio [21]

4 Effect of Ethanol Fraction on Engine Performance and Economy Parameters

4.1 Advantages of Ethanol Heat of Vaporisation

The higher latent heat on vaporisation of ethanol causes an increased cooling of the mixture during fuel evaporation than gasoline leading to lower engine knock. The lower cylinder temperatures also constitute lower exhaust gases, decreasing NOx emissions [18, 29, 71]. To illustrate this outcome, Fig. 10.1 depicts the effect of E0 and E10 on NOx emissions [21, 70].

Figure 10.1 indicates that at lean conditions with equivalents ratios < 0.9, the NOx emissions of gasoline and ethanol are comparable, and it can be concluded that the cooling effect of ethanol is less significant than the air/fuel ratio. So, based on the data presented by El-Faroug et al. [21], the reduction of NOx emissions depends on the engine settings like ignition timing and air/fuel ratio. Thus, the positive effect of ethanol's high heat of vaporisation demands for precise engine tuning.

The cooling effect of the higher latent heat of evaporation of ethanol has a positive effect on the volumetric efficiency during port fuel injection, as the cooling effect allows more air to enter the cylinder. An increase in volumetric efficiency can be observed with increased ethanol rates. Some effects are observed even with direct injection engines [57]. Moreover, there is a direct link between the high heat or vaporisation of ethanol and the rise in thermal efficiency of up to 20% due to the energy used to vaporise ethanol [29, 68].

4.2 Advantages of Ethanol Octane Number

Ethanol displays good anti knock properties due to an octane number in RON 108/ MON 90 [47]. Consequently, ethanol is used as an octane number improver during gasoline and ethanol blending. So, ethanol fuels can be used in engines with higher compression ratios and boost pressures to increase engine efficiency. Hydrous ethanol has an even higher-octane number than anhydrous ethanol, but blending hydrous ethanol with gasoline requires additives to mitigate fuel stratification [21].

4.3 Disadvantages of Low Heating Value

The lower heating value of ethanol is two times lower than that of gasoline. As a result, the fuel quantity and the consequent fuel consumption must be higher to sustain the same engine work and similar loading conditions. The rise in fuel consumption can depend on engine tuning, geometry, etc. However, hydrous ethanol with 40% water content can increase BSFC up to 75% compared to anhydrous ethanol due to the reduced heating value of hydrous ethanol that prolongs combustion duration [21].

The rise in fuel consumption is proportional to ethanol content in fuel blends and makes engine modification for flex fuel applications difficult. Converting a conventional gasoline engine to ethanol fuel requires not only the adjustment of engine settings but, in some cases, fitting higher flow injectors to accommodate the higher fuel flow rates.

For port-injected fuel systems, the lower calorific value of ethanol is not beneficial to the volumetric efficiency. Large fuel volumes disturb the cylinder filling because the heating value is lower, and the injected fuel quantity is higher. This means less air is inducted as a substantial portion of the cylinder is filled with fuel vapours. To illustrate this, Fig. 10.2 represents the relationship between volumetric efficiency and ethanol fraction in the fuel [60].

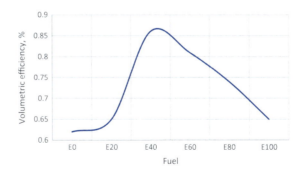

Fig. 10.2 Change in volumetric efficiency with different ethanol fuel blend in a naturally aspirated engine at 100% loading (based on Sadiq et al. [60])

4.4 Disadvantages of Ethanol During Cold Start

Due to higher latent heat of vaporisation values of ethanol, engine cold starting is problematic because lower cylinder temperatures are insufficient for ethanol vaporisation. Moreover, the lower heating value of ethanol requires higher volumes of fuel to be injected into the cylinder than gasoline. This mandates that the initial cylinder temperature be even higher to ensure fuel vaporisation. Based on literary sources, cold starting ethanol fuelled engines below 13 °C is difficult [7, 49]. Other sources suggest difficulties arise at temperatures below 5 °C [29]. Fuels with hydrous ethanol require temperatures above 13 °C as the heat of vaporisation increases with water content [21]. Various solutions have been developed to combat cold start issues, ranging from heated fuel lines to pre-heating the intake air. Furthermore, additives are blended with ethanol to ease cold starting [49, 62].

4.5 Effect of Ethanol Content on Efficiency and Fuel Economy Data

Due to ethanol having a heating value two times lower than gasoline, fuel consumption should presumably rise by a factor of two as well [60]. To describe engine efficiency and fuel consumption adequately between different fuels like gasoline and ethanol brake specific fuel consumption (BSFC) is the reasonable metric. Figure 10.3 describes the BSFC of different fuel blends where for E5 and E10, the BSFC is similar to E0 within a 5% margin but for fuel blends above E50, the BSFC can rise to 70–75% [21, 29].

The thermal brake efficiency of an engine increases with ethanol, and the rise in ethanol content of fuel blends also constitutes a rise in thermal efficiency [29, 60, 68].

Figure 10.4 describes the relation between ethanol-gasoline fuelling at different equivalence ratios. It is important to note that the increase in thermal efficiency largely depends on the engine tuning, ignition timing and air/fuel ratio.

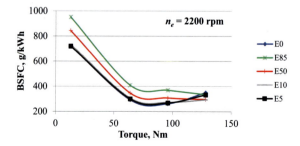

Fig. 10.3 Break specific fuel consumption of various ethanol-gasoline blends

Fig. 10.4 Brake thermal efficiency at different equivalence ratios with E0 and E10 fuels [21]

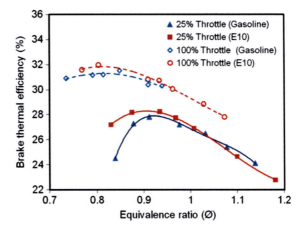

4.6 *Engine Design for High Ethanol Fraction Fuel*

Engines fuelled with ethanol can largely be divided into spark- and compression-ignition engines, where the SI engines are modified to be fuelled by ethanol, which is named flex-fuel engines (FFE). These engines are built to run on fuels from E10 to E85 with greater efficiency than conventional gasoline engines. They have higher compression ratios to utilise the benefits of the higher-octane number and high latent heat of vaporisation of ethanol, by which they also display higher thermal efficiency. Although higher CR values have been mentioned in scientific works, the compression ratios of FFE have remained between 11 and 13 throughout various generations of development [7]. This can be explained by the benefits of implementing forced induction in commercial engines that allows for higher thermal efficiency while maintaining optimum engine design. As there are various blends of ethanol fuel commercially available, it is essential that we can determine the ethanol content of the fuel in the tank. Modern FFE has specialised sensors for determining the alcohol content and can adjust engine settings like injection and ignition timing, EGR valve position etc. [7]. One such FFE management system is depicted in Fig. 10.5.

Fuels with high ethanol content require corrosion-resistant material and lubricants. The engine components most susceptible to corrosion in the fuel supply system are the fuel tank, fuel pump, fuel lines, common rail, and injectors. In addition, the piston assembly, combustion chamber and valves are prone to corrosion. The main source of corrosion is the water in the ethanol. Moreover, additives with high chloride ion content cause galvanic and stress corrosion in fuels with high polarity. As a countermeasure, various surface coating technologies are used to protect FFE components from corrosion. For example, aluminium alloys are coated with nickel to reduce the oxidising effects of ethanol [7].

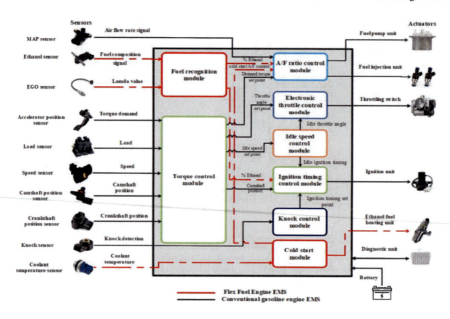

Fig. 10.5 Engine management system and components of an FFE [7]

5 Effect of Ethanol Fraction on Engine Out Emissions

5.1 C/H/O Ratios

The main elements involved in fuel combustion are carbon, oxygen, and hydrogen. For this reason, their ratio within fuel is an important property. The ratio determines not only the combustion products of these elements but also the combustion characteristics and the amount of emissions produced in the process. Ethanol has a higher oxygen content than gasoline, whereas gasoline has higher carbon and almost equal hydrogen content [1, 29, 38, 43, 49].

The ratios and the percentage content of elements for fuels are given in Table 10.4.

5.2 Flame Temperatures

In ideal stoichiometric conditions, ethanol burns with a lower adiabatic flame temperature than gasoline [45]. This contributes to better thermal efficiency as less heat is dissipated through the cylinder walls. In conditions where ethanol or gasoline and ethanol blends are used without appropriate fuelling, a very lean mixture can increase combustion temperatures due to an abundance of oxygen. In addition, as the flame speed of ethanol is higher, the resulting increase in combustion speed can also induce higher flame temperatures [44].

5.3 Effect of Ethanol Fraction on NOx

With ethanol fuels, the effect on increasing or reducing NOx formation is strongly influenced by engine loading and speed, mixture strength and combustion chamber temperature and pressure, as well as initial charge temperature and pressure; therefore, a conclusion is not easy to draw [8, 38, 44, 74].

Effects contributing to the reduction of NOx are mainly linked to the cooling effect of ethanol mixture preparation due to the high latent heat of evaporation and lower adiabatic flame temperature that leads to lower combustion temperatures and hinders thermal NOx formation [61]. Moreover, increased ethanol rates bring additional oxygen radicals into the reaction that inhibit thermal NOx products [74]. However, the properties of ethanol may contribute to an increase in NOx under different combustion conditions. The main factor in increasing NOx emissions is an increase in cylinder pressure and increased cylinder temperature [8, 44]. A higher oxygen concentration within the cylinder can cause such a situation since fuels burn hotter under a surplus of oxygen. Excess oxygen is introduced into the charge by a leaner mixture or the oxygen molecules within ethanol. As seen in Table 10.3, ethanol has a lower air–fuel ratio. Using ethanol and ethanol blends in gasoline engines without proper modifications can lead to lean and very lean conditions and increased NOx formation due to thermal effects. Furthermore, the oxygen in the fuel can react with nitrogen from the air, resulting in N_2O formation [38, 63, 74].

5.4 Effect of Ethanol Fraction on PM

Ethanol is seen as a way to reduce particle matter or soot during combustion as the –OH bonds contribute to soot oxidation [38, 74]. The oxygen content and easier evaporation of ethanol show lower PM values than gasoline when comparing directly injected fuels. Blended fuels like E30 also have lower PM values. However, they display an increase in particle size as the ethanol lowers charge temperature due to a high latent heat of evaporation that disrupts the evaporation of gasoline. Due to this property, a dual-fuel solution is favourable to fuel blending [11].

5.5 Effect of Ethanol Fraction on CO

Engines fuelled with ethanol fuels have a higher fuel consumption due to the lower air/fuel ratio of ethanol compared to gasoline. As a result, higher concentrations of

Table 10.3 Engine-relevant properties of ethanol and gasoline

	Neat Ethanol	Hydrous Ethanol	Gasoline
Chemical formula	C_2H_5OH	–	$C_nH_{1.87n}$
Molar weight (g/mol)	46.07	–	~ 100
Density (kg/dm^3)	0.79^d	–	0.72–0.78
RON	108.4^a	110.3^c	91–99
MON	90.0^a	92.1^c	82–89
Octane sensitivity (RON-MON)	18.4	–	–
H/C Ratio	3	–	1.87
Water (%vol)	–	5.8^c	–
Lower heating value (MJ/kg)	26.95^d	–	43.0
Lower heating value (MJ/L)			
Stoichiometric air–fuel ratio	8.98^d	–	
Heat of vaporisation (kJ/kg)	885^b	1020^c	305^b
Lower heating value/mass stoichiometric mixture (MJ/kg)	2.69^e	–	2.76^e
Heat of vaporisation/mass stoichiometric mixture (kJ/kg)	88.5^b	–	19.5^b
Max. charge cooling delta T (K)	81^b	–	19^b
Stoichiometric A/F ratio	9.0^e	–	14.6^e

[a] average of values provided by Foong [24]
[b] Kasseris and Heywood [36]
[c] average of values provided by Anderson et al. [6]
[d] Fan et al. [23]
[e] Heywood [17]

Table 10.4 Properties of ethanol, gasoline and their blends

	Ethanol	Gasoline	78% Gasoline + 22% anhydrous ethanol
Chemical formula	C_2H_5OH	C2-C12c	C6.39H13.60O0.16
H/C ratio	3	1.795	–
O/C ratio	0.5	0	–
Carbon %	50…52	85…88	76.7
Hydrogen %	13	12…15	13.6
Oxygen %	34…37	0…4	9.7
Stoichiometric air–fuel ratio	9:1	14.7:1	13.07:1
Laminar flame speed (100 kPa, 100 °C) m/s	0.62	0.49	
Lower heating value (MJ/kg)	26.9	42.9	39.22
Mixture energy density (MJ/kg)	2.61	2.67	–

Source [1, 38, 43, 47, 49]

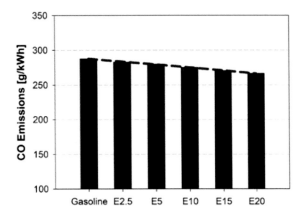

Fig. 10.6 Variation of specific CO emissions with increased ethanol content [38]

carbon-based molecules like CO and CO_2 are produced. At the same time, more oxygen is introduced into the cylinder due to chemically bound molecules and the air/fuel ratio. This effectively reduces CO formation as more oxygen is available for combustion, and the complete oxidation of carbon to CO_2 is promoted [44, 61].

Even lower ethanol blends of E10 contribute to lower CO emissions (Fig. 10.6). A noticeable effect is seen until E30, after which only minor improvements or increases in emissions arise (Fig. 10.6) [8, 38, 44, 61].

5.6 Effect of Ethanol Fraction on UHC

As discussed with CO emissions, the oxygen content of ethanol promotes complete combustion, and with an increased ethanol ratio, a decrease in HC emissions can be observed (Fig. 10.7). Moreover, the faster laminar flame speed of ethanol promotes higher cylinder pressures and temperatures, providing better conditions for combustion [8, 38, 63].

Due to a change in air/fuel ratio when switching from gasoline to ethanol, an increased amount of carbon and hydrogen is involved in the combustion reaction, thus giving way to higher concentrations of carbon-based emissions, especially with unmodified fuel supply systems. The resulting lean mixture decreases flame speed due to a diluted charge, resulting in a misfire and incomplete combustion.

Furthermore, ethanol blends above E30 also contribute to a leaner mixture and increased HC emissions, as in some cases studied (Fig. 10.8). The positive effect of increased ethanol content diminished and even increased HC emissions. The lower combustion temperatures resulting from lower combustion efficiency can also lead to wall quenching and produce HC emissions. Higher engine speeds increase combustion temperatures and pressures, thus providing better conditions for combustion [44, 61, 63].

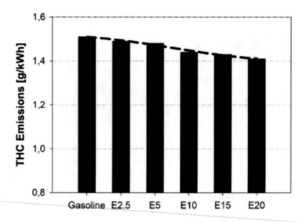

Fig. 10.7 Variation of specific HC emissions with increased ethanol content [38]

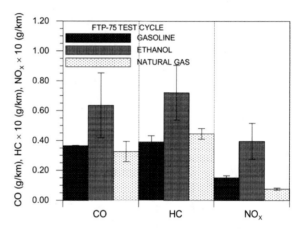

Fig. 10.8 Average CO, HC and NOx emissions of E22, hydrous ethanol and CNG gas during a full FTP-75 test cycle [44]

5.7 Effect of Ethanol Fraction on Cold Start Emissions

The drawbacks of ICE are their long warm-up period and the need to idle even without loading to reach operational temperatures. Moreover, the engines are sensitive to lower ambient temperatures and produce multiple times more emissions during cold starting than at optimum temperatures. The lower temperatures prolong combustion due to fuel vaporisation difficulties and wall wetting inhibiting proper mixture building. As a result, incomplete combustion and misfire occur, and unburned combustion products are expelled into the exhaust. Ethanol displays a higher heat of vaporisation and vapour pressure than neat gasoline. Thus, it is more difficult to ignite in colder conditions and produces higher CO and HC emissions during cold starting with increasing ethanol content (Fig. 10.9) [27, 45, 63].

Ambient temperatures lower than 15 °C make it near impossible to start an engine fuelled with higher ethanol blends without an additional gasoline injection of

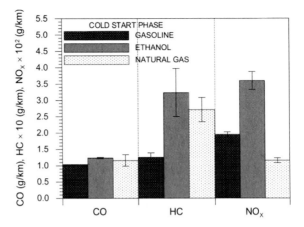

Fig. 10.9 Average CO, HC and NOx emissions of E22, hydrous ethanol and CNG during cold start phase operation [44]

heating systems for fuel or intake air implemented [45, 49]. Colder engine components draw heat from the mixture and combustion process, increasing heat loss and inhibiting mixing during combustion, contributing to increased PM emissions due to rich combustion zones [27].

6 Conclusion

Ethanol as a motor fuel has been around for a long time, and several ethanol blends have been introduced. In modern times ethanol mixtures such as E5, E10, E15, E70, and E85 are used worldwide. Ethanol fuel blends like E5 and E10 can be used in a conventional gasoline engine, whereas engine tuning is required for higher ethanol content fuels. When blended into gasoline-range motor fuels, the physicochemical properties of ethanol impact various aspects of the combustion engine such as performance, emissions, and efficiency. Characteristics including oxygen content, heating value, liquid density, heat of vaporization, surface tension, viscosity, vapour pressure and flame speed are well known to significantly affect engine behaviours such mileage, cold-starting ability, emissions and knock tendency. Using ethanol has many benefits to the combustion process of the SI-engine like the high-octane number of ethanol that allows to rise engine compression ratio and thereby increase thermal efficiency. Therefore, the usage of ethanol as a motor fuel should be more widespread, as it is seen as an alternative fuel for polluting fossil fuels. To reduce the negative effect of internal combustion engines on the environment the usage of fuels derived from renewable, and carbon neutral sources is most welcome and future production engines should be made capable of running on alcohol-based fuels and their blends from the factory.

References

1. Al-Muhsen NFO, Hong G (2017) Effect of spark timing on performance and emissions of a small spark ignition engine with dual ethanol fuel injection. SAE Technical Papers.https://doi.org/10.4271/2017-01-2230
2. AlRamadan AS, Sarathy SM, Khurshid M, Badra J (2016) A blending rule for octane numbers of PRFs and TPRFs with ethanol. Fuel 180:175–186. https://doi.org/10.1016/j.fuel.2016.04.032
3. AlRamadan AS, Sarathy SM, Badra J (2021) Unraveling the octane response of gasoline/ethanol blends: paving the way to formulating gasoline surrogates. Fuel 299. https://doi.org/10.1016/j.fuel.2021.120882
4. Anderson JE, Kramer U, Mueller SA, Wallington TJ (2010) Octane numbers of ethanol- and methanol-gasoline blends estimated from molar concentrations. Energy Fuels 24(12):6576–6585. https://doi.org/10.1021/ef101125c
5. Anderson JE, Wallington TJ (2020) Novel method to estimate the octane ratings of ethanol-gasoline mixtures using base fuel properties. Energy Fuels 34(4):4632–4642. https://doi.org/10.1021/acs.energyfuels.9b04204
6. Anderson JE, Thomas GL, Michael HS, Wallington TJ, Bizub JJ, Foster M, Lynskey MG, Polovina D (2012) Octane numbers of ethanol-gasoline blends: measurements and novel estimation method from molar composition. SAE Technical Papers. https://doi.org/10.4271/2012-01-1274
7. Azhaganathan G, Bragadeshwaran A (2022) Critical review on recent progress of ethanol fuelled flex-fuel engine characteristics. Int J Energy Res 46(5):5646–5677. https://doi.org/10.1002/er.7610
8. Barboza ABV, Mohan S, Dinesha P (2022) On reducing the emissions of CO, HC, and NOx from gasoline blended with hydrogen peroxide and ethanol: optimization study aided with ANN-PSO. Environ Pollut 310(October):119866. https://doi.org/10.1016/j.envpol.2022.119866
9. Bielaczyc P, Woodburn J, Klimkiewicz D, Pajdowski P, Szczotka A (2013) An examination of the effect of ethanol–gasoline blends' physicochemical properties on emissions from a light-duty spark ignition engine. Fuel Process Technol 107:50–63. https://doi.org/10.1016/j.fuproc.2012.07.030
10. Bureshaid K, Shimura R, Feng D, Zhao H, Bunce M (2019) Experimental studies of the effect of ethanol auxiliary fuelled turbulent jet ignition in an optical engine. SAE Int J Engines 12(4):387–399. https://doi.org/10.4271/03-12-04-0026
11. Catapano F, Di Iorio S, Luise L, Sementa P, Vaglieco BM (2019) Influence of ethanol blended and dual fueled with gasoline on soot formation and particulate matter emissions in a small displacement spark ignition engine. Fuel 245(June):253–262. https://doi.org/10.1016/j.fuel.2019.01.173
12. Chow EW, Heywood JB, Speth RL (2014) Benefits of a higher octane standard gasoline for the U.S. light-duty vehicle fleet. SAE Technical Papers 1. https://doi.org/10.4271/2014-01-1961
13. Chupka GM (2015) Heat of vaporization measurements for ethanol blends up to 50 volume percent in several hydrocarbon blendstocks and implications for knock in SI engines. SAE Int J Fuels Lubr 8(2):251–263. https://doi.org/10.4271/2015-01-0763
14. Cornell JA (2011) A primer on experiments with mixtures. A primer on experiments with mixtures, no. Hoboken, NJ, USA, pp 1–351. https://doi.org/10.1002/9780470907443
15. Costanzo V, Yu X, Chapman E, Davis R (2021) Fuel effects on the propensity to establish propagating flames at SPI-relevant engine conditions. SAE Technical Papers, no. 2021. https://doi.org/10.4271/2021-01-0488
16. Dahmen M, Marquardt W (2017) Model-based formulation of biofuel blends by simultaneous product and pathway design. Energy Fuels 31(4):4096–4121. https://doi.org/10.1021/acs.energyfuels.7b00118
17. Davis SC, Boundy RG (2022) Transportation energy data book edition 40
18. De Melo TCC, Machado GB, Belchior CR, Colaço MJ, Barros JE, de Oliveira EJ, de Oliveira DG (2012) Hydrous ethanol-gasoline blends-combustion and emission investigations on a flex-fuel engine. Fuel 97:796–804

19. Dempsey AB, Das Adhikary B, Viswanathan S, Reitz RD (2012) Reactivity controlled compression ignition using premixed hydrated ethanol and direct injection diesel. J Eng Gas Turbines Power 134(8)
20. Energy API (2010) Determination of the potential property ranges of mid-level ethanol blends. American Petroleum Institute, pp 1–108
21. El-Faroug MO, Yan F, Luo M, Turkson RF (2016) Spark ignition engine combustion, performance and emission products from hydrous ethanol and its blends with gasoline. Energies 2016:9
22. EPA (2020) The 2020 EPA automotive trends report. Epa
23. Fan Q, Wang Z, Qi Y, Wang Y (2019) Investigating auto-ignition behavior of n-Heptane/Iso-Octane/Ethanol mixtures for gasoline surrogates through rapid compression machine measurement and chemical kinetics analysis. Fuel 241:1095–1108. https://doi.org/10.1016/j.fuel.2018.12.112
24. Foong TM, Morganti KJ, Brear MJ, Da Silva G, Yang Y, Dryer FL (2014) The octane numbers of ethanol blended with gasoline and its surrogates. Fuel 115:727–739. https://doi.org/10.1016/j.fuel.2013.07.105
25. Haenel P, Kleeberg H, de Bruijn R, Tomazic D (2017) Influence of ethanol blends on low speed pre-ignition in turbocharged. Dir Injection Gasoline Engines. https://doi.org/10.4271/2017-01-0687
26. Hanson R, Curran S, Wagner R, Reitz R (2013) Effects of biofuel blends on RCCI combustion in a light-duty, multi-cylinder diesel engine. SAE Int J Engines 6(1):488–503. https://doi.org/10.4271/2013-01-1653
27. He X, Zhou Y, Liu Z, Yang Q, Sjöberg M, Vuilleumier D, Ding CP, Liu F (2022) Impact of coolant temperature on the combustion characteristics and emissions of a stratified-charge direct-injection spark-ignition engine fueled with E30. Fuel 309(February):121913. https://doi.org/10.1016/j.fuel.2021.121913
28. Hunwartzen I (1982) Modification of CFR test engine unit to determine octane numbers of pure alcohols and gasoline-alcohol blends. SAE technical papers 820002. https://doi.org/10.4271/820002
29. Ilves R, Küüt A, Olt J (2019). Ethanol as internal combustion engine fuel. In: Angelo Basile Adolfo Iulianelli Francesco Dalena T. Nejat Veziroglu (ed) Ethanol, 1st edn. Science and Engineering. Elsevier, pp 215–229
30. Kalghatgi GT (2015) Developments in internal combustion engines and implications for combustion science and future transport fuels. Proc Combust Inst 35(1):101–115. https://doi.org/10.1016/j.proci.2014.10.002
31. Kalghatgi GT (2001a) Fuel anti-knock quality-part I. Engine studies. SAE technical papers, no. 724. https://doi.org/10.4271/2001-01-3584
32. Kalghatgi GT (2001b) Fuel anti-knock quality-part II. Vehicle studies-how relevant is motor octane number (MON) in modern engines? SAE technical papers, no. 724. https://doi.org/10.4271/2001-01-3585
33. Kalghatgi G, Risberg P, Ångstrom HE (2003) A method of defining ignition quality of fuels in HCCI engines. SAE Technical Papers. https://doi.org/10.4271/2003-01-1816
34. Kalghatgi G (2013) Fuel/engine interactions. SAE International
35. Kar K (2009) Measurement of vapor pressures and enthalpies of vaporization of gasoline and ethanol blends and their effects on mixture preparation in an SI engine. SAE Int J Fuels Lubr 1(1):132–144. https://doi.org/10.4271/2008-01-0317
36. Kasseris E, Heywood JB (2012b) Charge cooling effects on knock limits in SI DI engines using gasoline/ethanol blends: part 2-effective octane numbers. SAE Int J Fuels Lubr 5(2):844–854. https://doi.org/10.4271/2012-01-1284
37. Kasseris E, Heywood JB (2012a) Charge cooling effects on knock limits in SI Di engines using gasoline/ethanol blends: part 1-quantifying charge cooling. SAE Technical Papers. https://doi.org/10.4271/2012-01-1275
38. Köten H, Karagöz Y, Balcı Ö (2020) Effect of different levels of ethanol addition on performance, emission, and combustion characteristics of a gasoline engine. Adv Mech Eng 12(7):1–13. https://doi.org/10.1177/1687814020943356

39. Leone TG (2014) Effects of fuel octane rating and ethanol content on knock, fuel economy, and CO2 for a turbocharged DI engine. SAE Int J Fuels Lubr 7(1):9–28. https://doi.org/10.4271/2014-01-1228
40. Leone TG (2015) The effect of compression ratio, fuel octane rating, and ethanol content on spark-ignition engine efficiency. Environ Sci Technol 49(18):10778–10789. https://doi.org/10.1021/acs.est.5b01420
41. Leppard WR (1990) The chemical origin of fuel octane sensitivity. SAE Technical Papers. https://doi.org/10.4271/902137
42. MON, ASTM D2700-16. n.d. ASTM D2700-16 (2016) Standard test method for motor octane number of spark-ignition engine fuel. ASTM Int. https://doi.org/10.1520/D2700-16
43. Malaquias ACT, Netto NAD, Filho FAR, da Costa RBR, Langeani M, Baêta JGC (2019) The misleading total replacement of internal combustion engines by electric motors and a study of the brazilian ethanol importance for the sustainable future of mobility: a review. J Braz Soc Mech Sci Eng. https://doi.org/10.1007/s40430-019-2076-1
44. Martins AA, Rocha RAD, Sodré JR (2014) Cold start and full cycle emissions from a flexible fuel vehicle operating with natural gas, ethanol and gasoline. J Nat Gas Sci Eng 17(March):94–98. https://doi.org/10.1016/j.jngse.2014.01.004
45. Martins J, Brito FP (2020) Alternative fuels for internal combustion engines. Energies 13(15). https://doi.org/10.3390/en13164086
46. Mehl M, Faravelli T, Giavazzi F, Ranzi E, Scorletti P, Tardani A, Terna D (2006) Detailed chemistry promotes understanding of octane numbers and gasoline sensitivity. Energy Fuels 20(6):2391–2398. https://doi.org/10.1021/ef060339s
47. Meng L (2019) Ethanol in automotive applications. In: Angelo Basile Adolfo Iulianelli Francesco Dalena T. Nejat Veziroglu (ed) Ethanol, 1st edn. Science and Engineering. Elsevier, pp 289–302
48. Mittal V, Heywood JB (2008) The relevance of fuel RON and MON to knock onset in modern SI engines. SAE Technical Papers 2(2):1–10. https://doi.org/10.4271/2008-01-2414
49. Monteiro Sales LC, Sodré JR (2012) Cold start characteristics of an ethanol-fuelled engine with heated intake air and fuel. Appl Therm Eng 40(July):198–201. https://doi.org/10.1016/j.applthermaleng.2012.01.057
50. Morganti K, Viollet Y, Head R, Kalghatgi G, Al-Abdullah M, Alzubail A (2017) Maximizing the benefits of high octane fuels in spark-ignition engines. Fuel 207:470–487. https://doi.org/10.1016/j.fuel.2017.06.066
51. Pan J, Ding Y, Tang R, Wang L, Wei H, Shu G (2022) Ethanol blending effects on auto-ignition and reaction wave propagation under engine-relevant conditions. SSRN Electron J. https://doi.org/10.2139/ssrn.4126116
52. Patrick H, Kleeberg H, de Bruijn R, Tomazic D (2017) Influence of ethanol blends on low speed pre-ignition in turbocharged, direct-injection gasoline engines. SAE Int J Fuels Lubr 10(1):95–105. https://doi.org/10.4271/2017-01-0687
53. Pawlowski A, Splitter D (2015) SI engine trends: a historical analysis with future projections. In: SAE 2015 world congress & exhibition. SAE International. https://doi.org/10.4271/2015-01-0972
54. Prakash A, Cracknell R, Natarajan V, Doyle D, Jones A, Jo YS, Hinojosa M, Lobato P (2016) Understanding the octane appetite of modern vehicles. SAE Int J Fuels Lubr 9(2):345–357. https://doi.org/10.4271/2016-01-0834
55. Prakash A, Wang C, Janssen A, Aradi A, Cracknell R (2017) Impact of fuel sensitivity (RON-MON) on engine efficiency. SAE Int J Fuels Lubr 10(1):115–125. https://doi.org/10.4271/2017-01-0799
56. RON, ASTM D2699–19. n.d. ASTM D2699–19 (2016) Standard test method for research octane number of spark-ignition engine fuel. ASTM International. https://doi.org/10.1520/D2699-19
57. Ratcliff MA, Windom B, Fioroni G M, St John P, Burke S, Burton J, Christensen ED, Sindler P, McCormick RL (2019) Impact of ethanol blending into gasoline on aromatic compound evaporation and particle emissions from a gasoline direct injection engine. Appl Energy 250:1618–1631

58. Risberg P, Kalghatgi G, Ångstrom HE (2003) Auto-ignition quality of gasoline-like fuels in HCCI engines. SAE Technical Papers, no. January. https://doi.org/10.4271/2003-01-3215
59. Robertson D, Prucka R (2019) A review of spark-assisted compression ignition (SACI) Research in the context of realizing production control strategies. SAE Technical Papers 2019-Septe (September). https://doi.org/10.4271/2019-24-0027
60. Sadiq MAR, Ali YK, Noor AR (2011) Effects of ethanol-gasoline blends on exhaust and noise emissions from 4-stroke S.I. Engine. Eng Technol J Univ Technol Iraq 29:1438–1450
61. Sakthivel P, Subramanian KA, Mathai R (2020) Experimental study on unregulated emission characteristics of a two-wheeler with ethanol-gasoline blends (E0 to E50). Fuel 262(February):116504. https://doi.org/10.1016/j.fuel.2019.116504
62. Savelenko VD, Ershov MA, Kapustin VM, Chernysheva EA, Abdellatief TM, Makhova UA, Makhmudova AE, Abdelkareem MA, Olabi A (2022) Pathways resilient future for developing a sustainable E85 fuel and prospects towards its applications. Sci Total Environ 844:157069
63. Schifter IU, González L, Díaz R, Rodríguez I, Mejía-Centeno MCG (2018) From actual ethanol contents in gasoline to mid-blends and E-85 in conventional technology vehicles. Emission control issues and consequences. Fuel 219(May):239–247. https://doi.org/10.1016/j.fuel.2018.01.118
64. Sjöberg M, John E (2011) Smoothing HCCI heat release with vaporization-cooling-induced thermal stratification using ethanol. SAE Technical Papers 5(1)
65. Stein RA (2012) Effect of heat of vaporization, chemical octane, and sensitivity on knock limit for ethanol—gasoline blends. SAE Int J Fuels Lubr 5(2):823–843. https://doi.org/10.4271/2012-01-1277
66. Stein RA, Anderson JE, Wallington TJ (2013) An overview of the effects of ethanol-gasoline blends on SI engine performance, fuel efficiency, and emissions. SAE Int J Engines 6(1):470–487. https://doi.org/10.4271/2013-01-1635
67. Szybist JP, Splitter DA (2018) Understanding chemistry-specific fuel differences at a constant RON in a boosted SI engine. Fuel 217:370–381. https://doi.org/10.1016/j.fuel.2017.12.100
68. Thangavelu SK, Ahmed AS, Ani FN (2016) Review on bioethanol as alternative fuel for spark ignition engines. Renew Sustain Energy Rev 56:820e835
69. Thrän D, Naumann K, Billig E, Millinger M, Oehmichen K, Pfeiffer D, Zech K (2018) Data on biofuels production, trade and demand. In: Riazi MR, Chiaramonti D (eds) Biofuels production and processing technology. CRS Press, Taylor and Francis Group, pp 55–95
70. Venugopal T, Sharma A, Satapathy S, Ramesh A, Gajendra Babu M (2013) Experimental study of hydrous ethanol gasoline blend (E10) in a four stroke port fuel-injected spark ignition engine. Int J Energy Res 37:638–644
71. Wang X, Chen Z, Ni J, Liu S, Zhou H (2015) The effects of hydrous ethanol gasoline on combustion and emission characteristics of a port injection gasoline engine. Case Stud Ther Eng 6:147–154
72. Waqas M, Naser N, Sarathy M, Morganti K, Al-Qurashi K, Johansson B (2016) Blending octane number of ethanol in HCCI, SI and CI combustion modes. SAE Int J Fuels Lubr 9(3):659–682. https://doi.org/10.4271/2016-01-2298
73. Westbrook CK, Mehl M, Pitz WJ, Sjöberg M (2017) Chemical kinetics of octane sensitivity in a spark-ignition engine. Combust Flame 175:2–15. https://doi.org/10.1016/j.combustflame.2016.05.022
74. Xie M, Li Q, Fu J, Yang H, Wang X, Liu J (2022) Chemical kinetic investigation on NOx emission of SI engine fueled with gasoline-ethanol fuel blends. Sci Total Environ 831(July):154870. https://doi.org/10.1016/j.scitotenv.2022.154870
75. Yates ADB, Swarts A, Viljoen CL (2005) Correlating auto-ignition delays and knock-limited spark-advance data for different types of fuel. SAE Technical Papers. https://doi.org/10.4271/2005-01-2083

Overview of Commercial Bioethanol Production Plants

Bárbara P. Moreira, William G. Sganzerla, Paulo C. Torres-Mayanga, Héctor A. Ruiz, and Daniel Lachos-Perez

Abstract Compared to fossil fuels, biofuel production has been growing progressively due to much-reduced greenhouse gas emissions. Currently, bioethanol is the biggest biofuel used worldwide, and its production has been increasingly extensive. This chapter discusses different generations of bioethanol processing production, such as the first, second, third and futuristic fourth generations. Also, an overview of commercial bioethanol production shows the United States and Brazil are the countries with dominant industries, highly influenced by their national energy policies. However, the EU and China intend to increase bioethanol production driven by government support in the next few years. Then, bioethanol production is promising due to the growing expansion of countries such as China and within the EU and the demand for renewable fuels.

Keywords Bioethanol · First-generation · Second-generation · Third-generation · Commercial plants

---

B. P. Moreira · D. Lachos-Perez (✉)
Department of Chemical Engineering, Federal University of Santa Maria, 1000 Roraima Avenue, Santa Maria, RS 97105-900, Brazil
e-mail: daniel.perez@ufsm.br; daniel_lachosperez@uml.edu

W. G. Sganzerla
Food Engineering School, University of Campinas (UNICAMP), São Paulo, Brazil

P. C. Torres-Mayanga
Universidad Tecnológica del Perú, Lima, Peru

Innovative Technology, Food and Health Research Group, Departamento de Ingeniería de Alimentos Y Productos Agropecuarios, Facultad de Industrias Alimentarias, Universidad Nacional Agraria La Molina, Av. La Molina S/N—Lima 12, Lima, Peru

H. A. Ruiz
Biorefinery Group, Food Research Department, Faculty of Chemistry Sciences, Autonomous University of Coahuila, 25280 Saltillo, Coahuila, Mexico

© The Author(s), under exclusive license to Springer Nature Switzerland AG 2023
E. Betiku and M. M. Ishola (eds.), *Bioethanol: A Green Energy Substitute for Fossil Fuels*, Green Energy and Technology,
https://doi.org/10.1007/978-3-031-36542-3_11

1 Introduction

Petroleum-based energy consumed in the transport sector accounted for 41.5% of global CO_2 emissions in 2016; on the other hand, plants capture 56×10^9 tons of CO_2 per year and produce $170-200 \times 10^9$ tons of biomass [73]. Thus, replacing of fossil fuels such as diesel and gasoline is an ecologically correct and promising way to reduce dependence on crude oil in the transport sector. Ethanol produced from biomass is termed bioethanol, and its use to fuel an internal combustion engine has been around for a long time, so it is the most used biofuel in the transport sector in the world. Brazil and the USA are the largest bioethanol producers worldwide (7.43 and 15.015 B Gal of ethanol per year, respectively) [73]. However, several countries have adopted new renewable energy policies to introduce bioethanol as a biofuel in the transport sector.

Bioethanol production can be categorized into three generations, the most studied and a fourth a futuristic generation [5]. First-generation ethanol (E1G) is produced by fermenting monomeric sugars extracted from raw materials derived from sugar or starch, such as cassava, corn, sugarcane, or grains, followed by centrifugation and subsequent distillation. The E1G process is relatively simple, provides high yields compared to other generations, and its industrial production is largely consolidated. However, there are some disadvantages related to the production, such as the high costs of feedstock (sugar/starch), representing 40–70% of the total cost, and ethical and social concerns due to the use of food or feed [64, 79].

Ethanol produced from lignocellulosic biomass, called Second Generation Ethanol (E2G), is highly available, low cost, does not involve food consumption (sugar/starch) and can reduce CO_2 emissions produced by up to 50% when compared to E1G, is now considered the most promising alternative for biofuel production [36]. Research and companies have been trying to develop E2G production on a commercial scale. However, the main barrier has been the technological limitations, with only 20 companies worldwide producing at this level. Two companies in Brazil, GranBio and Raízen use sugarcane bagasse and sugarcane straw as feedstock. The United States has seven companies and the rest are in Europe and China [5].

Ethanol produced from micro/macroalgae biomass is named Third Generation Ethanol (E3G). E3G is the latest technology studied in the literature. Among the main merits of this technology is that the growth yield of micro/macroalgae is greater than the growth yield of E1G and E2G feedstock. Also, micro/macroalgae do not contain lignin favoring the bioethanol production compared to E2G. However, culture media requires nutrients and harvesting constitutes an expensive process (represents 20–30% of the total cost). In addition, it is necessary to rupture the cell wall of micro/macroalgae to release the compounds of interest, which requires greater investments in different complex pretreatment methods that also can lead to formation of by-products [68]. Thus, to the best of our knowledge, there are no E3G industrial-scale production [64]. Fourth-generation ethanol (E4G) is obtained from genetically-modified micro/macroalgae and cyanobacteria. But there is limited information about

E4G, and international patents protect it. Thus, there is no real prospect for industrial scale [36].

Several governments around the world have been working extensively to promote novel and clean knowledge for bioethanol production. Therefore, this chapter reviews updates and a detailed overview of the main processes of bioethanol production 1G, 2G, and 3G, as well as the factors that attract and limit commercial bioethanol production. As aforementioned, E1G and E2G are already produced on commercial scales, even though most of the commercial bioethanol is E1G. Therefore, E1G and E2G commercial bioethanol production worldwide is discussed, and economic feasibility and sustainability approach proposed. The main objective is to indicate the importance of producing this biofuel in the current environmental scenario and the possibility of expanding its production.

2 Bioethanol Production: an Overview

Bioethanol production employs several feedstocks composed of carbohydrates or fermentable sugars to obtain the product through fermentation. Most bioethanol in the world belongs to the first-generation food crops such as corn, sugarcane, rice, wheat, and barley using as feedstock [47, 64]. Feedstock like sugarcane juice needs only a fermentation process, and biomass starch-based needs a hydrolysis process before the fermentation [36].

The number of steps required to produce bioethanol changes based on the feedstock used. Figure 11.1 illustrates the process steps according to feedstock.

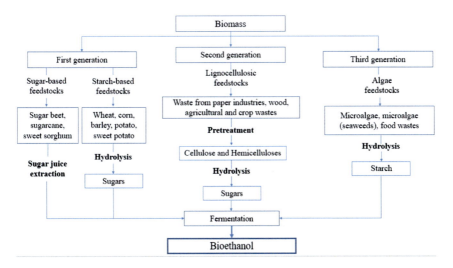

Fig. 11.1 First, second and third ethanol based on different feedstock. Adapted from Melendez et al. [64] and Halder et al. [36]

2.1 First-Generation Bioethanol Production

To produce E1G from sugar-based feedstock, firstly, the sugar juice is extracted and then washed with whitewash to neutralize organic acids and reduce colorants [14]. Subsequently, the purified sugars are fermented using the appropriate microorganisms to obtain bioethanol, which is then subjected to a distillation process. On the contrary, bioethanol production from starch-based feedstock requires a hydrolysis process prior to fermentation to produce sugars converted to bioethanol [52]. Two conventional technologies are applied as the first step in bioethanol production from starchy crops: wet and dry milling. In the wet milling, the corn kernel is subjected to a steeping process to separate its components, and the starch obtained is then transferred to subsequent steps. Dry milling eliminates the steeping process. Thus, the entire corn kernel is ground, and the process follows the same steps as the wet milling. Until now, most commercial bioethanol plants employ the dry milling route since wet milling demands a high capital cost due to extensive and complex equipment [102].

2.2 Second-Generation Bioethanol Production

E2G requires a multi-step process. The presence of lignin in lignocellulosic biomass makes the structure rigid, promoting recalcitrance against hydrolyzing agents [14]. Therefore, before the hydrolysis and fermentation steps, a pretreatment step is required to convert lignocellulosic biomass to bioethanol.

2.2.1 Pretreatment

The conditioning of feedstock by pretreatment improves the yield in any process. Pretreatment of biomasses for biofuel production aims to make the biopolymers present in the biomass more accessible for subsequent degradation by other methods. Conventional pretreatments for this purpose are chemical, biological, mechanical and physicochemical [38, 62, 85, 90].

Subjecting biomasses to pretreatment is beneficial to obtain better results in subsequent stages (Table 11.1). For example, in the hydrolysis stage, regardless of the method used (biological or chemical), the pretreatment allows greater access to the carbohydrate structures and facilitates the release of sugars due to reduced crystallinity.

Table 11.1 Main characteristics of the pretreatments used in the bioethanol process

Pretreatment	Characteristics	References
Chemical	- Use of chemical reagents, acids, and alkalis - Different reaction mechanisms - by-products - Easy handling and availability, and economical - Low yields in longer times with alkali pretreatment (NaOH, lime, NH_4OH) - High yields in short times with acid pretreatment - Provides higher porosity of the feedstock and improves the digestibility of the microorganisms in the hydrolysis stage	[62, 90, 107]
Biological	- Uses microorganisms - Degrade lignocellulosic components - Requires mild process conditions - Low energy consumption - No use of chemical reagents	[85]
Mechanical	- Applies different mechanical techniques to reduce size, by compression, hammers, balls, vibrating balls, and rollers - Reducing the size of the raw material improves degradation	[62]
Physic-chemist	- Application of methods such as steam explosion and wet oxidation efficiently affects fibrous biomasses - Biomasses are subjected to steam explosion and are accompanied by high temperatures and pressures in short periods (1 min) - Higher xylose concentrations are recovered from the hemicellulose fraction by steam explosion - Wet oxidation applies high pressures and temperatures in relative times (10–15 min), degrading lignocellulose biomass and benefiting the removal of the lignin fraction while reducing cellulose crystallinity	[1, 26, 63]

2.2.2 Hydrolysis

The critical factor in bioethanol production is hydrolysis, which is carried out by three methods to obtain sugars from feedstock before fermentation, such as dilute acid hydrolysis, concentrated acid hydrolysis, and enzymatic hydrolysis [102].

Hydrolysis enzymatic: This biological method of fractionating the lignocellulosic matrix has advantages in releasing sugars from hemicellulose and cellulose through enzymes. Because of the action of enzymes, such as cellulases, cellobiohydrolases, and 1,4-β-D-glucosidases, they develop a mechanism that converts cellulose into glucose [13]. Similarly, it happens with hemicellulose, xylanases (endo-β-1,4-xylanases and β-xylosidases) that act on the bonds of the hemicellulose structure releasing xylooligosaccharides and xylose [13, 42]. In summary, the removed components, such as glucans and xylan, mannans, and galactans, present in biopolymers, finally cluster pentose sugars and hexoses as the product of significant interest in developing fermentation [26].

Hydrolysis acid: Concentrated and dilute acids can be applied, which generates the release of sugars for the production of bioethanol; this originated through a process of reactions, firstly, the hydrolysis of hemicellulose and cellulose through a reaction catalyzed by acid under temperature conditions in a range of 150–230 °C [102]. According to research, the acid medium destroys the covalent bonds of the lignocellulosic matrix. Likewise, treatment with dilute acid advantages the solubilization of hemicellulose, generating greater accessibility to microorganisms to the cellulosic fraction and decreasing cellulose's crystallinity. The most commonly used reagents are hydrochloric acid, sulfuric acid, and nitric acid [104]. Under acidic conditions, it does not influence lignin, which is present in the entire production process line. However, this depends on the type of acid used in the pretreatment. According to Sindhu et al. [87], sulfuric acid facilitates the elimination of lignin and the low cost of the process because it is cheap, which is essential for industrial-scale applications.

Commercial companies such as BlueFire Renewables have centralized and improved their bioethanol production technologies by integrating methods such as concentrated acid hydrolysis into their conversion processes [11]. The concentrated acid hydrolysis process allows biopolymers such as cellulose and hemicellulose to be converted into economically viable sugars and feed the fermentation stage and its conversion into bioethanol. In the processes designed, the focus is on obtaining the sugar-acid solution from the process by intensified hydrolysis with concentrated acid. Subsequently, the acid is separated from the sugar by chromatography and taking vital care to maintain the sugar concentrations using ion exchange resins. Then, the operation line continues to the fermentation process.

Some researchers have studied designs with various stages, such as pretreatment, production, purification, hydrolysis, saccharification and simultaneous fermentation, aiming to produce commercial cellulosic bioethanol and by-products of the process [8]. Other studies propose using solvents, such as γ-valerolactone, in combination with acid catalysts for the simultaneous conversion of the cellulose-hemicellulose matrix, which has good bioethanol yields from corn residues [37, 59]. All studies are essentially related to the production of bioethanol and the analysis of different biomass recovery scenarios through mechanical, biological and chemical pre-treatments. These scenarios lead us to infer that many of these process configurations are promising and should be incorporated and designed on an industrial scale for the development of distillery, seeking a viable projection from a technical and economic perspective.

2.2.3 Fermentation

Bioethanol production is generated by developing integrated technologies based on hydrolysis (Table 11.2), such as separated hydrolysis and fermentation, simultaneous saccharification and fermentation, simultaneous saccharification and co-fermentation, and consolidated bioprocessing [1, 20, 25, 61, 67, 92].

Knowledge of pre-treatments and fermentation methods allows defining a competitive technological process in the production of biofuels, aiming to generate benefits in the reduction of environmental impacts, thus achieving the development of a sustainable process on a commercial scale. However, criteria related to infrastructure and production techniques should be evaluated first.

Table 11.2 Main characteristics of the processes

Pretreatment	Characteristics	References
Simultaneous saccharification and fermentation	- Parallel hydrolysis and fermentation in a single reactor - High hydrolysis rates and lower concentrations of accumulated sugars due to the fast conversion of sugars into bioethanol - Reduced inhibitor formation - High bioethanol yields - Economical design and process costs—reaction-conversion is performed in a single reactor - Uses microorganisms suitable for saccharification and fermentation temperature: *Kluyveromyces marxianus* (42–48 °C)—Applicable to various sugars	[16, 21, 30, 39]
Consolidated bioprocessing	- Develops direct microbial conversion - A group of microorganisms produces enzymes (celluloses) and bioethanol in a single step in a compatible way - Uses thermophilic microorganisms such as cellulolytic anaerobic bacteria due to their ability to be applied to various raw materials and withstand high temperatures - Eliminates costs associated with enzyme production and the purchase of additional equipment	[17, 102]
Simultaneous saccharification and co-fermentation	- The integrated process used in microorganisms' total conversion of hexoses and pentoses - Requires genetic modification of yeast or bacteria for pentose conversion - Generates higher ethanol productivity - Favorable for thermotolerant microorganisms tolerant to temperature ranges of 34–37 °C	[7, 16, 34, 82]

2.3 Third-Generation Bioethanol

The process of E3G production follows the same steps as for lignocellulosic materials, including pretreatment, hydrolysis, fermentation, and product recovery [64]. The first step is processing fresh seaweed collected from the sea. The algal paste and powder obtained by a particle reduction process are used for the subsequent hydrolysis step. The most commonly used method to hydrolyze algae polysaccharide is chemical hydrolysis. Researchers found that the classes of macroalgae such as red, green and brown can be hydrolyzed to monosaccharides by treatment with dilute H_2SO_4 at high temperatures [46]. Rapid and complete hydrolysis of the hydrolyzate to monosaccharide can be provided with addition or dilution with water at moderate temperatures. Comparatively, the growth yield of microalgae is higher than that of first and second-generation feedstock, producing 30 times more energy per acre than land crops. Another advantage is the capacity of microalgae to grow in different modes of cultivation, while the feedstock of other generations only grow in autotrophic mode [45].

3 Processing technologies for commercial scale bioethanol production

Bioethanol production has increased worldwide, and a productive approach has positioned some countries, such as the United States and China, to produce biofuels [2, 54]. This can be noticed in some companies and industrial plants that have focused their efforts primarily on processing cellulosic raw materials for bioethanol production [22, 66], highlighting a common good of an emerging economy, waste utilization policy and improvement of production technologies [77, 81].

The technological processes in bioethanol production depend on the methods applied in its pretreatment, fermentation, hydrolysis, and distillation. The various steps involved in bioethanol production are shown in Fig. 11.2 [38]. The first step is conditioning the raw material and applying physical treatments for particle size reduction. Then, the addition of microorganisms and chemicals for the degradation of complex structures in the cell matrix, through biological or chemical hydrolysis, fractionates the components of higher molecular weights, such as starch, cellulose, and hemicellulose to lower molecular weight molecules (monomers). The latter are converted into bioethanol through the fermentation process using microorganisms or biocatalysts, and finally, the broth obtained is subjected to distillation. It should be noted that, depending on the raw material processed, the biofuel is classified as E1G, E2G, and E3G. In these terms, bioethanol production will also depend on the proper development of the stages and conditions involved in the process [69, 80, 85]. Likewise, it should be mentioned that bioethanol production has been widely focused on processes on lignocellulosic biomasses, which do not involve food consumption, as it is a subject of worldwide controversy. For this reason, E2G production is positioned as an eco-friendly and sustainable alternative. Thus, in Fig. 11.3, industries and commercial companies worldwide that produce bioethanol as a biofuel have focused their efforts on processing feedstock such as corn, sugarcane residues, rice, wheat straw, wood residues, corn stover and leaves [31].

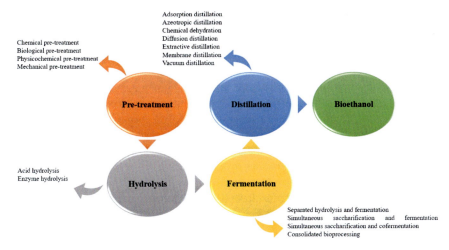

Fig. 11.2 Bioethanol process and main technologies applied [1, 6, 20, 25, 61, 62, 67, 85, 90–92]

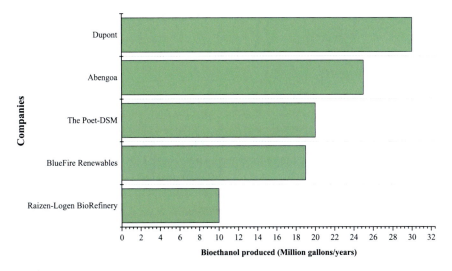

Fig. 11.3 Commercial industries in bioethanol production, adapted from Chilakamarry et al. [22]

4 Main Factors that Attract and Limit Commercial Bioethanol Production

4.1 Composition

One of the limiting factors for bioethanol commercialization is the composition and recalcitrance of the lignocellulosic biomass. This resistance to matrix degradation affects the extraction of monomeric sugars [40]. The recalcitrance of the material is the first reason that limits biomass processing since this implies implementing an entire engineering process to obtain bioethanol. In addition, due to controversial issues of competition with raw materials for direct human consumption, it is more convenient to use residual food matrices (lignocellulosic) in bioethanol production, providing greater sustainability to the process [33, 58, 76]. On the other hand, biomass characterized by relatively low energy and mass densities, which can only be efficient for ethanol production at small scales, should be avoided [58].

Biomass composition is essential for bioethanol commercialization, which implies considering raw materials with relatively low lignin content. According to the composition, the biomass used must have high cellulose, starch, and sugar contents to be efficient in this process [50, 58, 69]. Unfortunately, the lignin fraction is often an interfering component in ethanol yield. It is non-reactive, blocks enzyme access to carbohydrates, and provides a dead volume when it comes to enzymatic carbohydrate hydrolysis reaction [49, 58, 89]. For this reason, it is necessary to separate hemicellulose and cellulose by applying pretreatments (chemical or biological), achieving their degradation into sugars, and fermenting them into ethanol [24, 86].

4.2 Pretreatment

Pretreatments are necessary to overcome the challenges related to the usability of biomass, and this stage would represent a considerable cost (30–40%) of the production process [26, 35, 55, 100]. For this reason, the commercial viability of the process should be sought to reduce production costs. However, some industries dedicated to commercializing bioethanol and its by-products from different agricultural biomasses have focused on implementing and designing bioethanol production processes according to the nature of the biomass.

The reaction mechanism in bioethanol conversion becomes complex due to the composition of the sugar cocktail (pentoses and hexoses) previously released in the pretreatment. Therefore, attention must be paid to developing studies on the affinity of the microorganisms used and their efficient development in the bioprocess. Consequently, this generates a higher cost of the process due to acquiring genetically modified enzymes and yeasts that can metabolize pentoses and hexoses into bioethanol [13, 37, 64, 67]. From an environmental perspective, biological pretreatment in bioethanol production would be an acceptable way to go because no chemicals are used, and it is ecologically safe. However, the low bioethanol production rate on an industrial scale is the main drawback [85]. At the same time, research shows that enzymatic pretreatment is more expensive than chemical pretreatment [48]. However, acid pretreatment requires adequate attention because the reaction system must be developed in a reactor resistant to operational conditions, leading to high process costs. For this reason, it is necessary to perform detailed evaluations of the environmental impacts, risks, and sustainability assessments for the acid pretreatment to determine and ensure the process scenarios that are feasible for the technical-operational parameters, seeking economic and competitive profitability.

When pretreatments are applied, generating other chemical compounds as by-products is a drawback for bioethanol yield. This is related to the affinity and favorable conditions of the microorganisms and chemical reagents added to the pretreatment. The addition of acids acts on the biopolymer structures (cellulose and hemicellulose). When operating on hemicellulose, acetyl groups are released, intensifying the acidic environment. Under these conditions, hemicellulosic sugars from the dehydration reaction mechanism generate other compounds, such as furanic aldehydes (5-hydroxymethylfurfural and furfural) and organic acids [26, p. 17, 98]. In addition, these compounds tend to provide a toxic medium for the microorganisms causing them to inhibit fermentation. In addition, the generation of these compounds can also enhance the commercialization of the process since, from an economic point of view, it improves the viability of the process because the by-products can be used to produce other chemical compounds of industrial interest [98].

Complete compression is imperative from the interaction of the different parameters involved in pretreatment and fermentation, as they can become critical to improve the bioethanol yield, which would affect technical and economic feasibility. The improvement of the production of any product is based on the optimization of the applied process, specifically, its operating conditions and all the components involved in the process. First, there is the need to determine the potential feedstock to produce ethanol, determine different pretreatment technologies to increase biomass accessibility, and improve the fermentation process to obtain high bioethanol yields. In addition, a relative amount of research reports the profitability of the process and its projection on an industrial scale. For example, researchers have highlighted the efficiency of the steam explosion process on plantain waste, where the optimal pretreatment conditions are defined at 2.2% sulfuric acid, 5 min, and 177 °C. The optimal pretreatment conditions are reported as 2.2% sulfuric acid, 5 min, and 177 °C with an extract of 91% glucose [32]. This pretreatment showed promising results, also the combination of temperature conditions and dilute acid concentrations favored the release of hemicellulosic sugars and part of the starch by steam explosion, leaving the cellulose exposed and as a strategy to perform a sequential process to increase the concentration of fermentative sugars and bioethanol yields [32]. Operational conditions determine whether the use of commercial and conventional feedstock and microorganisms impact bioethanol production. For example, studies predicted high bioethanol yields of approximately 82% from the hemicellulosic fraction of wood and sorghum biomass with yeasts in fermentation [12]. In addition, they determined that operating costs were reduced due to the strategic use of secondary effluents with nutrient sources, which favored the reduction of growth time requirements, yeast inoculum, and dry yeast content. Within this context of the reuse of effluents with unique characteristics for microorganisms, this would be attractive to develop a process applied on an industrial scale.

From a technological and scalable point of view, these studies seek to develop a cost-effective process with high bioethanol yields, which turns out to be more promising among the different processes, considering that they must present high hydrolysis rates and lower concentrations of by-products [30, 51]. Therefore, for efficient development of the fermentative pathway and the microorganisms acting in the process, research details that the process conditions must be adequate. The reaction mechanism of the fermentation depends on the conditions of each process, pH between 3.5 and 7.5, and temperature ranging from 25 to 30 °C in which yeasts present good microbial activity. For example, the fermentative affinity of the microorganisms is selective on certain sugars; on the microbial side, *Zymomonas mobilis* generates bioethanol from the conversion of glucose, sucrose and fructose [22, 102]. In the same way, to develop a highly competitive technology at the commercial level, it is vital to highlight the critical effort made in innovative research on processes and technologies applied in ethanol production. To develop this process efficiently and obtain high yields, and in parallel, seeking to reduce the environmental impacts that the operation at an industrial scale can bring [37, 51, 60, 70, 72, 102]. These investigations have discussed topics on the design, simulation, and evaluation of a wide range of industrial chemical processes and integrated process configurations using software such as Aspen Hysys, Aspen Plus® V11 and SuperPro Designer with various bioethanol production models and focusing on the valorization of different biomass [8, 23, 53, 102].

5 Worldwide Commercial Companies for Bioethanol

According to the International Energy Agency, fuel from renewable sources is expected to become about 90% for the transport sector by 2023 due to different reasons (armed conflicts, volatile fossil fuel prices, and global warming). Among the main biofuels, bioethanol achieved 119 billion liters at an annual growth rate of 2%, primarily helped by the increase in bioethanol production from the United States, European Union, and China [5].

Bioethanol world production in 2021 was 27.3 billion Gallons. The United States and Brazil accounted for 82% of global bioethanol production, driven by their governments' support (Table 11.3). The United States is the world's largest producer of ethanol, followed by Brazil, using corn and sugarcane as biomasses, respectively. Over the next ten years, European Union (EU) member countries and China are expected to continue growing [2]. Other countries, including Columbia, Peru, Cuba,

Table 11.3 Bioethanol production around the world (Mil. Gal.)

Region	2021	% Production (2021)
United States	15.015	55
Brazil	7.43	27
European Union	1.35	5
China	860	3
India	860	3
Canada	434	2
Thailand	350	1
Argentina	260	1
Rest of World	711	3
Total	27.27	

Sources from Renewable Fuels Association (RFA)

India, Australia, Ethiopia, Vietnam, and Zimbabwe, will increase their production in the next few years.

In 2021, the United States was the largest, the United States was the largest producer of gasoline blend, with nearly 15,000 million gallons per year (equivalent to 357 million barrels) of total capacity, recovering to a pre-pandemic level of 55% of global output. However, due to the COVID-19 pandemic, there was a decrease in Brazilian bioethanol production. It was about half the volume of the US, with a total annual production capacity of 7,430 million gallons. In addition to other issues, such as adverse conditions for sugarcane cultivation, the sector's economy, and the persistent effects of the pandemic. No other country had a share of more than 3%, although India has seen a notable increase in production.

Due to the fuel against food (FvsF) dispute, E2G companies are expanding in the Brazil, United States, Asia, EU and other regions. The bioethanol worldwide market is ready for significant growth in the next few years. E1G is an established technology with fully developed markets and infrastructure. However, it is criticized for its implications involving land use in the FvsF relationship due to the use of substrates such as corn (United States), sugarcane (Brazil) and wheat and sugar beet (EU) and also for its price production (generally 50–80% of total costs). E2G technology has been widely expanded in recent years around the world due to different aspects, such as reduction of greenhouse gas emissions and global energy security. A large number of commercial scale bioethanol companies have been equipped, however many of them remain inactive or idle due to economic issues or lack of suitable E2G technology as shown in Table 11.4. Furthermore, the production of biofuels from algae has aroused extensive interest. However, several obstacles in the processing of algae fuels must be overcome to commercialize E3G. The huge capital cost of the facilities appears as one of the main limitations in producing E3G for the public market. This is a critical issue to reduce the vulnerability of algae to contamination, and to increase productivity. Some other critical issues that limited the production of E3G on an industrial scale are algae composition (high lipid content), production

Table 11.4 Pioneer global commercial scale E2G plants and their current status

Country output	Company (status)	Start-up year	Raw material	Capacity (ktons)
USA	Abengoa Bioenergy Biomass of Kansas, LLC (Idle)	2014	Corn stover, straw	75
	Aemetis (Planned)	2019	Corn stover	35
	Beta Renewables (On hold)	2018	Wood, Agri-waste	60
	DuPont VERBIO (Idle)	2016	Corn stover	83
	POET-DSM Advanced Biofuels (On hold)	2014	Corn stover, corn cob	25 M Gal
	Fiberight LLC (Under construction)	2019	Paper/Card Fibre	18
	Ineos Bio (Idle)	NI	Vegetative and wood waste	24
Brazil	Raízen Energia (Planned)	2015	Bagasse, straw	1 B liters
	Granbio (Operational)	2014	Bagasse, straw	65
China	Beta Renewables (On hold)	2018	Wood, Agri-waste	90
	Henan Tianguan Group (Idle)	2011	Agri-waste	30
	COFCO Zhaodong Co (Planned)	2018	Corn, tapioca and cellulose	50
	Longlive Bio-technology Co. Ltd (Idle)	2012	NA	60
Norway	Borregaard Industries AS ChemCell Ethanol (Operational)	1938	Wood pulping	16
Sweden	Domsjoe Fabriker (Operational)	1940	Wood pulping	10
Finland	Chempolis Ltd. (Operational)	2008	Wheat straw, corn stover	5
Denmark	Maabjerg Energy Concept Consortium (On hold)	2018	Agri-waste	50
France	Futurol Procethol 2G (Under construction)	2018	Forest crops, Agri-waste	180 kL/year
Italy	Beta Renewables 1 (Idle)	2018	Agri-waste	40
	Beta Renewables (On hold)	2018	Wheat & rice straw	60
	Versalis (Planned to restart)	2020	Corn stover, wheat straw	40
Belgium	ArcelorMittal, (Under construction)	2020	Agricultural waste	16
Spain	Sainc Energy Limited, Cordoba (Planned)	2020	Agricultural waste and wood biomass	25
Austria	AustroCel Hallein (Under construction)	2020	Wood pulping	50
Norway	St1 Biofuels Oy (Planned)	2024	Organic rich waste, agricultural waste	40

Adapted from Raj et al. [75]

cost (which depicts the biofuel price), reduced tolerance to temperature and salinity fluctuations and vulnerability to contamination, and capital costs for photobioreactors production.

6 Economic Feasibility and Sustainability of Commercial Bioethanol: Challenges and Future Perspectives

For the industrial implementation of ethanol plants, it is imperative to understand the economic viability of bioethanol production technologies. This would help implement a fuel ethanol policy [18]. Bioethanol production from lignocellulosic feedstock is expensive compared to sugar-based and starch-based feedstock. Ethanol production costs by corn and sugarcane have been calculated to range from US$9–20 GJ–1 and US$5–9, respectively. Production with lignocellulosic raw materials has an estimated cost of between US$19 and $ 62 [74]. There is a need to overcome several technical barriers to produce profitable bioethanol from lignocellulosic biomass, such as the exploitation of suitable and economic pretreatment technology. Beyond that, the recalcitrant structure of lignocellulosic feedstock is a hurdle to bioethanol production, impacting the costs of the process [28]. Up till the moment, enzymatic hydrolysis of biomass is the most applied pretreatment on an industrial scale to produce fermentable sugars [101, 105].

Some companies employed enzymatic hydrolysis to ethanol, such as Abengoa (capacity: 25 MGY; feedstock: corn stover; capital cost: $350 M), Beta Renewables (capacity: 20 MGY; feedstock: switchgrass; capital cost: $170 M), DuPont Biofuel Solutions (capacity: 25 MGY; feedstock: corn stover; capital cost: $276 M), and POET (capacity: 20 MGY; feedstock: corn stover; capital cost: $250 M) [15]. In addition, two Brazilian E2G plants have been using cellulose derived from sugarcane straw and bagasse as feedstock. The first company (GranBio) presented an initial production capacity of 82 M liters of ethanol per year (21.6 M US Gal), which was necessary for an initial invested US$190 M to build the plant. The second plant (Raízen) invested US$130 M for the production capacity of 40 M liters of E2G per year [70].

Several studies described the possibilities of E2G production following techno-economic studies. Hossain et al. [41] studied the techno-economic analysis of microalgae as a commercial feedstock for bioethanol production. The total cost of bioethanol production from microalgae was US$ 2.22 M, while the selling price was US$ 2.87 M. An advantage of the project is that the by-product selling price was equivalent to US$1.6 M. In addition to the sale price of the by-product, the total production cost for bioethanol from microalgae was US$ 10,666 year^{-1} where the total bioethanol production was 57,087.62 gal year^{-1}. The fixed capital investment of this project was 78% of the total capital investment, which is lower than the other commercial biofuel plant derived from algae [27]. The total plant profit was

US$ 591,333 and even with sensitivity analysis comprising variable ranges of all influencing factors, the project is still feasible for industrial application.

Littlewood et al. [56] investigated the technical–economic of bioethanol production from bamboo. Several pretreatments and different saccharification conditions were investigated in order to identify the optimal conditions to maximize sugar release from bamboo. The liquid hot water pretreatment (190 °C for 10 min) was selected as the optimal condition for maximizing sugar release, which reached 69% of the theoretical maximum after 72 h of saccharification. The enzymatic saccharification with five loadings (10–140 FPU/g glucan) of Cellic CTec2 led to a total sugar release ranging from 59–76% of the theoretical maximum. The economic analysis revealed that the lowest enzyme loading had the most commercially viable scenario with a minimum ethanol selling price (MESP) of $0.484/L. Besides, various pretreatment technologies (steam explosion with and without acid catalyst, liquid hot water, dilute acid, and wet oxidation) was evaluated to produce ethanol from wheat straw [56]. The economic analysis revealed that wet oxidation pretreatment had the lowest MESP of $2.032/gal. The main cost contributors demonstrated through the analysis were the prices of wheat straw (37–56%) and the enzyme costs (17–41%) of the MESP. The focus of research should be directed towards methods aimed at increasing conversion efficiency and decreasing the raw material price of wheat straw in order to allow further development of a profitable E2G industry [56]. Bioethanol production from biomass can supply renewable energy and help the world meet its greenhouse gas emissions [94, 95].

According to Melin et al. [65], E2G production from lignocellulosic material generally has a lower carbon footprint, but its production cost is higher than that of first-generation fuels. Recently, a case study regarding the application of sugarcane by-products (bagasse and straw) for ethanol production was conducted [84]. The costs of by-products by subcritical water pretreatment from sugarcane were analyzed considering an industrial scheme linked to a Brazilian sugarcane plant. The investment of the industrial process in fixed capital was approximately US$ 27 M, and 35% was associated with subcritical water pretreatment. The lowest manufacturing cost of the process reached 5.45 US$ L^{-1} [84]. In addition, Zhao et al. [106] evaluated the bioethanol production from corn stover using dilute-acid pretreatment and enzymatic hydrolysis.

Eco-friendly energy sources that are irreproachable, inexhaustible, efficient, economical, and have low greenhouse gas emissions are necessary to meet future energy demands [10]. Integrating E1G and E2G production processes can create a significant synergy that can culminate in a lower environmental impact. Bioethanol production using first-generation feedstock has contributed mainly to the emission of greenhouse gases that invariably have a significant environmental impact. Bioethanol production using first-generation feedstock has contributed mainly to the emission of greenhouse gases that invariably have a significant environmental impact [3]. The major causes of greenhouse gas emissions from the ethanol life cycle are straw transportation, uses of chemicals and enzymes, and pretreatment [93]. Reducing greenhouse gas emissions depends on gases emitted throughout the chain, which can be substantial relative to end-use emissions [88]. The savings of $CO_{2\text{-eq}}$ emissions using E1G and E2G in 2018–2030 are expected to be 325-fold higher than those observed between 2008 and 2017 (160 versus 0.49 billion tons) [96].

Finally, there is a consensus that E2G feedstock integrated into E1G plants could have positive techno-economic and environmental impacts. However, according to Ferreira et al. [29], E2G from lignocelluloses is unfeasible as a stand-alone process. Moreover, Chandel et al. [19] described that large-scale production of biofuels and biochemicals from second-generation biomass (i.e., agricultural residues) is yet to be competitive compared to chemical routes currently used on a large scale. Novel studies should be addressed to decrease the cost of manufacturing of ethanol, especially applying low-cost technologies to the management of lignocellulosic feedstock. Recent advances in the use of agricultural residues composed of polysaccharides, algal polysaccharides and genetic manipulation to develop high-carbon crops have opened a new extent in bioethanol production [4, 71]. Besides, further research is required to reduce the price costs of harvesting processes. In addition, it is necessary to decrease the recalcitrance of the lignocellulosic biomass to aid downstream bioconversion [97]. The optimization of distillation costs and the solids content in the hydrolysis bioreactor are some major factors that should be optimized for industrial operations [83].

7 Current Status of Global Commercial-Scale Bioethanol Production

The demand for transport biofuels had a reduction of 8% in previous years, largely due to the situation of the COVID-19 pandemic, but demand is expected to increase in the coming years and, based on the net zero emissions by 2050, is requires much higher growth [43, p. 105].

Currently, the production of bioethanol at an industrial level has grown for more than 10 years. According to the Renewable fuels association (RFA), in 2021, global ethanol production was 27 B Gal. The USA is the leading producer with 55% (15 B Gal), followed by Brazil with 27% (7.5 B Gal), the European Union with 5% (1.3 B Gal), China with 3% (0.86 B Gal), India 3% (0.820 B Gal), Canada 2% (0.44 B Gal), Thailand 1% (0.39 B Gal), Argentina 1% (0.26 B Gal) and the rest of the World 1% (0.740 B Gal). Figure 11.4 shows the fuel ethanol production in the world and the U.S. Ethanol production in the USA is based on the following raw materials: 93.8% (Corn starch), 3.9% (cellulosic biomass/starch), 2.1% (corn, sorghum, and wheat), 0.2% (waste sugars/alcohol/starch) using two types of technologies: dry mill with 91.4% and wet mill 8.6% in the use of operation for ethanol production. There are 208 commercial-scale ethanol biorefineries installed in the USA, with an average capacity per biorefinery of 85 M Gal. According to the U.S. Department of Energy,

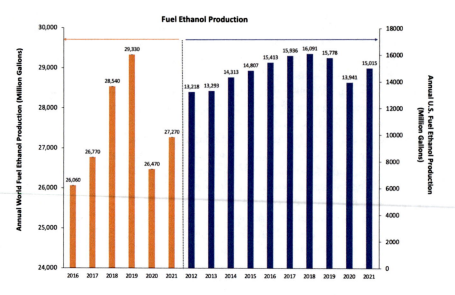

Fig. 11.4 Trends in world and US fuel ethanol production (Adapted from US Department of Energy [99])

4232 E85 filling stations are available. Also, the national average prices in the USA for the end of the year 2021 (December) were: US$2.79 and US$2.97/gallon for E85 and E15, respectively [99].

On the other hand, the ethanol industry has had an economic, social, and environmental impact, specifically in the USA. For example, the production of 15 B Gal has generated more than US$52 B directly to the Gross Domestic Product, generating 407,000 direct and indirect. Moreover, greenhouse gas emissions were reduced in the U.S. transportation sector by 980 M metric tons, equivalent to removing 12 M cars on the road for a year in the USA.

The estimated scenarios are that by 2030 and 2050, the use of ethanol will reduce at least 70% of the greenhouse gas emissions compared directly with gasoline and will reach a net zero lifecycle of GE emissions [2].

In Brazil, sugarcane is a major source of ethanol production, and the regular gasoline in Brazil already contains a 27% ethanol blend (E27). In a recent work, Barros [9] reported that in Brazil, there are 269 sugarcane ethanol plants, six flex plants of corn and sugarcane, and five corn-ethanol plants.

Six E2G production plants are operating in the United States, and one is in Brazil. One of the important limitations in developing the E2G at a large scale is the high production costs. Therefore, seeking and implementing technologies could help mitigate this situation. One of the main alternatives is to integrate the first- and second-generation to reduce costs in unit operations, equipment, transportation of raw materials, and the cost of enzymes. Figure 11.5 shows a scenario of integrating first- and second-generation processes in bioethanol production.

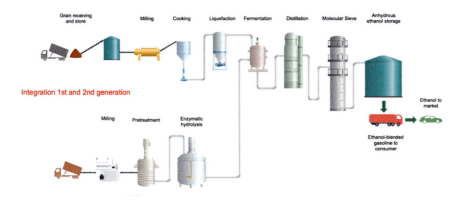

Fig. 11.5 Integration process for ethanol production (Adapted and modified from RFA [2]

The production of new advanced biofuels will develop in the coming years on an industrial scale. For this, the implementation of biofuels requires continuous support to overcome financial, technical, and market barriers and public policy frameworks will be of utmost importance for the implementation. According to (IRENE) (2019) the development of advanced biofuels will be important for the low-carbon path for the transport sector.

8 Conclusions

Commercial interest in bioethanol production has grown year after year. First-generation ethanol is a quite consolidated technology all over the world, producing billions of gallons per year. However, integrated biorefineries, including first and second-generation bioethanol, have great potential for low-cost production blended with gasoline reducing greenhouse gas emissions. Third-generation bioethanol is a promising technology with no plants at a commercial scale at present. Raw material unpredictability, research investment, expensive pretreatment, and solid government policies are the major concerns to commercializing second, third, and even fourth-generation bioethanol. Moreover, energetic, economic, and LCA are the main aspects of effective bioethanol production.

References

1. Aditiya H, Mahlia T, Chong W, Nur H, Sebayang A (2016) Second generation bioethanol production: a critical review. Renew Sustain Energy Rev 66:631–653
2. Annual Ethanol Production [Internet] (2022) @EthanolRFA. https://ethanolrfa.org/markets-and-statistics/annual-ethanol-production
3. Ayodele BV, Alsaffar MA, Mustapa SI (2020) An overview of integration opportunities for sustainable bioethanol production from first- and second-generation sugar-based feedstocks. J Clean Prod 245:118857
4. Baig KS, Wu J, Turcotte G (2019) Future prospects of delignification pretreatments for the lignocellulosic materials to produce second generation bioethanol. Int J Energy Res 43(4):1411–1427
5. Bajpai P (2021) General background and introduction. In: Bajpai P (ed) Developments in bioethanol. Springer Singapore, Singapore, pp 1–13
6. Balakrishnan D, Kumar SS, Sugathan S (2019) Amylases for food applications—updated information. Springer, Green Bio-processes, pp 199–227
7. Banerjee S, Mudliar S, Sen R, Giri B, Satpute D, Chakrabarti T et al (2010) Commercializing lignocellulosic bioethanol: technology bottlenecks and possible remedies. Biofuels Bioprod Biorefin Innov Sustain Econ 4(1):77–93
8. Bariani M, Cebreiros F, Guigou M, Cabrera MN (2022) Integrated production of furfural and second-generation bioethanol from Eucalyptus wood residues: experimental results and process simulation. Wood Sci Technol 56(4):1149–1173
9. Barros S (2022) Biofuels annual
10. Bibi R, Ahmad Z, Imran M, Hussain S, Ditta A, Mahmood S et al (2017) Algal bioethanol production technology: a trend towards sustainable development. Renew Sustain Energy Rev 71:976–985
11. BlueFire. BlueFire Renewables | Production Plant. 2022.
12. Boboescu I-Z, Gélinas M, Beigbeder J-B, Lavoie J-M (2018) High-efficiency second generation ethanol from the hemicellulosic fraction of softwood chips mixed with construction and demolition residues. Biores Technol 266:421–430
13. Bornscheuer U, Buchholz K, Seibel J (2014) Enzymatic degradation of (ligno) cellulose. Angew Chem Int Ed 53(41):10876–10893
14. Brodeur G, Yau E, Badal K, Collier J, Ramachandran KB, Ramakrishnan S (2011) Chemical and physicochemical pretreatment of lignocellulosic biomass: a review. Enzyme Res 2011:787532
15. Brown TR, Brown RC (2013) A review of cellulosic biofuel commercial-scale projects in the United States. Biofuels, Bioprod Biorefin 7(3):235–245
16. Cardona CA, Sánchez ÓJ (2007) Fuel ethanol production: process design trends and integration opportunities. Biores Technol 98(12):2415–2457
17. Carere CR, Sparling R, Cicek N, Levin DB (2008) Third generation biofuels via direct cellulose fermentation. Int J Mol Sci 9(7):1342–1360
18. Chandel AK, Chan E, Rudravaram R, Narasu ML, Rao LV, Ravindra P (2007) Economics and environmental impact of bioethanol production technologies: an appraisal. Biotechnol Mol Biol Rev 2(1):14–32
19. Chandel AK, Forte MBS, Gonçalves IS, Milessi TS, Arruda PV, Carvalho W et al (2021) Brazilian biorefineries from second generation biomass: critical insights from industry and future perspectives. Biofuels Bioprod Biorefin 15(4):1190–1208
20. Chen W-C, Lin Y-C, Ciou Y-L, Chu I-M, Tsai S-L, Lan JC-W et al (2017) Producing bioethanol from pretreated-wood dust by simultaneous saccharification and co-fermentation process. J Taiwan Inst Chem Eng 79:43–48
21. Chiaramonti D (2007) Bioethanol: role and production technologies. Improvement of crop plants for industrial end uses. Springer, pp 209–251

22. Chilakamarry CR, Sakinah AM, Zularisam A, Pandey A, Vo D-VN (2021) Technological perspectives for utilisation of waste glycerol for the production of biofuels: a review. Environ Technol Innov 24:101902
23. Choo B, Ismail K, Ma'Radzi A (eds) (2021) Scaling-up and techno-economics of ethanol production from cassava starch via separate hydrolysis and fermentation. In: IOP conference series: earth and environmental science. IOP Publishing
24. Chundawat SP, Beckham GT, Himmel ME, Dale BE (2011) Deconstruction of lignocellulosic biomass to fuels and chemicals. Annu Rev Chem Biomol Eng 2(1):121–145
25. Cripwell RA, Favaro L, Viljoen-Bloom M, van Zyl WH (2020) Consolidated bioprocessing of raw starch to ethanol by Saccharomyces cerevisiae: Achievements and challenges. Biotechnol Adv 42:107579
26. Das N, Jena PK, Padhi D, Kumar Mohanty M, Sahoo G (2021) A comprehensive review of characterization, pretreatment and its applications on different lignocellulosic biomass for bioethanol production. Biomass Convers Biorefin 1–25
27. Davis R, Markham J, Kinchin C, Grundl N, Tan EC, Humbird D (2016) Process design and economics for the production of algal biomass: algal biomass production in open pond systems and processing through dewatering for downstream conversion. National Renewable Energy Lab (NREL), Golden, CO (United States)
28. de Souza NRD, Fracarolli JA, Junqueira TL, Chagas MF, Cardoso TF, Watanabe MDB et al (2019) Sugarcane ethanol and beef cattle integration in Brazil. Biomass Bioenerg 120:448–457
29. Ferreira JA, Brancoli P, Agnihotri S, Bolton K, Taherzadeh MJ (2018) A review of integration strategies of lignocelluloses and other wastes in 1st generation bioethanol processes. Process Biochem 75:173–186
30. Foust TD, Aden A, Dutta A, Phillips S (2009) An economic and environmental comparison of a biochemical and a thermochemical lignocellulosic ethanol conversion processes. Cellulose 16(4):547–565
31. Gavahian M, Munekata PE, Eş I, Lorenzo JM, Khaneghah AM, Barba FJ (2019) Emerging techniques in bioethanol production: from distillation to waste valorization. Green Chem 21(6):1171–1185
32. Guerrero AB, Ballesteros I, Ballesteros M (2017) Optimal conditions of acid-catalysed steam explosion pretreatment of banana lignocellulosic biomass for fermentable sugar production. J Chem Technol Biotechnol 92(9):2351–2359
33. Gupta A, Verma JP (2015) Sustainable bio-ethanol production from agro-residues: a review. Renew Sustain Energy Rev 41:550–567
34. Ha S-J, Galazka JM, Rin Kim S, Choi J-H, Yang X, Seo J-H et al (2011) Engineered Saccharomyces cerevisiae capable of simultaneous cellobiose and xylose fermentation. Proc Natl Acad Sci 108(2):504–509
35. Haldar D, Purkait MK (2021) A review on the environment-friendly emerging techniques for pretreatment of lignocellulosic biomass: Mechanistic insight and advancements. Chemosphere 264:128523
36. Halder P, Azad K, Shah S, Sarker E (2019) 8—Prospects and technological advancement of cellulosic bioethanol ecofuel production. In: Azad K (ed) Advances in eco-fuels for a sustainable environment: Woodhead Publishing, pp 211–236
37. Han J, Luterbacher JS, Alonso DM, Dumesic JA, Maravelias CT (2015) A lignocellulosic ethanol strategy via nonenzymatic sugar production: Process synthesis and analysis. Biores Technol 182:258–266
38. Hassan SS, Williams GA, Jaiswal AK (2018) Emerging technologies for the pretreatment of lignocellulosic biomass. Biores Technol 262:310–318
39. Hasunuma T, Kondo A (2012) Consolidated bioprocessing and simultaneous saccharification and fermentation of lignocellulose to ethanol with thermotolerant yeast strains. Process Biochem 47(9):1287–1294
40. Ho DP, Ngo HH, Guo W (2014) A mini review on renewable sources for biofuel. Biores Technol 169:742–749

41. Hossain N, Mahlia TMI, Zaini J, Saidur R (2019) Techno-economics and sensitivity analysis of microalgae as commercial feedstock for bioethanol production. Environ Prog Sustain Energy 38(5):13157
42. Hu J, Arantes V, Saddler JN (2011) The enhancement of enzymatic hydrolysis of lignocellulosic substrates by the addition of accessory enzymes such as xylanase: is it an additive or synergistic effect? Biotechnol Biofuels 4(1):1–14
43. IEA (2022) International energy agency transport biofuels 2022 [cited 2022 12/09/2022]. https://www.iea.org/reports/transport-biofuels
44. (IRENE) IREA. Advanced biofuels. What Holds Them Back? : International Renewable Energy Agency Abu Dhabi, United Arab Emirates; 2019.
45. Jambo SA, Abdulla R, Mohd Azhar SH, Marbawi H, Gansau JA, Ravindra P (2016) A review on third generation bioethanol feedstock. Renew Sustain Energy Rev 65:756–769
46. Jang SS (2010) Production of mono sugar from acid hydrolysis of seaweed. African J Biotechnol
47. Kang Q, Appels L, Tan T, Dewil R (2014) Bioethanol from lignocellulosic biomass: current findings determine research priorities. Sci World J 2014:298153
48. Kazi FK, Fortman JA, Anex RP, Hsu DD, Aden A, Dutta A et al (2010) Techno-economic comparison of process technologies for biochemical ethanol production from corn stover. Fuel 89:S20–S28
49. Koppram R, Tomás-Pejó E, Xiros C, Olsson L (2014) Lignocellulosic ethanol production at high-gravity: challenges and perspectives. Trends Biotechnol 32(1):46–53
50. Koutinas A, Wang R, Webb C (2004) Evaluation of wheat as generic feedstock for chemical production. Ind Crops Prod 20(1):75–88
51. Kumar D, Juneja A, Singh V (2018) Fermentation technology to improve productivity in dry grind corn process for bioethanol production. Fuel Process Technol 173:66–74
52. Kumar S, Singh N, Prasad R (2010) Anhydrous ethanol: A renewable source of energy. Renew Sustain Energy Rev 14(7):1830–1844
53. Le PK, Le TD, Nguyen QD, Tran VT, Mai PT (eds) (2020) Process simulation of the pilot scale bioethanol production from rice straw by Aspen Hysys. In: IOP conference series: materials science and engineering. IOP Publishing
54. Leading countries based on biofuel production worldwide in 2021 [Internet] (2022) Statista https://www.statista.com/aboutus/our-research-commitment
55. Limayem A, Ricke SC (2012) Lignocellulosic biomass for bioethanol production: current perspectives, potential issues and future prospects. Prog Energy Combust Sci 38(4):449–467
56. Littlewood J, Wang L, Turnbull C, Murphy RJ (2013) Techno-economic potential of bioethanol from bamboo in China. Biotechnol Biofuels 6(1):173
57. Littlewood J, Murphy RJ, Wang L (2013) Importance of policy support and feedstock prices on economic feasibility of bioethanol production from wheat straw in the UK. Renew Sustain Energy Rev 17:291–300
58. Liu C-G, Xiao Y, Xia X-X, Zhao X-Q, Peng L, Srinophakun P et al (2019) Cellulosic ethanol production: progress, challenges and strategies for solutions. Biotechnol Adv 37(3):491–504
59. Luterbacher JS, Rand JM, Alonso DM, Han J, Youngquist JT, Maravelias CT et al (2014) Nonenzymatic sugar production from biomass using biomass-derived γ-valerolactone. Science 343(6168):277–280
60. Mabrouki J, Abbassi MA, Khiari B, Jellali S, Zorpas AA, Jeguirim M (2022) The dairy biorefinery: Integrating treatment process for Tunisian cheese whey valorization. Chemosphere 293:133567
61. Mahmoodi P, Karimi K, Taherzadeh MJ (2018) Efficient conversion of municipal solid waste to biofuel by simultaneous dilute-acid hydrolysis of starch and pretreatment of lignocelluloses. Energy Convers Manage 166:569–578
62. Mankar AR, Pandey A, Modak A, Pant K (2021) Pretreatment of lignocellulosic biomass: a review on recent advances. Biores Technol 334:125235
63. Martín C, Thomsen MH, Hauggaard-Nielsen H, BelindaThomsen A (2008) Wet oxidation pretreatment, enzymatic hydrolysis and simultaneous saccharification and fermentation of clover–ryegrass mixtures. Biores Technol 99(18):8777–8782

64. Melendez JR, Mátyás B, Hena S, Lowy DA, El Salous A (2022) Perspectives in the production of bioethanol: a review of sustainable methods, technologies, and bioprocesses. Renew Sustain Energy Rev 160:112260
65. Melin K, Nieminen H, Klüh D, Laari A, Koiranen T, Gaderer M (2022) Techno-economic evaluation of novel hybrid biomass and electricity-based ethanol fuel production. Front Energy Res 10
66. Mohanty SK, Swain MR (2019) Bioethanol production from corn and wheat: food, fuel, and future. In: Bioethanol production from food crops. Elsevier, pp 45–59
67. Muhamad A, Noor SFM, Syahirah N, Hamid AI, Zaidi MAHM (2021) Physical factors optimization of saccharomyces cerevisiae fermentation to enhance production of bioethanol: a review. Multi Appl Res Innov 2(2):266–277
68. Müller et al (2023) Challenges and opportunities for third-generation ethanol production: a critical review. Eng Microbiol 3:100056
69. Naik SN, Goud VV, Rout PK, Dalai AK (2010) Production of first and second generation biofuels: a comprehensive review. Renew Sustain Energy Rev 14(2):578–597
70. Neto AC, Guimarães MJO, Freire E (2018) Business models for commercial scale second-generation bioethanol production. J Clean Prod 184:168–178
71. Niphadkar S, Bagade P, Ahmed S (2018) Bioethanol production: insight into past, present and future perspectives. Biofuels 9(2):229–238
72. Oliveira T, Interlandi M, Hanlon K, Torres-Mayanga P, Silvello M, Timko M et al (2022) Integration of subcritical water and enzymatic hydrolysis to obtain fermentable sugars and second-generation ethanol from sugarcane straw. BioEnergy Res 15(2):1071–1082
73. Production AE (2021) Renewable fuels association 2021. https://ethanolrfa.org/markets-and-statistics/annual-ethanol-production
74. Qiao Z, Lü X (2021) Chapter 10—Industrial bioethanol production: status and bottlenecks. In: Lü X (ed) Advances in 2nd generation of bioethanol production. Woodhead Publishing, pp 213–227
75. Raj T, Chandrasekhar K, Naresh Kumar A, Rajesh Banu J, Yoon J-J, Kant Bhatia S et al (2022) Recent advances in commercial biorefineries for lignocellulosic ethanol production: current status, challenges and future perspectives. Biores Technol 344:126292
76. Ramos JL, Valdivia M, García-Lorente F, Segura A (2016) Benefits and perspectives on the use of biofuels. Microb Biotechnol 9(4):436–440
77. Rehan M, Nizami A-S, Rashid U, Naqvi MR (2019) Waste biorefineries: future energy, green products and waste treatment. Front Energy Res 7:55
78. Robak K, Balcerek M (2018) Review of second generation bioethanol production from residual biomass. Food Technol Biotechnol 56(2):174
79. Rosales-Calderon O, Arantes V (2019) A review on commercial-scale high-value products that can be produced alongside cellulosic ethanol. Biotechnol Biofuels 12(1):240
80. Rusco F (ed) (2012) Biofuels infrastructure in the united states: current status and future challenges 2012. In: The IFP/IEA/ITF workshop on: developing infrastructure for alternative.
81. Sadh PK, Duhan S, Duhan JS (2018) Agro-industrial wastes and their utilization using solid state fermentation: a review. Bioresour Bioprocess 5(1):1–15
82. Sanchez OJ, Cardona CA (2008) Trends in biotechnological production of fuel ethanol from different feedstocks. Biores Technol 99(13):5270–5295
83. Senthil Rathi B, Senthil Kumar P, Vinoth Kumar V (2022) Chapter 24—Sustainability assessment of third-generation biofuels: a life cycle perspective. In: Gurunathan B, Sahadevan R, Zakaria ZA (eds) Biofuels and bioenergy. Elsevier, pp 523–534
84. Sganzerla WG, Lachos-Perez D, Buller LS, Zabot GL, Forster-Carneiro T (2022) Cost analysis of subcritical water pretreatment of sugarcane straw and bagasse for second-generation bioethanol production: a case study in a sugarcane mill. Biofuels Bioprod Biorefin 16(2):435–450
85. Sharma HK, Xu C, Qin W (2019) Biological pretreatment of lignocellulosic biomass for biofuels and bioproducts: an overview. Waste Biomass Valorization 10(2):235–251
86. Shelley M (2006) Alcoholic fuels. FL, Boca Raton, pp 33487–42742

87. Sindhu R, Kuttiraja M, Binod P, Janu KU, Sukumaran RK, Pandey A (2011) Dilute acid pretreatment and enzymatic saccharification of sugarcane tops for bioethanol production. Biores Technol 102(23):10915–10921
88. Singh R, Srivastava M, Shukla A (2016) Environmental sustainability of bioethanol production from rice straw in India: a review. Renew Sustain Energy Rev 54:202–216
89. Siqueira G, Arantes V, Saddler JN, Ferraz A, Milagres AM (2017) Limitation of cellulose accessibility and unproductive binding of cellulases by pretreated sugarcane bagasse lignin. Biotechnol Biofuels 10(1):1–12
90. Solarte-Toro JC, Romero-García JM, Martínez-Patiño JC, Ruiz-Ramos E, Castro-Galiano E, Cardona-Alzate CA (2019) Acid pretreatment of lignocellulosic biomass for energy vectors production: a review focused on operational conditions and techno-economic assessment for bioethanol production. Renew Sustain Energy Rev 107:587–601
91. Srivastava N, Srivastava M, Mishra P, Gupta VK, Molina G, Rodriguez-Couto S et al (2018) Applications of fungal cellulases in biofuel production: advances and limitations. Renew Sustain Energy Rev 82:2379–2386
92. Szambelan K, Nowak J, Szwengiel A, Jeleń H, Łukaszewski G (2018) Separate hydrolysis and fermentation and simultaneous saccharification and fermentation methods in bioethanol production and formation of volatile by-products from selected corn cultivars. Ind Crops Prod 118:355–361
93. Sun Y, Cheng J (2002) Hydrolysis of lignocellulosic materials for ethanol production: a review. Biores Technol 83(1):1–11
94. Szulczyk KR, Tan Y-M (2022) Economic feasibility and sustainability of commercial bioethanol from microalgal biomass: the case of Malaysia. Energy 253:124151
95. Szulczyk KR, Yap CS, Ho P (2021) The economic feasibility and environmental ramifications of biodiesel, bioelectricity, and bioethanol in Malaysia. Energy Sustain Dev 61:206–216
96. Sydney EB, Letti LAJ, Karp SG, Sydney ACN, Vandenberghe LPdS, de Carvalho JC et al (2019) Current analysis and future perspective of reduction in worldwide greenhouse gases emissions by using first and second generation bioethanol in the transportation sector. Bioresour Technol Rep 7:100234
97. Terán Hilares R, Sanchez Muñoz S, Alba EM, Prado CA, Ramos L, Ahmed MA et al (2022) 7—Recent technical advancements in first, second and third generation ethanol production. In: Chandel AK, Segato F (eds) Production of top 12 biochemicals selected by USDOE from renewable resources. Elsevier, pp 203–232
98. Torres-Mayanga PC, Lachos-Perez D, Mudhoo A, Kumar S, Brown AB, Tyufekchiev M et al (2019) Production of biofuel precursors and value-added chemicals from hydrolysates resulting from hydrothermal processing of biomass: a review. Biomass Bioenerg 130:105397
99. US Department of Energy (2022)
100. Vanneste J, Ennaert T, Vanhulsel A, Sels B (2017) Unconventional pretreatment of lignocellulose with low-temperature plasma. Chemsuschem 10(1):14–31
101. Vasić K, Knez Ž, Leitgeb M (2021) Bioethanol production by enzymatic hydrolysis from different lignocellulosic sources. Molecules 26(3):753
102. Vohra M, Manwar J, Manmode R, Padgilwar S, Patil S (2014) Bioethanol production: feedstock and current technologies. J Environ Chem Eng 2(1):573–584
103. Wong KH, Tan IS, Foo HCY, Chin LM, Cheah JRN, Sia JK et al (2022) Third-generation bioethanol and L-lactic acid production from red macroalgae cellulosic residue: prospects of industry 5.0 algae. Energy Convers Manag 253:115155
104. Xu Z, Huang F (2014) Pretreatment methods for bioethanol production. Appl Biochem Biotechnol 174(1):43–62
105. Yu H-T, Chen B-Y, Li B-Y, Tseng M-C, Han C-C, Shyu S-G (2018) Efficient pretreatment of lignocellulosic biomass with high recovery of solid lignin and fermentable sugars using Fenton reaction in a mixed solvent. Biotechnol Biofuels 11(1):287

106. Zhao L, Zhang X, Xu J, Ou X, Chang S, Wu M (2015) Techno-economic analysis of bioethanol production from lignocellulosic biomass in China: dilute-acid pretreatment and enzymatic hydrolysis of corn stover. Energies 8(5):4096–4117
107. Zhao X, Li S, Wu R, Liu D (2017) Organosolv fractionating pre-treatment of lignocellulosic biomass for efficient enzymatic saccharification: chemistry, kinetics, and substrate structures. Biofuels Bioprod Biorefin 11(3):567–590

Techno-Economic and Life Cycle Analysis of Bioethanol Production

Ana Belén Guerrero and Edmundo Muñoz

Abstract Nowadays, there is a wide variety of pathways to transform biomass into biofuels. There are economically viable pathways that have been escalated to an industrial level and others still at the pilot and/or research and development scale. As well, considering a life cycle perspective for the environmental analysis, there are pros and cons that should be considered regarding the bioethanol environmental performance, since it has been considered that the use of bioethanol could help to reduce fossil fuels depletion and greenhouse gas emissions. In this chapter, some insights about techno-economic and environmental analyses of different bioethanol production methods developed in the last five years are determined. The techno-economic analysis details the potential economic viability of a selected technology at the research and development stage. The analysis considers the maturity level of the technologies available today: first, second, and third-generation bioethanol. Regarding the environmental performance of bioethanol from a life cycle perspective, it is clear that no biofuel is "climate neutral" because of the inputs of fossil fuels needed, in the agricultural and industrial processes, before the use stage. Given this background, a life cycle approach in a bioethanol environmental analysis facilitates an environmental profile containing all inputs and outputs that entailed in the bioethanol system, from the extraction of raw materials to the use of bioethanol in the engine of a vehicle. Life cycle assessment results allow informed decision-making and avoiding shifting environmental impacts.

Keywords Biofuels · Ethanol · Costs · Policies · Sustainability · Environment

---

A. B. Guerrero
Faculty of Environmental and Agricultural Sciences, Universidad Rafael Landivar, Guatemala, Guatemala

Trisquel Consulting Group, Quito, Ecuador

E. Muñoz (✉)
Center for Sustainability Research, Universidad Andres Bello, Santiago, Chile
e-mail: edmundo.munoz@unab.cl

1 Introduction

Energy demand is expected to increase by 1.3% annually from 2020 to 2030 [24]. This trend is contrary to the International Energy Agency's landmark net zero emissions by 2050 scenario, which establishes an ambitious roadmap to a 1.5 °C maximum for rising global temperatures. In 2020, fossil fuels accounted for almost 80% of the global energy mix [24]. While 2023 is when countries will review progress towards the Paris Agreement goals, the Intergovernmental Panel on Climate Change (IPCC) announced in its annual climate change report that the global surface temperature has already warmed by 1.1 °C, compared to the pre-industrial period, and that is set to pass the 1.5 °C threshold by 2040 [28]. This highlights the key role of renewable energies. Thus, countries should act quickly and decisively to develop sustainable, affordable, and environmentally friendly energy sources.

Nowadays, a vast quantity of technologies exists that could help to achieve the Paris Agreement goals. One of them is the inclusion of biofuels in countries' energy mix. Bioethanol is an oxygenated biofuel containing 35% oxygen, reducing particulate matter and greenhouse gas (GHG) emissions from combustion. This biofuel can be used in road transport vehicles, blended with gasoline, with no engine adaptation. The nearly pure bioethanol fuel demonstrates a clean burning characteristic, reducing GHG emissions into the atmosphere [44]. Bioethanol is the most-used liquid biofuel in the world, and is largely produced from corn, sugar cane or cereals (first-generation). Its benefits in the energy balance, emissions reduction, and cost-effectiveness depend on technology, raw material, facility location, and the products and co-products obtained in the transformation [4].

Besides techno-economic data of the different production pathways, environmental aspects are a key factor in evaluating bioethanol sustainability. Thus, this chapter aims to provide information about the techno-economic and environmental performance of bioethanol, considering its main transformation pathways.

2 Techno-Economic Analysis

Bioethanol can be obtained by several technologies that are at different levels of maturity. It is important to highlight that having one specific method to produce bioethanol is not possible. In fact, biomass could be considered as a heterogeneous raw material as it comes from different sources (agro-food crops, agro-energy crops, residues, wastes, genetically modified organisms, etc.). In addition, bioethanol yields depend on biomass physicochemical composition [11] and the transformation technology used to produce it. Thus, the bioethanol production process must be tailored to each kind of biomass. In this context, the techno-economic analysis is focused on each maturity level of the technologies available today: first, second, and third-generation bioethanol (Fig. 1).

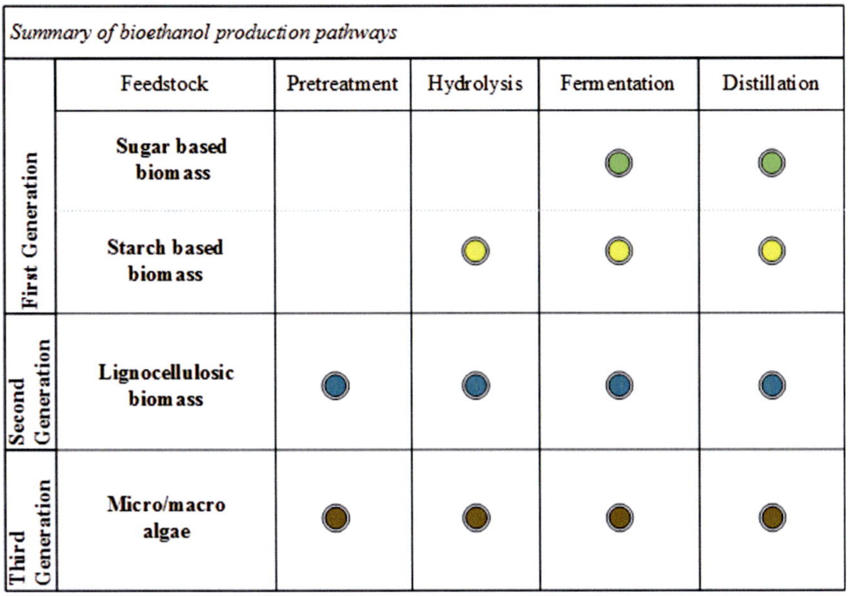

Summary of bioethanol production pathways					
	Feedstock	Pretreatment	Hydrolysis	Fermentation	Distillation
First Generation	Sugar based biomass			●	●
First Generation	Starch based biomass		●	●	●
Second Generation	Lignocellulosic biomass	●	●	●	●
Third Generation	Micro/macro algae	●	●	●	●

Fig. 1 Summary of bioethanol production pathways, considering its main stages of the transformation process

The main objective of a techno-economic analysis (TEA) is to determine the potential economic viability of a selected technology at the research and development (R&D) stage. Evaluating the production costs of a given technology facilitates assessing the potential economic viability of a production process. Results obtained from these analyses are useful to determine which emerging technologies have the highest potential for deployment success [43]. Considering that the main objective of bioethanol is to transition to a bioeconomy, TEA of bioethanol production is crucial.

For the analysis, the following indicators were considered:

- *Scale of technology*: It is important to consider whether laboratory development of the technology is still in process. If this is the case, modeling a theoretical potential is vital to determine process efficiency.
- *Minimum ethanol selling price (MESP)*: This indicator is influenced by manufacturing costs, overhead costs, financing, discount rates and incentives. The MESP allows the costs of different technologies to be compared and is considered a good metric for setting policy and cost targets for a technology [43].
- *Net present value (NPV):* This is the best tool to determine whether a project is potentially profitable (NPV > 0) or should be rejected (NPV < 0). The NPV is the cumulative sum of the discounted cash flows calculated by the expected future cash flows to the present using a discount rate.
- *Operational and maintenance costs (O&M)*: These are daily expenses incurred to run a business, which can be labor, costs of chemicals, equipment regular

maintenance and major repair and replacement, site and building maintenance, utilities, etc.
- *Internal rate of return* (*IRR*): This is an annualized effective compounded return rate that could be earned on a project.
- *Payback time or period*: This is the time required to recover the initial investment. A desired payback time should be under 5 years [13].
- *Ethanol yield*: Volume unit of ethanol obtained from a mass unit of biomass.

2.1 Bioethanol Market Price and Regulations

The main bioethanol producers are the United States of America (USA) and Brazil, contributing more than 80% of worldwide production (Fig. 2) [1]. Therefore, any bioethanol price analysis must include these two countries. Both countries produce mainly first-generation bioethanol from corn and sugarcane, respectively. Sugarcane ethanol is cheaper to produce than corn ethanol, and the former is seven times more efficient than the latter in terms of the energy required to produce it, compared to the energy delivered in its consumption [17].

Bioethanol has achieved economic viability due to the incentives or subsidies each country mentioned earlier applied over the years. This is not surprising since both countries have had a love-hate history with bioethanol for more than 100 years. In addition, it has been demonstrated that the key role of government institutions is vital to ensure the economic sustainability of biofuels. A recent study of governance systems of biofuels demonstrated that systematic legislations and sustainable and robust renewable energy policies of governments helped the USA and Brazil to boost their bio-based economies [61].

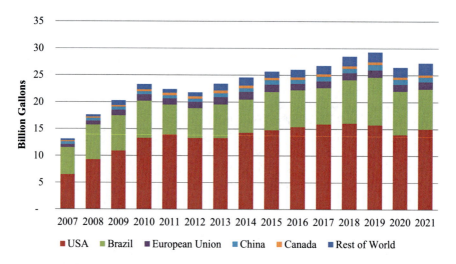

Fig. 2 Global ethanol production by country/region [1]

Brazil started to increase the production and use of ethanol from sugarcane in the 1970s. This country adopted an E11 blending mandate (11% ethanol, 89% regular gasoline) in 1976, and this percentage has been increasing ever since [39]. The initial governmental drivers that allowed the ethanol industry to become economically sustainable are: (a) guaranteed purchases by the state-owned oil company Petrobras, (b) low-interest loans for agro-industrial ethanol firms; and (c) fixed gasoline and ethanol prices where hydrous ethanol sold for 59% of the government-set gasoline price at the pump. Nowadays, Brazil has an E27 blending mandate [61], but as the Brazilian mandate has periodically been increased, the country ceased to be self-sufficient at some point. In 2010, Brazil withdrew the import tax on ethanol, and since then, it has been importing ethanol from the USA [39].

For its part, the USA has a long history with ethanol, but its glory started in the 1970s when the Environmental Protection Agency (EPA) was established. This government institution was responsible for implementing biofuel policy. In 1999, E5.7 blending became mandatory, though this mandate varied depending on the state. In order to accomplish the stipulated blend, various loans and subsidies were provided to support research and development activities in biofuel [61]. USA ethanol production and consumption skyrocketed with the introduction of the Renewable Fuel Standard program in 2005 [17]. Nowadays, the USA has a blending federal mandate of at least 10% ethanol mixed with regular gasoline. Bioethanol produced in the USA is currently under strong criticism with respect to its sustainability and environmental impact due to the high participation of coal in the country's electricity mix, which is needed for bioethanol production.

Additionally, in the last decade, a sustained ethanol price increase was experienced until 2020, after which the COVID-19 pandemic caused a drop of 8.5% in global transport fuel use, compared to 2019, due to restrictions on people's movements and disruption in global trade logistics [48]. As can be seen in Fig. 3, bioethanol demand in hand with ethanol price recovered in 2021 from Covid-19 lows, to near 2019 levels, and it is expected to grow by 5% in 2022 and by 3% in 2023 [25].

It is difficult to make a good forecast for 2023 prices due to the possible impact of the Russian invasion of Ukraine and the preponderant role that these two countries have in the worldwide energy and agricultural sector. In fact, the prices plotted in Fig. 3 are from early 2022, and in the third trimester of 2022, these prices skyrocketed. In August 2022, the worldwide average ethanol price was 1.23 USD/L, and gasoline was 1.41 USD/L [14]. In addition, increasing feedstock prices and policy reaction from multiple countries to the war has put upward pressure on an already high-price environment for bioethanol feedstocks [26].

In short, credit, gasoline prices, and ethanol blending policies seem to have increased ethanol consumption, creating dependence on imports [39]. With this background, it is possible to foresee bioethanol market behavior and its selling price. In some cases, this price is considered the minimum ethanol selling price, which corresponds to the ethanol selling price, rendering NPV equal to zero. Now, let us discuss how each transformation technology and feedstock can affect the main economic indicators considered in this techno-economic analysis.

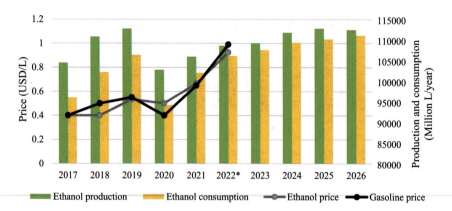

Fig. 3 Historical gasoline price, bioethanol price, consumption and production. *Source* [26, 48]. * Forecast data for ethanol demand

2.2 First Generation

First-generation bioethanol or conventional ethanol includes well-established processes that use as feedstock sugar and starch biomass which are staple crops (corn, sugarcane, wheat, etc.), and have already achieved a commercial scale [27]. It is the largest contributor to total bioethanol production thus, represents the largest renewable contribution to transport.

A literature review was carried out for the writing of this chapter, considering the latest research papers published in journals indexed in Scopus in the last 5 years. Interestingly, only 3% of the articles focus on TEA of first-generation bioethanol. There are two possible explanations for the lack of interest in developing techno-economic analyses of first generation bioethanol: (a) the main objective of a TEA is to determine the potential economic viability of a selected technology at the research and development stage, and first-generation ethanol has passed the R&D stage and is scaled up to an industrial level because it is an economically viable technology.; or (b) all scientific efforts are focused on developing a biofuel that is economically viable, environmentally friendly and will not contribute to food insecurity. It is important to highlight that increasing first-generation biofuel production has generated concern about the potential for food resource competition [38].

The cost of producing first-generation ethanol is expectedly to be high because of the elevated raw material price [63]. Feedstock costs account for 60–80% of conventional ethanol production costs [29]. Corn prices are continually rising, thus increasing the MESP for corn ethanol. Furthermore, the operation and maintenance costs account for 3–6% of the investment cost [45].

First-generation biofuels are currently the preferred option for commercial biofuel production [66]. However, some studies have demonstrated that while first-generation bioethanol has the highest production and energy efficiency, it has the worst environmental performance [65], as discussed in Sect. 3.4. Over the years and with scientific evidence, it has been possible to demonstrate various negative effects of expanding the production of first-generation bioethanol. Although, these problems can be overcome by transitioning to second-generation bioethanol. In fact, integrating both first- and second-generation bioethanol production can maximize ethanol yields and revenue, and would require less capital investment to integrate lignocellulose processes into an industrial design [66]. For example, the government offers various subsidies, grants, loans, and cash prizes to support biofuel research to ensure the sustainable use of bioethanol in the US economy. It provides incentives for bio-refineries that help to replace 80% of fossil fuels and use second-generation biofuel [61].

Subsidies of waste and residues derived fuels deserve more policy attention to rely on non-food biomass and avoid food versus fuel competition. The European Union has already achieved this; since 2010, it has renewable energy legislation, which requires that at least 10% of transport fuels come from renewable sources. In 2018, it was agreed to increase the biofuels target to 14%, limiting first-generation biofuels from exceeding 7%. This new framework was adopted in 2018 and will be implemented in 2030 [48].

2.3 Second Generation

Nearly 98% of liquid fuel demand today is met by petroleum; the remainder is met mainly by conventional liquid biofuels, and second-generation biofuels have a very small share [24]. Second-generation bioethanol, also known as cellulosic bioethanol, is produced from lignocellulosic biomass such as sugarcane bagasse, cereal straw, corn stover, wood residues, and some energy crops such as miscanthus [59]. The origin of the feedstock used in this technology leads to significantly fewer greenhouse gas emissions than those generated by fossil fuels, does not represent a fuel versus food competition, and the revalorization of agricultural and forestry residues helps to avoid environmental problems caused by uncontrolled biomass decomposition [38]. Ethanol production from cellulosic material is considered a highly promising option, allowing the energy diversification and decarbonization of the transport sector [50].

This technology reached profitability in 2013 when the first commercial plant began operations. A handful of commercial-scale projects are currently in operation and several new plants are under construction. These plants are mainly located in the USA, Brazil, and India [59]. To ensure economic viability, R&D is essential for the development of this technology. Scientific articles on bioethanol production using lignocellulosic biomass have been published since the 1990s [73]. Thus, 30 years ago, this idea started to blossom, and 10 years ago, it began to bear fruit.

Cellulosic bioethanol production involves several processes: biomass production/acquiring, pre-treatment, enzyme production/acquiring, hydrolysis, fermentation, and distillation. Each of these processes contributes differently to production costs. Feedstock costs depend significantly on the type of biomass used, whether it has to be specifically collected or produced, or if it is already on-site (such as sugar cane bagasse). The cost of enzymes also plays a significant role in overall production costs. For example, in the case of olive tree pruning, the most representative costs can be attributed to the raw materials (olive tree pruning, enzymes, and water), which together accumulate more than 86% of the costs [62]. This is also the case with wheat straw, where about 43% of total manufacturing cost comes from raw materials, attributed mainly to the cost of wheat straw (22.47%) and enzymes (18.47%) [19]. Regarding corn stover, utility (hot and cold utility and electricity) encompasses 57% of production costs, and raw material 35% [22]. Fortunately, enzyme costs have been significantly reduced by about a factor of 10 since 2000 through improving enzyme efficiency and better matching of enzymes to specific feedstock types [59].

The capital cost of a cellulosic ethanol production plant represents a significant part of the overall costs. It depends on various specifications (plant size and location, technology complexity, and evolution of the learning curve). For example, considering the use of waste biomass, high capital and operational costs were attributed to the requirement of some equipment and chemicals for pre-treatment, conversion of biomass, and waste generation [11]. On the other hand, the operation and maintenance costs are considered to fluctuate between 5 and 10% [45].

To facilitate the fast screening of lignocellulosic biomass options for bioethanol production in an integrated biorefinery, Sadhukhan et al. [60] determined an interesting correlation to estimate ethanol yield, net electricity production, MESP, and global warming potential (GWP) based on biomass physicochemical composition. Correlations showed that a ratio of lignin (wet basis) to the sum of cellulose and hemicellulose (dry basis), $xL > 0.22$ is required, for both an energy-balanced process and net global warming potential impact savings.

Table 1 summarizes techno-economic data from which it was possible to determine that the average second-generation ethanol yield is 200 L/t, and the average production cost of ethanol is around 0.70 USD/L. However, the results of these techno-economic models vary significantly from one to another. Every study analyzes different scenarios with unique peculiarities, which is why Table 1 shows economic indicators marked according to the results offered in the article analyzed. Still, no data is included on the table because of the great range of values presented by each scientific article due to the details of each study such as analysis of different scenarios, biorefinery approach, scale of technology, different raw materials, products and coproducts, and different conversion pathways. For additional data, please refer to the research articles cited in Table 1.

Unfortunately, despite incentive policies to support targets in some countries (as discussed in Sect. 2.1) and significant R&D funding in the last decade, cellulosic ethanol is still not competitive compared to conventional ethanol and efforts are still needed to reduce costs [50]. Advanced biofuels are hindered by feedstock availability and price, efficiency of conversion, technological reliability, stability of national policies, subsidy levels and mixing requirements [33]. Further process developments will lower production costs and increase coproduct utilization [66]. Some studies suggest that to decrease production costs, it is necessary to change the bioethanol single production to a biorefinery approach [62, 67]. Another option to reduce the

Table 1 Techno-economic indicators of different raw materials used for the production of second-generation bioethanol

Reference	Geographical relevance	Raw material	Scale of technology/ Technical assessment	Ethanol yield	MESP	NPV	O&M	IRR	Payback time
Hasanly et al. [19]	Iran	Wheat straw	Simulation of an industrial biorefinery (SuperPro Designer® 8.5)	191.58 L/t	1.23 USD/L	✓	✓		✓
Vaskan et al. [67]	Brazil	Palm empty fruit bunches	Simulation of a pilot plant (Aspen Plus V8.8)		0.8–1.0 USD/L		✓		
Solarte-Toro et al. [62]	Spain	Olive tree pruning biomass	Simulation of a small-scale process (Aspen Plus v9.0)	139.45–122.43 L/t	2.4 USD/L		✓		
Hossain et al. [22]		Corn stover	Simulation of a profitable bioethanol and furfural coproduction (Aspen Plus™ V7.1)		0.68 USD/L		✓		

(continued)

Table 1 (continued)

Reference	Geographical relevance	Raw material	Scale of technology/Technical assessment	Ethanol yield	MESP	NPV	O&M	IRR	Payback time
Sadhukhan et al. [60]	Mexico	Wood and grass species, bagasse and crop residues	Biochemical conversion technology studied by NREL has been taken as the basis to model the process yields	184 - 340 L/t	0.54 USD/L (wheat straw) and 0.75 USD/L (coffee pulp)		✓		✓
Demichelis et al. [11]	European Union	Sugarcane for sugar-based, potatoes for starch-based and rice straw, cattle manure and organic fraction of municipal solid waste (OFMSW) for lignocellulosic biomasses	The bioethanol conversion process was defined according to (1) physico-chemical composition of the selected biomasses and (2) conversion process available in the scientific literature	- Sugarcane 160 L/t, - Potatoes 170 L/t, - Rice straw 220 L/t, - Cattle manure 190 L/t, - OFMSW 140 L/t	0.91 USD/L	✓		✓	✓
Sondhi et al. [63]	India	Kitchen waste (peels of potato (50%), onion (20%) and seasonal vegetables (30%))	Laboratory scale (500 ml Erlenmeyer flasks)	0.32 g per g biomass ethanol production	0.143 USD/L		✓		

(continued)

Table 1 (continued)

Reference	Geographical relevance	Raw material	Scale of technology/Technical assessment	Ethanol yield	MESP	NPV	O&M	IRR	Payback time
Pratto et al. [53]	Brazil	Sugarcane straw	Laboratory scale (200 mL bench-scale reactors with 50 mL of the total reaction medium)	290 L/t	0.51 USD/L	✓	✓	✓	
Ntimbani et al. [47]	South Africa	Mixture of sugarcane bagasse and harvest residues	Simulation of an industrial process (Aspen Plus® V8.8)	95.69% based on initial glucan & xylan	0.59 USD/L		✓	✓	
Manhongo et al. [40]	South Africa	Mango waste	Simulation of an industrial process (Aspen Plus)	–	3.29 USD/kg	✓	✓	✓	
Wang et al. [68]	China	*Arundo donax*	Simulation of an industrial process (Aspen Plus)	–	0.44 USD/L	✓	✓	✓	
Ngigi et al. [46]	Kenya	Sila sorghum stalks	Simulation of a large-scale plant (Aspen Plus)	–	0.8–1.6 USD/L	✓		✓	

high bioethanol costs is to cogenerate the solid residues obtained to decrease the utility costs [62].

Finally, the literature review done for this chapter revealed that 70% of the revised articles were about second-generation bioethanol. There is a large amount of research, not only in quantity, but also in time. It has been 30 years since the first publications about this topic saw the light. Pilot plants have opened and closed their doors worldwide since then. It is time to step back and reconsider whether research needs to change its focus in order to achieve the much-desired economic viability of second-generation bioethanol.

2.4 Third Generation

The third-generation ethanol production process has yet to reach the commercial stage; it is still at laboratory scale. For feedstock, this technology uses algal biomass (micro and macroalgae) and genetically modified organisms (GMO), which can give several distinct sorts of sustainable biofuels, including bioethanol and biodiesel [9].

The main advantage of using microalgae as a feedstock is the low cultivation cost, as it can be produced in a moist marginal area or wastewater [65]. Moreover, microalgae have been considered as a suitable biomass for bioethanol production since several types of them are rich in carbohydrates which can be transformed to bioethanol [21]. On the other hand, macroalgae have proved to be viable bioethanol feedstock as it avoids fresh water, food crops, or arable land competition [56]. However, the energy required for processing third-generation bioethanol is high, making it less environmentally friendly as fossil fuels are used to generate the electricity needed for its transformation. In addition, the aquatic nature of the biomass "farming" and processing logistics, as well as inherent seasonality fluctuations in biomass cultivation are challenges that third-generation ethanol faces.

According to Tse et al. [66], ethanol yield presents a great range of variability between microalgae (167–501 L/t) and macroalgae classification based on thallus pigment: green (72–608 L/t), brown (12–1128 L/t) and red (12–595 L/t). These values are higher than the ones obtained in the second-generation pathway. Even though, the analyzed studies concluded that ethanol production from algal biomass is not economically viable. A constant loss of value over the years has been demonstrated, as the revenue from ethanol has never been able to cover operating costs [13, 36]. Ethanol production becomes economically feasible when it has a biorefinery approach; i.e., algal biomass ethanol production must not be the only intended product of the process; there must be other value-added products. The same conclusion was drawn for second-generation ethanol observed in the previous section.

Alternative approaches can be considered in further studies in order to identify paths towards economic viability and environmentally friendly. For example, to increase ethanol productivity of GMO cyanobacteria leading to higher concentrations of ethanol, subsequently increasing the energy efficiency of ethanol extraction [36] or that environmental feasibility study of third-generation bioethanol should be emphasized to ensure green production [8].

In the literature review (Table 2), only some articles intended to develop a techno-economic analysis were found, perhaps due to the novelty of this technology. These research papers did not present production costs in a disaggregated manner. In fact, the results of Ranjbari et al. [58] demonstrate that techno-economic analysis of algal biofuels is still lacking. As detailed in the second-generation bioethanol section, values are not included in Table 2 due to the great variability of results presented by each study, starting with the type of biomass, transformation route, scenarios, and assumptions for simulation, etc. Therefore, no values are included in order to avoid confusion and misinterpretation of data.

Overall, using these types of raw materials requires improvements to reduce costs and increase economic viability [66]. Moreover, payback periods > 10 years [36] have been calculated for third-generation ethanol, making this an unsustainable technology with current conditions. In any case, there is still a need for more in-depth research into TEA of algae biofuels to determine commercial improvements and increase economic feasibility, and more R&D is needed to scale up this type of biofuel to an industrial level and enable bioethanol production at more competitive prices.

3 Life Cycle Assessment

3.1 Overview

From an environmental point of view, bioethanol is presented as a potential solution to minimize the environmental impact of fossil fuels, as well as their depletion. However, bioethanol production can generate various environmental impacts, depending on the type of raw material, local conditions, the design and implementation of the conversion process [2], and pollutant control systems, among others. A comprehensive assessment of the entire production chain is necessary to determine the environmental burdens of bioethanol production systems. In this sense, Life Cycle Assessment (LCA), defined according to ISO standards 14,040 [30] and 14,044 [31], has been widely used to identify and measure environmental impacts on sustainability throughout the entire life cycle of a product [30]. An LCA allows determining the environmental burdens from the extraction of raw materials and inputs, production processes, transport, use, and final disposal of a product at the end of its useful life. Several LCA studies have been conducted on bioethanol systems to, among other applications, identify environmental hotspots [75], ascertain energy balances [23], determine carbon footprints [6], support public policy formulation [64], establish social [69] and economic impacts [6]. Various bioethanol LCA studies have considered different technical and methodological elements, such as feedstock type, the scope of the

Table 2 Techno-economic indicators of different raw materials used for the production of third generation bioethanol

Reference	Geographical relevance	Raw material	Scale of technology/Technical assessment	Ethanol yield	MESP	NPV	O&M	IRR	Payback time
Wu et al. [72]		Chlorella	Simulation of a commercial-scale process (Aspen Plus)	–	2.06 USD/L	✓	✓	✓	✓
Lopes et al. [36]	Portugal	Synechocystis GMO	Laboratory analysis and simulation of an industrial process (SuperPro Designer)	–	–	✓	✓	✓	✓
Hossain et al. [21]	Borneo	*Chlorella vulgaris*	Large-scale bioethanol production plant from microalgae simulated with integrated process design	58.90 m³/ha/year	–		✓		✓
Chong et al. [8]	Malaysia	Macroalgae waste	Simulation of an industrial process (Aspen Plus V10)	85.5%	0.54 USD/L	✓		✓	✓
Ferreira da Silva et al. [13]	Portugal	Synechocystis GMO	Optimization of a biorefinery process, using as a starting point and information source the results from a European Project (DEMA)	–	–	✓	✓	✓	

(continued)

Table 2 (continued)

Reference	Geographical relevance	Raw material	Scale of technology/ Technical assessment	Ethanol yield	MESP	NPV	O&M	IRR	Payback time
Wong et al. [71]	Malaysia	Macroalgae cellulosic residue	Simulation of a large-scale biorefinery (Aspen Plus V10)	–	–				✓

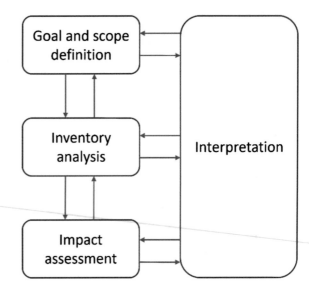

Fig. 4 Phases of an LCA (adapted from ISO 14040)

study, environmental impact assessment methodologies, and conversion technologies, among others. From a methodological point of view, LCA has several technical elements that can confuse LCA practitioners. For this reason, the objective of this section is to describe the main methodological steps of an LCA, highlighting the different criteria considered in bioethanol research and their relationship to the study goals.

According to the methodology proposed by ISO 14044 [31], there are four phases in a life cycle assessment: (i) goal and scope definition, (ii) inventory analysis, (iii) impact assessment, and (iv) interpretation (Fig. 4). These phases are related and interconnected with the interpretation phase, which is used transversally throughout the LCA. The methodological phases and their connections are presented next.

3.2 Goal and Scope Definition

The objective of an LCA study should unambiguously state the intended application, the reasons for conducting the study, and to whom the results will be communicated [30]. In bioethanol LCA studies, the objectives have mainly focused on determining environmental impacts, identifying environmental hotspots, determining environmental benefits of bioethanol, comparing impacts of different feedstock types, assessing the effect of bioethanol blending with gasoline, and determining the carbon footprint. The study's objective is fundamental for the following definitions because it conditions the type of information (data) needed to construct the inventory. In this sense, the depth and breadth of an LCA can differ considerably depending on the goals of the study. Examples of bioethanol LCA study objectives are presented in Table 3.

The scope of an LCA should clearly consider and describe the product system under scrutiny along with its functions. The product system consists of unit processes with elementary product flows, performing one or more defined functions and modelling a product life cycle [30]. Figure 5 presents a general outline of the life cycle of bioethanol. Studies have used different scopes depending on the objective(s) of the study, with the well-to-wheel scope being the most popular [5, 6, 12, 18]. This scope considers everything from feedstock extraction required for bioethanol production to bioethanol consumption or its use to fuel the engine in a vehicle. The well-to-wheel scope is equivalent to the cradle-to-grave scope. However, the latter is also used as a scope in several bioethanol studies [35, 37]. The cradle-to-gate approach is widely used in studies that focus on the environmental impacts of the production stage [11, 34, 64, 74]. In this case, the studies compare different biomass feedstock, approaches for bioethanol production, and the identification of environmental hotspots, among other issues associated with the production stage.

The system boundaries must be established to delimit the product system. These correspond to the criteria specifying which unit processes are part of a product system [31] and, therefore, which should be included in the LCA. The system boundaries also depend on the goal and scope, as these will determine which life cycle stages are to be included in the study. All processes and flows to be modeled will be inside the system boundaries, while outside are the unitary processes excluded from the study. For example, in Fig. 5, it can be observed that the production/acquisition of raw material is plotted in the middle of the system boundaries, which is because there are three types of allocation of the environmental burden associated with the raw material:

1. Some studies allocate environmental burdens (concerning input and output flows of farming activities) to raw material production, such as studies related to first-generation bioethanol. In this case, the feedstock production process is plotted inside the system boundaries.
2. Other studies, such as those focused on second-generation ethanol, use feedstock residues or wastes (seen as having no value) to which environmental burdens are not usually allocated. Therefore, the raw material is plotted outside the system boundaries.
3. When the feedstock used for bioethanol production is a co-product, meaning it is not a primary product of the main production process but has gained value, the environmental burdens are allocated according to the criteria selected.

For more details about allocation, see Sect. 3.3.

Various criteria can be applied to determine which processes will be considered as part of the product system, and which will not. The main cut-off criteria for the inclusion of inputs and outputs are mass, energy, and environmental significance. However, data availability is generally a constraint for the inclusion of inputs and outputs in most studies.

Table 3 Examples of LCA goals and functional units

Unit	Goals	Functional unit	Reference
Volume	To determine the environmental benefits of producing bioethanol	1 gal bioethanol	Byun et al. [6]
	To compare the environmental impacts of two approaches for bioethanol production from sugar beet	1 L bioethanol	Zaky et al. [74]
	To compare the environmental impacts of three bioethanol feedstocks	1000 L of bioethanol	Lyu et al. [37]
Mass	To scale up the best type of fermentation process in a bioreactor and its life cycle analysis	50 kg bioethanol	Mohapatra et al. [41]
	To evaluate the environmental impacts of the ethanol production and supply chain with different scenarios and identify the most economical and environmentally-friendly solution	1 t bioethanol	Sujata and Kaushal [64]
	To maximize environmental benefits in the EU by upscaling second-generation ethanol production and the respective substitution of petrol and first-generation bioethanol	t bioethanol per year	Wietschel et al. [70]
Distance	To analyze the feasibility of different cardoon- to-bioethanol processes and the environmental performance of the obtained biofuel compared to gasoline	1 km travelled in a fuel-driven vehicle	Espada et al. [12]
	To compare the current E10 blend situation in Jamaica and an alternative scenario where bioethanol comes from a local enterprise	To drive 6.67 × 10⁷ km	Batuecas et al. [5]
	To investigate the environmental performance of corncob-based bioethanol-blended E10 and E85 fuels and compare it with a fossil gasoline reference system	1 km distance driven	Liu et al. [35]
Feedstocks	To investigate the contribution of each unit in bioethanol production to the environmental impacts	The amount of raw material treated annually (875 kt/year)	Liu et al. [34]

(continued)

Table 3 (continued)

Unit	Goals	Functional unit	Reference
	1. To evaluate the potential environmental impacts of bioethanol production from different biomass categories and biorefineries 2. To define the process outline with the lowest impacts and to underline the pros, cons and bottlenecks	1 t of biomass and 1000 L of bioethanol	Demichelis et al. [11]
Energy	To compare the life cycle greenhouse gas emissions of ethanol production using loose and pelleted biomass	Energy (MJ)	Pandey et al. [51]
	To provide information on the environmental impacts of the use of different gasoline blends containing second-generation ethanol derived from banana agricultural waste	1 MJ	Guerrero and Muñoz [18]

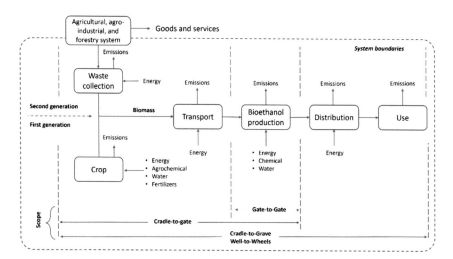

Fig. 5 Schematic representation of a generic bioethanol product system

Once the objectives and functions of the product system are established, it is necessary to define the functional unit (FU), which is the quantified performance of a product system for use as a reference unit [30]. As the FU needs to be consistent with the goal and scope of the study, it must be clearly defined and measurable. The most commonly used functional units in bioethanol studies are energy (MJ), mass (kg, t) or volume (L, gal) of bioethanol produced, distance traveled by a vehicle (km), and amount of feedstock or biomass waste treated (kg, t). Table 3 presents examples of functional units and goals used in different bioethanol LCA studies.

In general, there are common objectives in several FUs, such as determining the environmental impacts of bioethanol production. However, it is possible to observe specific characteristics in the objectives associated with each FU. When the objective is focused on the environmental impacts of bioethanol production, mass or volume is generally used as the FU. If the objective is to compare the performance of bioethanol versus gasoline or gasoline-bioethanol blends, the FUs generally relate to the distance traveled by the vehicle. In addition, this type of study uses the energy of bioethanol as a FU. When the study focuses on waste management, bioethanol production is a strategy for recycling biomass waste, and the FUs are associated with a mass quantity of the waste managed. The selection of the FU is key to the study's results, and its selection may generate different or contradictory findings. Consequently, González-García et al. [16] analyzed impact categories for two different functional units, comparing E10 with E85 fuels. According to their findings, LCA results are changed by choosing FUs for similar system boundaries.

Once the FU is established, the reference flow must be determined. The reference flow corresponds to the amount of product required to fulfill the function expressed in the FU. The reference flow is used as a basis for calculating the inputs and outputs of the product system [30]. In general, all bioethanol LCA studies present a FU appropriate to the goal and scope of each study. However, the reference flow is usually not stated, which can be complex in bioethanol product systems due to the significant variations in reference flows depending on the FU. For example, in comparative studies of different bioethanol feedstocks, the reference flow will depend on different factors. The bioethanol yield depends on the production technologies and on the type of lignocellulosic biomass [57]. Both factors will influence the amount of feedstock needed for bioethanol production. This amount is the reference flow. Figure 6 schematizes reference flow changes as a function of 1 kg of bioethanol (FU). This figure shows that the reference flows change significantly, hence also the data associated with their production, collection, transport, and processing. Although explicitly incorporating the reference flow does not condition the results, its inclusion makes it possible to visualize the material flows and highlights the importance of the function in the products. Additionally, this contributes to the completeness of data, which is necessary for the reproducibility and reliability of scientific studies and ensuring the quality of the results.

3.3 Inventory Analysis

A life cycle inventory (LCI) corresponds to the collection of input and output data of the product system under study throughout its whole life cycle. LCI is related to data collection and calculation procedures (ISO 2006a). In general, different foreground and background data sources can be used to construct a life cycle inventory. Secondary data are the most commonly used in LCA studies and generally account for approximately 80% of the data in an LCI [15].

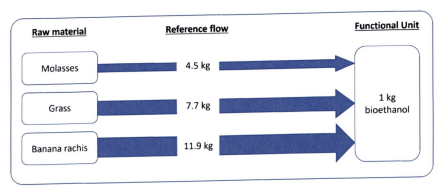

Fig. 6 Relationship between functional unit and reference flow (flows calculated from Batuecas et al. [5] for molasses, Mohapatra et al. [41] for grass, and Guerrero and Muñoz [18] for banana rachis; a bioethanol density of 790 kg/m^3 was used to estimate the reference flow and establish the same functional unit)

In LCA studies of bioethanol, literature is widely used in the agricultural production stage [5, 12, 70] and complements information throughout all stages of the bioethanol life cycle. Data for the biomass harvesting and transport stages are generally obtained from a database using LCA software [5], mixed with literature data. The bioethanol production stage presents several ways to obtain the LCI, and simulation software is one of the most frequently used. Aspen Plus [32, 52] and SuperPro Designer [12] have been used to model bioethanol production plants. However, data from laboratory studies [18, 41] and databases such as Ecoinvent [5] are often used to model the production stage. The distribution and use of bioethanol are stages generally not included in many studies. However, those that model this stage use the data from Ecoinvent through LCA software [12] and literature [18, 35]. In the literature, scientific articles, national statistics, and specific guidelines for determining emissions (e.g., IPCC and EMEP) stand out.

In short, no data source is more appropriate than another is. It is essential to be clear that regardless of the data source, the quality requirements that the data meet will reduce the uncertainty of the results. In general, the most used data quality criteria for selection from the literature and commercial databases are temporality, technology, geography, and reliability. Undoubtedly, all data are important for the completeness of an LCI. However, it is not necessary to spend resources to quantify inputs and outputs that will not produce significant changes in the overall conclusions of a study [30]. For this reason, LCI is considered an iterative process, which may constantly require updating and improvement, and cut-off rules.

A relevant issue in bioethanol LCA studies has to do with allocation procedures. Allocation corresponds to partitioning the input or output flows of a process or product system between the product system under study and one or more additional product systems [30]. Industrial processes generally produce more than one product and recycle intermediate products or waste. Most agricultural and industrial processes provide more than one output. Bioethanol production inevitably generates co-/by-products such as animal feed, electricity, and heat [3]. Choosing a suitable allocation method for sharing environmental loads among final outputs is still the most debated aspect of LCA methodology [7]. Allocation methods based on mass, energy, and environmental relevance are widely used and applied at different life cycle stages. In second-generation bioethanol, it is required to allocate part of the inputs and outputs of agricultural production to biomass residues. Some studies do not allocate environmental burdens to biomass waste, mainly when biomass waste is managed as waste. However, in many countries, biomass waste has agricultural, energy or building material uses and, therefore, economic value. In this case, part of the agricultural production flows has to be distributed among the wastes, with the economic allocation being the most commonly used method. This allocation method is also frequently applied in agro-industrial processes. For example, molasses is generated as a by-product of the sugar production process and can represent up to 20% of the inputs and outputs of the sugar production process, thus being considered an economic allocation [5]. The co-products of the bioethanol production process have been allocated through energy [74] and economics, while mass allocation has not been used as a criterion at this stage of the bioethanol life cycle. Despite the importance of allocation in bioethanol studies, many do not consider its use: in an extensive literature review by Angili et al. [3], only 56% of bioethanol studies considered allocation procedures.

3.4 Life Cycle Impact Assessment

Hundreds or thousands of consumption and emission data are generated for many compounds in the inventory phase, making their interpretation very difficult. The life cycle impact assessment (LCIA) aims to assess the inventory analysis results by quantifying the potential impacts of the product system. This stage consists of a technical phase that considers mandatory elements and an optional phase [30]. The mandatory elements include (i) the selection of impact categories, category indicators, and characterization models, (ii) the classification stage, and (iii) the characterization stage. The optional elements correspond to normalization, grouping, and weighting. Normalization transforms the result of an impact category indicator through its division by a reference value. This value can be obtained at the local, regional or global level, and allows us to see the relative importance of each impact category. Grouping consists of organizing and classifying impact categories into one or more pre-established groups. On the other hand, weighting consists of converting different impact category indicators into standard units through numerical factors based on value judgments. Weighting is one of the most controversial steps in life cycle impact assessment due to the difficulty of concentrating a wide range of impact categories into a single number.

The impact categories represent the environmental issues of interest within which the life cycle inventory data can be assigned, and their selection depends on the goal of the study. Table 4 presents examples of impact categories used in LCA studies, several of which vary according to the analysis method. LCA software such as SimaPro, Humberto, GaBi, and OpenLCA, among others, include several impact assessment methods (CML2001, ReCiPe, IPCC2013, EN 15,804, EPD, USEtox, Cumulative Energy Demand and AWARE, among others), which have different impact categories, many of which are equivalent. In order to transform the results of the LCI into indicators of an impact category, substances that generate equivalent environmental impacts are classified in the same group. For example, carbon dioxide, methane, and nitrous oxide can be classified as global warming substances (IPCC 2013 impact methodology). The substances classified can be transformed into common indicators that represent an impact category. This transformation is done through characterization factors. These represent the potential of a substance to generate a specific environmental impact, and their determination responds to environmental mechanisms (physical, chemical or biological) that have been scientifically determined. For example, the characterization factor for CO_2 in the global warming impact category can be equal to one, while the characterization factor of methane can be 34. The latter means that releasing 1 kg of methane causes the same amount of climate change as 34 kg of CO_2. Figure 7 schematically represents the LCA methodology, from the LCI to the derivation of impact category indicators. These impact categories are generally referred to as midpoints because they are oriented toward environmental problems.

The impact categories can be classified according to the scale of impact (global, regional or local) or in different hierarchies based on value judgments (high, medium and low importance).

Environmental damage-oriented methodologies are called endpoint methodologies, which use these optional elements. Endpoint methods represent impacts in terms of damage, mainly to human health, ecosystems, and natural resource depletion. These methods allow weighting across impact categories. This means the impact (or damage) category indicator results are multiplied by weighting factors, and are added to create a total or single score [49]. Examples of this type of methods are Ecoindicator 99 and ReCiPe 2016 Endpoint. In general, the impact assessment methods used in bioethanol research comprise both midpoint (problem-oriented) and endpoint (damage-oriented) levels as well as combined methods, in some cases [3].

Table 4 presents the main impact categories and methods used in bioethanol LCA studies in the last five years. As can be seen, the ReCiPe 2016 methodology has been one of the most widely used, which includes both midpoint and endpoint impact categories. The characterization factors are represented globally instead of the European scale employed in the previous version. For this reason, the method moved from the European category to the global category, which probably increased its use in recent years. The methodology comprises 18 impact categories at the midpoint level, which aggregate into three impact categories at the endpoint level (human health, ecosystem, and resource scarcity). Finally, the damage categories can be weighted to obtain a single score. The hierarchist version of ReCiPe with average weighting is chosen as default. In general, value choices made in the hierarchist version are scientifically and politically accepted [49].

Table 4 Main software, impact methodologies, and impact categories used in the environmental impact assessment in LCA studies

Authors	Software	Impact methodology	Impact categories													
			CC, GWP	HTP, HTPc, HTPnc	ETP, TE, FET, MEC	WD, WC	ADP, MD, FFD, MRS, FFS, FFP	AP, TAP	EP, FEP, MEP	ODP, SOD, OFhh, OFte	POF, POFPh, POFPe	PMF, FPMF	IR	ALO, ULO, NLT, LO	CED*	CP, NCP	RO, RI	Endpoint (HH, EQ, RS, ME)
Mohapatra et al. [41]	GaBi	ReCiPe endpoint	X	X		X											
Jayasundara et al. [32]	SimaPro	IPCC 2007 GWP100	X														
Espada et al. [12]	SimaPro	CML2001 Midpoint—Ecoinvent*	X	X	X		X	X	X						X		
Batuecas et al. [5]	SimaPro	ReCiPe midpoint (H)	X	X	X	X	X	X	X	X	X	X	X	X			
Parascanu et al. [52]	SimaPro	ReCiPe midpoint (H)	X	X		X	X	X	X	X	X						
Wietschel et al. [70]	SimaPro	ReCiPe (H) midpoint and endpoint	X	X	X	X	X	X	X	X	X	X	X	X			X
Sujata and Kaushal [64]	SimaPro	Impact 2002 + midpoint and endpoint	X		X		X	X	X	X			X	X		X	X	X
Demichelis et al. [11]	Open LCA	ReCiPe midpoint	X		X		X	X		X	X	X						

(continued)

Table 4 (continued)

Authors	Software	Impact methodology	Impact categories															
			CC, GWP	HTP, HTPc, HTPnc	ETP, TE, FET, MEC	WD, WC	ADP, MD, FFD, MRS, FFS, FFP	AP, TAP	EP, FEP, MEP	ODP, SOD, OFhh, OFte	POF, POFPh, POFPe	PMF, FPMF	IR	ALO, ULO, NLT, LO	CED*	CP, NCP	RO, RI	Endpoint (HH, EQ, RS, ME)
Mohd Yusof et al. [42]	SimaPro	CML 2 baseline 2000	X	X	X		X	X	X	X	X							
Sadhukhan et al. [60]	–	CML baseline V3.03	X					X	X		X							
Vaskan et al. [67]	SimaPro	ReCiPe	X	X			X		X									

Impact categories

CC: climate change; GWP: global warming potential—HTP: human toxicity potential; HTPc: human toxicity potential, cancer; HTPnc: human toxicity potential, non-cancer; ETP: ecotoxicity potential; TE: terrestrial ecotoxicity; FET: freshwater ecotoxicity; MEC: marine ecotoxicity—WD: water depletion; WC: water consumption—ADP: abiotic depletion potential; MD: mineral depletion; FFP: fossil fuel depletion; MRS: mineral resource scarcity; FFS: fossil fuel scarcity; FFP: fossil fuel potential—AP: acidification potential; TAP: terrestrial acidification potential—EP: eutrophication potential; FEP: freshwater eutrophication potential; MEP: marine eutrophication potential—ODP: ozone depletion potential; SOD: stratospheric ozone depletion; OFhh: ozone formation, human health; OFte: ozone formation, terrestrial ecosystems—POF: photochemical oxidation formation; POFPh: photochemical oxidation formation potential, humans; POFPe: photochemical oxidation formation potential, ecosystems—PMF: particulate matter formation; FPMF: Fine particulate matter formation—IR: ionizing radiation—ALO: agricultural land occupation; ULO: urban land occupation; NLT: natural land transformation; LO: land occupation—CED: cumulative energy demand—CP: carcinogens potential; NCP: non-carcinogens potential—RO: respiratory organic; RI: respiratory inorganic—HH: human health, EQ: ecosystem quality, RS: resource scarcity, ME: mineral extraction

*impact category from Ecoinvent methodology

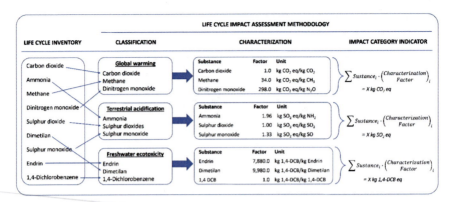

Fig. 7 Schematic representation of life cycle impact assessment methodology

Regardless of the environmental impact assessment methodology, GWP is the most used impact category in bioethanol LCA studies. In fact, most bioethanol LCAs are limited to GWP as a single impact category [3]. This is because biofuels in the transportation sector have generally been considered much better for the climate than fossil fuels. This reasoning is because the CO_2 emitted from biofuel combustion has a "neutral" effect on climate change since it belongs to the biogenic carbon cycle. However, from a life cycle perspective, it is clear that no biofuel can be "climate neutral" due to the fossil fuel inputs needed in the agricultural and industrial processes before the use stage [20]. Despite this, increased blending of bioethanol with fossil fuel can reduce the impact on global warming [18].

On the other hand, it is important to emphasize that decreasing impacts on global warming by substituting fossil fuels with biofuels has the potential to cause an increase in other environmental issues such as water scarcity, eutrophication, and land occupation and transformation [20]. Thus, many studies include other categories related to resource consumption, depletion and scarcity (ADP, MD, FFD, MRS, FFS, and FFP), eutrophication (EP, FEP, and MEP), acidification (AP and TAP), toxicity (HTP, HTPc, HTPnc), and ecotoxicity (ETP, TE, FET, MEC). This is because the bioethanol system influences the use of resources (such as fossil fuel, water and minerals) and agrochemicals in the agricultural stage, which can generate toxicity and ecotoxicity, and fertilizers that can generate acidification and eutrophication. The core reason for considering multiple environmental issues is to avoid burden shifting, which is also why a life cycle perspective is taken. Here burden shifting happens if efforts for lowering one type of environmental impact unintentionally increase other types of environmental impacts [20].

3.5 Interpretation

Interpretation represents the final phase of the LCA methodology. All LCA phases are summarized and discussed therein. Hence, conclusions and recommendations can be reached, and decisions can be made following the goal and scope of the study. It essentially describes several processes that must be performed to see if the conclusions are adequately supported by the data and procedures used. The processes to be considered include identifying significant issues as well as checking for completeness, sensitivity, and consistency [31]. The objectives of life cycle interpretation are to analyze the results, reach conclusions, explain limitations, provide recommendations based on the findings of the LCA and LCI study's preceding phases, and report the life cycle interpretation results in a transparent manner [30]. The interpretation also includes elements related to the purpose and scope of the study. In this regard, the proper definition of the functions of the system, the functional unit, and the system boundaries should be reviewed. Additionally, limitations associated with data quality must be identified to understand any uncertainty in the results.

Sensitivity analysis has been the most widely-used tool in bioethanol LCA studies. It seeks to determine the influence of assumptions, methods, and data variations on the results. Mainly, sensitivity analysis is performed on issues that have been identified as environmentally most significant (environmental hotspots), and in which contribution analysis is the primary tool for this identification. In practice, sensitivity analysis compares the results obtained by the study with certain assumptions or differences in parameters presented in other studies. Bioethanol LCA studies have conducted various sensitivity analyses, mainly due to the variability of options available throughout the life cycle. Some of the sensitized parameters have been: changes in biomass-to-bioethanol conversion yields [32, 35], percentages of bioethanol blended with regular gasoline [18], economic parameters [70], percentage changes of renewable energy in the electricity matrix used in the production process [52], process energy consumption and lignin composition in biomass [32], different allocation choices [10], and determining the influence of technical input parameters on the minimum ethanol selling price [51], among others.

Uncertainty analysis is another tool used in the interpretation of bioethanol LCA studies. It is a systematic procedure to quantify the uncertainty in the life cycle inventory results due to the cumulative effects of model inaccuracy, input uncertainties, and data variability [31]. Uncertainty analysis facilitates seeing how the quality of data and assumptions can influence environmental impact assessment reliability as well as the conclusions of the study. Thus, understanding uncertainty is essential in using LCA to support decisions. Quantitative uncertainty analysis has also been implemented in many LCA studies [54]. However, most bioethanol LCA studies do not include uncertainty analysis [3].

Monte Carlo simulation (MCS), a sampling method, is one of the most widely-used methods to characterize LCA variability [54]. Monte Carlo simulations require information on ranges or distributions of the underlying parameters. Experimental measurements offer the best source for such ranges and distributions, but unfortunately, this information is unavailable for all bioethanol life cycle stages. In some cases, specific laboratory or pilot plant parameters are available. However, these data require scaling, which increases their uncertainty. Other stages associated with agriculture and transport also require much data, often supplemented with literature and equivalent projects, increasing the uncertainty of the results. Absent such data, the pedigree method has often been used in LCAs to translate qualitative characteristics of underlying parameters into quantitative variability metrics [54]. The pedigree matrix includes criteria for data reliability, completeness, temporal correlation, geographic correlation, and technological correlation, criteria adopted by the Ecoinvent database since its version 2.0 [2]. The resulting uncertainty value is expressed as the geometric standard deviation (GSD) of a log-normal distribution. GSD measures the spread of log-normally distributed data points [55]. The uncertainty analysis by a Monte Carlo function is included in several LCA software such as SimaPro, Open LCA, and GaBi.

Although biofuels such as bioethanol are emerging as a solution to the depletion of fossil fuels and the impacts on global warming associated with their use, environmental criteria have yet to be a determining factor in developing a bioethanol project. Each feedstock type needs a particular bioconversion process, which involves water, energy, and chemical consumption, as well as the generation of emissions. Conversion technologies are, therefore, a significant source of environmental impacts in bioethanol production. However, in addition to bioethanol production processes, extraction, transport, and processing of feedstock and inputs are responsible for various environmental and social impacts. In this sense, first-generation bioethanol leads the international market despite the extensive bibliography highlighting the environmental and social effects generated throughout its life cycle [3].

Environmental analysis methodologies such as LCA have been an essential tool for identifying opportunities to improve the environmental performance of bioethanol throughout the life cycle by identifying environmental hotspots. LCA makes it possible to evaluate different configurations, technologies, feedstocks, and local realities of bioethanol production under the same reference unit (functional unit). The latter allows for comparing results and making decisions based on environmental criteria. Environmental analysis has strongly driven advances in clean technologies, the development of second and third-generation biofuels, and biorefineries around bioethanol production. These developments, together with the promotion of circular economy policies, make the joint analysis of techno-economic and environmental elements increasingly necessary, which has become more evident in recent research.

Finally, the current challenges associated with climate change, waste management, energy dependence, resource depletion, and environmental policies increasingly focused on the cyclical use and efficiency of natural resources make it necessary to address the development of bioethanol systems holistically. In this sense, a techno-economic and environmental approach that considers the life cycle of bioethanol systems is a fundamental requirement when analyzing the viability of biofuels.

4 Conclusions

The transportation sector is still heavily dependent on petroleum-based fuels. The shortage of fossil fuels announced for years is already palpable with skyrocketing gasoline prices, added to a global environmental awareness with several economic concerns. To face these setbacks, scientists have focused on obtaining bioethanol from renewable and low-cost bioresources, such as cellulosic biomass and algae, which are feedstock that avoids fuel versus food competition first-generation bioethanol leads the international market despite the extensive bibliography highlighting the environmental and social effects generated throughout its life cycle. It has had a sustained development through the years because of the subsidies and incentives that the governments of the producing countries have applied. Second- and third-generation bioethanol has better performance, considering ethanol yields and environmental performance, but these two bioethanol types are still at pilot or laboratory scale. These technologies have yet to blossom because of high processing costs. It has been demonstrated that biomass is a source of eco-friendly fuel, and it has earned its space in some biofuel regulations and mandates worldwide. Nonetheless, more R&D is needed to enable the production of bioethanol from non-food biomass at more competitive prices, accelerating the shift towards bioeconomy.

Regarding the environmental performance of bioethanol, LCA has been widely used as a reliable methodology to determine the environmental impact of the bioethanol system, identify environmental hotspots, and assess its contribution to global warming. Due to the multiple technical choices and decisions that must be made throughout the bioethanol life cycle, many impact categories should be assessed to avoid shifting environmental impacts between different sub-systems within system boundaries. Hence, the most frequently used impact categories are global warming potential, eutrophication, acidification, resource depletion, toxicity, and ecotoxicity, as they represent the cause-effect relationship between different system stages and their environmental impact. However, considering other impact categories would lead to identifying environmental effects that have not traditionally been observed but have gained more interest in recent years. In this regard, social impact assessments are almost absent in bioethanol LCA studies.

References

1. Alternative Fuels Data Center (2022) Maps and data—global ethanol production by country or region. https://afdc.energy.gov/data/. Accessed 11 Aug 2022
2. Althaus H-J, Doka G, Dones R et al (2007) Overview and methodology. Ecoinvent report No. 1. Dübendorf
3. Angili ST, Grzesik K, Rödl A, Kaltschmitt M (2021) Life cycle assessment of bioethanol production: a review of feedstock, technology and methodology. Energies 14:2939. https://doi.org/10.3390/EN14102939
4. Ballesteros M (2010) Producción de bioetanol. In: Nogués F, García-Galindo D, Rezeau A (eds) Energía de la Biomasa. Prensas Universitarias de Zaragoza, Zaragoza, pp 461–487
5. Batuecas E, Contreras-Lisperguer R, Mayo C et al (2021) Jamaican bioethanol: an environmental and economic life cycle assessment. Clean Technol Environ Policy 23:1415–1430. https://doi.org/10.1007/S10098-021-02037-8/TABLES/2
6. Byun J, Kwon O, Kim J, Han J (2022) Carbon-negative food waste-derived bioethanol: a hybrid model of life cycle assessment and optimization. ACS Sustain Chem Eng 4512–4521. https://doi.org/10.1021/acssuschemeng.1c08300
7. Cherubini F, Strømman AH, Ulgiati S (2011) Influence of allocation methods on the environmental performance of biorefinery products—a case study. Resour Conserv Recycl 55:1070–1077. https://doi.org/10.1016/j.resconrec.2011.06.001
8. Chong TY, Cheah SA, Ong CT et al (2020) Techno-economic evaluation of third-generation bioethanol production utilizing the macroalgae waste: a case study in Malaysia. Energy 210:118491. https://doi.org/10.1016/J.ENERGY.2020.118491
9. Chowdhury H, Loganathan B (2019) Third-generation biofuels from microalgae: a review. Curr Opin Green Sustain Chem 20:39–44. https://doi.org/10.1016/J.COGSC.2019.09.003
10. Costa D, Jesus J, Virgínio e Silva J, Silveira M (2018) Life cycle assessment of bioethanol production from sweet potato (Ipomoea batatas L.) in an experimental plant. Bioenergy Res 11:715–725. https://doi.org/10.1007/S12155-018-9932-1/TABLES/3
11. Demichelis F, Laghezza M, Chiappero M, Fiore S (2020) Technical, economic and environmental assessement of bioethanol biorefinery from waste biomass. J Clean Prod 277. https://doi.org/10.1016/J.JCLEPRO.2020.124111
12. Espada JJ, Villalobos H, Rodríguez R (2021) Environmental assessment of different technologies for bioethanol production from Cynara cardunculus: a life cycle assessment study. Biomass Bioenerg 144:105910. https://doi.org/10.1016/J.BIOMBIOE.2020.105910
13. Ferreira da Silva A, Brazinha C, Costa L, Caetano NS (2020) Techno-economic assessment of a Synechocystis based biorefinery through process optimization. Energy Rep 6:509–514. https://doi.org/10.1016/J.EGYR.2019.09.016
14. Global Petrol Prices (2022) Ethanol prices around the world. https://www.globalpetrolprices.com/ethanol_prices/. Accessed 26 Jul 2022
15. Goedkoop M, De Schryver A, Oele M et al (2010) Introduction to LCA with SimaPro 7. PRé Consultants
16. González-García S, Luo L, Moreira T et al (2009) Life cycle assessment of flax shives derived second generation ethanol fueled automobiles in Spain. Renew Sustain Energy Rev 113:1922–1933. https://doi.org/10.1016/j.rser.2009.02.003
17. Gro Intelligence (2019) How big ethanol plans will rock global corn and sugar markets. https://gro-intelligence.com/insights/how-big-ethanol-plans-will-rock-global-corn-and-sugar-markets. Accessed 15 Aug 2022
18. Guerrero AB, Muñoz E (2018) Life cycle assessment of second generation ethanol derived from banana agricultural waste: Environmental impacts and energy balance. J Clean Prod 174:710–717. https://doi.org/10.1016/J.JCLEPRO.2017.10.298
19. Hasanly A, Khajeh Talkhoncheh M, Karimi Alavijeh M (2018) Techno-economic assessment of bioethanol production from wheat straw: a case study of Iran. Clean Technol Environ Policy 20:357–377. https://doi.org/10.1007/S10098-017-1476-0/FIGURES/15

20. Hauschild MZ, Rosenbaum RK, Olsen SI (eds) (2018) Life cycle assessment. Springer, Cham, Switzerland
21. Hossain N, Mahlia TMI, Zaini J, Saidur R (2019) Techno-economics and sensitivity analysis of microalgae as commercial feedstock for bioethanol production. Environ Prog Sustain Energy 38:1–14. https://doi.org/10.1002/ep.13157
22. Hossain MS, Theodoropoulos C, Yousuf A (2019) Techno-economic evaluation of heat integrated second generation bioethanol and furfural coproduction. Biochem Eng J 144:89–103. https://doi.org/10.1016/j.bej.2019.01.017
23. Hossain N, Zaini J, Indra Mahlia TM (2019c) Life cycle assessment, energy balance and sensitivity analysis of bioethanol production from microalgae in a tropical country. Renew Sustain Energy Rev 115. https://doi.org/10.1016/J.RSER.2019.109371
24. IEA (2021) World energy outlook 2021. www.iea.org/weo. Accessed 12 May 2022
25. IEA (2022a) Transport biofuels. https://www.iea.org/reports/renewable-energy-market-update-may-2022/transport-biofuels#abstract. Accessed 13 May 2022
26. IEA (2022b) Renewables 2021 data explorer. https://www.iea.org/articles/renewables-2021-data-explorer?mode=transport&region=World&publication=2021&flow=Production&product=Ethanol. Accessed 9 Aug 2022
27. IEA (2017) Transport biofuels. https://www.iea.org/etp/tracking2017/transportbiofuels/. Accessed 1 Jun 2017
28. IPCC (2022) Synthesis report. https://www.ipcc.ch/ar6-syr/. Accessed 12 May 2022
29. IRENA (2012) Fast facts. In: Bioethanol/costs/presentations/fast-facts. Accessed 16 Aug 2022
30. ISO (2006a) Environmental management. Life cycle assessment. Principles and framework (ISO 14040:2006). Geneva, Switzerland
31. ISO (2006b) Environmental management. Life cycle assessment. Requirements and guidelines (ISO 14044:2006). Geneva, Switzerland
32. Jayasundara MP, Jayasinghe KT, Rathnayake M, Rathnayake mratnayake M (2022) Process simulation integrated life cycle net energy analysis and GHG assessment of fuel-grade bioethanol production from unutilized rice straw. Waste Biomass Valorization 1. https://doi.org/10.1007/s12649-022-01763-4
33. Larnaudie V, Ferrari MD, Lareo C (2022) Switchgrass as an alternative biomass for ethanol production in a biorefinery: perspectives on technology, economics and environmental sustainability. Renew Sustain Energy Rev 158. https://doi.org/10.1016/J.RSER.2022.112115
34. Liu F, Guo X, Wang Y et al (2021) Process simulation and economic and environmental evaluation of a corncob-based biorefinery system. J Clean Prod 329:129707. https://doi.org/10.1016/J.JCLEPRO.2021.129707
35. Liu F, Short MD, Alvarez-Gaitan JP, et al (2020) Environmental life cycle assessment of lignocellulosic ethanol-blended fuels: a case study. J Clean Prod 245. https://doi.org/10.1016/J.JCLEPRO.2019.118933
36. Lopes TF, Cabanas C, Silva A et al (2019) Process simulation and techno-economic assessment for direct production of advanced bioethanol using a genetically modified Synechocystis sp. Bioresour Technol Rep 6:113–122. https://doi.org/10.1016/J.BITEB.2019.02.010
37. Lyu H, Zhang J, Zhai Z et al (2020) Life cycle assessment for bioethanol production from whole plant cassava by integrated process. J Clean Prod 269. https://doi.org/10.1016/J.JCLEPRO.2020.121902
38. Ma T, Kosa M, Sun Q (2014) Fermentation to bioethanol/biobutanol. In: Ragauskas A (ed) Materials for biofuels. World Scientific Printers, pp 155–189
39. Machado Neto PA (2021) Why Brazil imports so much corn-based ethanol: the role of Brazilian and American ethanol blending mandates. Renew Sustain Energy Rev 152. https://doi.org/10.1016/J.RSER.2021.111706
40. Manhongo TT, Chimphango A, Thornley P, Röder M (2021) Techno-economic and environmental evaluation of integrated mango waste biorefineries. J Clean Prod 325. https://doi.org/10.1016/J.JCLEPRO.2021.129335

41. Mohapatra S, Behera BC, Acharya AN, Thatoi H (2022) Life cycle assessment of bioethanol production from Pennisetum sp. using fed-batch simultaneous saccharification and cofermentation at high solid loadings. Int J Energy Res 46:2904–2922. https://doi.org/10.1002/ER.7352
42. Mohd Yusof SJH, Roslan AM, Ibrahim KN et al (2019) Life cycle assessment for bioethanol production from oil palm frond juice in an oil palm based biorefinery. Sustain 11:6928. https://doi.org/10.3390/SU11246928
43. NREL (2021) Biochemical conversion techno-economic analysis. https://www.nrel.gov/bioenergy/biochemical-conversion-techno-economic-analysis.html. Accessed 7 May 2022
44. Nanda S, Mohammad J, Reddy SN et al (2013) Pathways of lignocellulosic biomass conversion to renewable fuels. Biomass Convers Biorefinery 4:157–191. https://doi.org/10.1007/s13399-013-0097-z
45. Navas-Anguita Z, García-Gusano D, Iribarren D (2019) A review of techno-economic data for road transportation fuels. Renew Sustain Energy Rev 112:11–26. https://doi.org/10.1016/J.RSER.2019.05.041
46. Ngigi W, Siagi Z, Anil Kumar ·, Arowo M (2022) Predicting the techno-economic performance of a large-scale second-generation bioethanol production plant: a case study for Kenya. Int J Energy Environ Eng. https://doi.org/10.1007/s40095-022-00517-1
47. Ntimbani RN, Farzad S, Görgens JF (2021) Techno-economic assessment of one-stage furfural and cellulosic ethanol co-production from sugarcane bagasse and harvest residues feedstock mixture. Ind Crops Prod 162. https://doi.org/10.1016/j.indcrop.2021.113272
48. OECD-FAO (2021) Agricultural outlook 2021–2030. In: 9. Biofuels. https://www.oecd-ilibrary.org/agriculture-and-food/oecd-fao-agricultural-outlook-2021-2030_19428846-en. Accessed 13 May 2022
49. PRé Sustainability (2021) Simapro database manual methods library
50. Padella M, O'Connell A, Prussi M (2019) What is still limiting the deployment of cellulosic ethanol? Analysis of the current status of the sector. Appl Sci 9:4523. https://doi.org/10.3390/APP9214523
51. Pandey R, Nahar N, Pryor SW, Pourhashem G (2021) Cost and environmental benefits of using pelleted corn stover for bioethanol production. Energies 14:2528. https://doi.org/10.3390/EN14092528
52. Parascanu MM, Sanchez N, Sandoval-Salas F et al (2021) Environmental and economic analysis of bioethanol production from sugarcane molasses and agave juice. Environ Sci Pollut Res 28:64374–64393. https://doi.org/10.1007/s11356-021-15471-4/Published
53. Pratto B, dos Santos-Rocha MSR, Longati AA, et al (2020) Experimental optimization and techno-economic analysis of bioethanol production by simultaneous saccharification and fermentation process using sugarcane straw. Bioresour Technol 297. https://doi.org/10.1016/J.BIORTECH.2019.122494
54. Qin Y, Suh S (2021) Method to decompose uncertainties in LCA results into contributing factors. Int J Life Cycle Assess 1:977–988. https://doi.org/10.1007/s11367-020-01850-5
55. Qin Y, Cucurachi S, Suh S (2020) Perceived uncertainties of characterization in LCA: a survey. Int J Life Cycle Assess 1846–1858. https://doi.org/10.1007/s11367-020-01787-9
56. Ramachandra TV, Hebbale D (2020) Bioethanol from macroalgae: prospects and challenges. Renew Sustain Energy Rev 117. https://doi.org/10.1016/J.RSER.2019.109479
57. Ramesh P, Arul V, Selvan M, Babu D (2022) Selection of sustainable lignocellulose biomass for second-generation bioethanol production for automobile vehicles using lifecycle indicators through fuzzy hybrid PyMCDM approach. Fuel 322:124240. https://doi.org/10.1016/J.FUEL.2022.124240
58. Ranjbari M, Shams Esfandabadi Z, Shevchenko T et al (2022) An inclusive trend study of techno-economic analysis of biofuel supply chains. Chemosphere 309. https://doi.org/10.1016/J.CHEMOSPHERE.2022.136755
59. Saddler J, Ebadian M, Mcmillan JD (2020) Advanced biofuels-potential for cost reduction

60. Sadhukhan J, Martinez-Hernandez E, Amezcua-Allieri MA et al (2019) Economic and environmental impact evaluation of various biomass feedstock for bioethanol production and correlations to lignocellulosic composition. Bioresour Technol Reports 7. https://doi.org/10.1016/J.BITEB.2019.100230
61. Sajid Z, Da Silva MAB, Danial SN (2021) Historical analysis of the role of governance systems in the sustainable development of biofuels in Brazil and the United States of America (USA). Sustain 13:6881. https://doi.org/10.3390/SU13126881
62. Solarte-Toro JC, Romero-García JM, Martínez-Patiño JC et al (2019) Acid pretreatment of lignocellulosic biomass for energy vectors production: a review focused on operational conditions and techno-economic assessment for bioethanol production. Renew Sustain Energy Rev 107:587–601. https://doi.org/10.1016/J.RSER.2019.02.024
63. Sondhi S, Kaur PS, Kaur M (2020) Techno-economic analysis of bioethanol production from microwave pretreated kitchen waste. SN Appl Sci 2:1–13. https://doi.org/10.1007/S42452-020-03362-1/TABLES/5
64. Sujata AA, Kaushal P (2021) Life cycle assessment of strategic locations to establish molasses based bioethanol production facility in India. Clean Environ Syst 3:100055. https://doi.org/10.1016/J.CESYS.2021.100055
65. Syahirah N, Aron M, Kuan I et al (2020) Sustainability of the four generations of biofuels—a review. Int J Energy Res 44:9266–9282. https://doi.org/10.1002/ER.5557
66. Tse TJ, Wiens DJ, Reaney MJT (2021) Production of bioethanol—a review of factors affecting ethanol yield. Ferment 7:268. https://doi.org/10.3390/FERMENTATION7040268
67. Vaskan P, Pachón ER, Gnansounou E (2018) Techno-economic and life-cycle assessments of biorefineries based on palm empty fruit bunches in Brazil. J Clean Prod 172:3655–3668. https://doi.org/10.1016/J.JCLEPRO.2017.07.218
68. Wang X, Yang J, Zhang Z et al (2022) Techno-economic assessment of poly-generation pathways of bioethanol and lignin-based products. Bioresour Technol Rep 17. https://doi.org/10.1016/J.BITEB.2021.100919
69. Wang C, Malik A, Wang Y et al (2020) The social, economic, and environmental implications of biomass ethanol production in China: a multi-regional input-output-based hybrid LCA model. J Clean Prod 249. https://doi.org/10.1016/J.JCLEPRO.2019.119326
70. Wietschel L, Messmann L, Thorenz A, Tuma A (2021) Environmental benefits of large-scale second-generation bioethanol production in the EU: an integrated supply chain network optimization and life cycle assessment approach. J Ind Ecol 25:677–692. https://doi.org/10.1111/JIEC.13083
71. Wong KH, Tan IS, Foo HCY et al (2022) Third-generation bioethanol and L-lactic acid production from red macroalgae cellulosic residue: prospects of industry 5.0 algae. Energy Convers Manag 253. https://doi.org/10.1016/J.ENCONMAN.2021.115155
72. Wu W, Lin KH, Chang JS (2018) Economic and life-cycle greenhouse gas optimization of microalgae-to-biofuels chains. Bioresour Technol 267:550–559. https://doi.org/10.1016/j.biortech.2018.07.083
73. Wyman CE (1994) Ethanol from lignocellulosic biomass: technology, economics, and opportunities. Bioresour Technol 50:3–15. https://doi.org/10.1016/0960-8524(94)90214-3
74. Zaky AS, Carter CE, Meng F, French CE (2021) A preliminary life cycle analysis of bioethanol production using seawater in a coastal biorefinery setting. Process 9:1399. https://doi.org/10.3390/PR9081399
75. Zucaro A, Forte A, Fierro A (2018) Life cycle assessment of wheat straw lignocellulosic bioethanol fuel in a local biorefinery prospective. J Clean Prod 194:138–149. https://doi.org/10.1016/J.JCLEPRO.2018.05.130

Concluding Remarks and Future Directions

Eriola Betiku and Mofoluwake M. Ishola

Abstract This chapter highlights the various topics covered in the book. The topics include benefits, prospects, and challenges of bioethanol production; new and cheap feedstock used for bioethanol production; lignocellulose biomass conditioning and pretreatment; novel starch sources; substrate hydrolysis to fermentable sugars; bioethanol production from lignocellulosic wastes and microalgae; bioethanol production through microbial fermentation including modeling and optimization of the process; ethanol recovery and dehydration procedures; ethanol application in internal combustion engines and emission characteristics; review of commercial bioethanol production plants; and techno-economic and life cycle analysis of bioethanol production. The chapter also suggests the direction of future research for bioethanol production to tackle the challenges identified in the book.

Keywords Bioethanol · Feedstock · Substrate · Lignocellulose biomass · Pretreatment · Hydrolysis · Fermentation · Microalgae · Techno-economic analysis · Life cycle

The importance of ethanol as a product, precursor, and/or solvent for manufacturing many other products, such as food beverages, cosmetics, perfumes, pharmaceutics (colognes, lotions, rubbing compounds, etc.), chemical compounds, polishes, plastics, aerosols, etc. formed the basis of this book. Its primary usage as a transportation fuel, either directly or as a blend with gasoline (e.g., E5, E10, and E25), was also emphasized.

E. Betiku (✉)
Department of Chemical Engineering, Obafemi Awolowo University, Ile-Ife 220005, Osun State, Nigeria
e-mail: eriola.betiku@famu.edu; ebetiku@oauife.edu.ng

Department of Biological Sciences, Florida Agricultural and Mechanical University, Tallahassee, FL 32307, USA

M. M. Ishola
Department of Research and Development, Scanacon, Stockholm, Sweden

© The Author(s), under exclusive license to Springer Nature Switzerland AG 2023
E. Betiku and M. M. Ishola (eds.), *Bioethanol: A Green Energy Substitute for Fossil Fuels*, Green Energy and Technology,
https://doi.org/10.1007/978-3-031-36542-3_13

Thus, the biological processes and unit operations involved in ethanol production were compiled. Although most bioethanol produced today comes from first-generation feedstocks such as starch sources (corn, wheat, cassava, barley, potato, sorghum, sugarcane, sugar beets, molasses, vinasse, etc.), the associated challenge of food security concerns led to the identification of second-generation substrates (corncobs, corn stover, corn leaves, wheat straw, rice husks, rice straw, empty oil palm fruit, sorghum residue, sugarcane bagasse, cotton stalk, wood and forest residues, cassava peels and potato peels) as alternative feedstocks. The main problem identified with this class of substrates is that they require major pretreatments to break down the inherent lignin-cellulose-hemicellulose matrix to access the sugars present. Some of the pretreatments suggested in this book include physical (e.g., ball milling, wet disk milling, and extrusion), chemical (acid or alkali), and biological (microbes) methods. However, some pretreatment methods may have to be combined for effectiveness. Some bacterial and fungal species have been demonstrated to be effective and eco-friendly in the pretreatment of lignocellulose biomass compared to other pretreatment methods but are not economically viable now. Expansion of this process can be achieved with microbe consortium, low operational costs, and efficient reactor configuration. Besides, the application of genetic engineering to manipulate the microbes for better performance and the biosynthesis of plant biomass are areas recommended for future studies. Also, some advanced pretreatment methods (e.g., ultrasound, pulsed electric field, and microwave irradiation) to enhance biomass degradation rate are still in their infancy. While the potential of these methods has been demonstrated, their successful application at an industrial-scale level at a comparative cost may constitute a major breakthrough for commercializing bioethanol production. Future research should be targeted toward the optimization of the pretreatment processes as well as the overall system economics.

The third-generation substrates discussed in this book are macroalgae and microalgae. Some of the benefits making algae biomass a promising feedstock for bioethanol include a high growth rate, all-year-round production, less pressure on freshwater, wastewater bioremediation, tolerance to harsh conditions, CO_2 sequestration, high lipid/carbohydrate content, relatively simple production processing, and non-competition with food sources are the merits that make algae biomass a promising source of biofuel. However, algae biomass requires pretreatment to obtain carbohydrate fractions, like lignocellulose biomass. Fourth-generation feedstock was developed due to the bottlenecks surrounding algae biomass utilization for bioethanol production. Some of the challenges are resolved through metabolic engineering on photosynthetic algae. More research is needed in this area to improve the rate and quantity of algae biomass output.

The carbohydrates obtained from the different categories of feedstocks are subjected to hydrolysis to convert them to simple sugars. This book reviewed the different hydrolysis (acid, enzymatic, and microbial) methods used to achieve this goal. Nowadays, both acid and enzymatic hydrolysis methods are assisted with microwave and ultrasound to make the process faster and obtain higher yields. Another major focus of this book is identifying factors affecting the hydrolysis process. Future research should aim to discover novel, efficient, and cheap enzymes that can act on starch, hemicellulose, and cellulose. This hydrolysis aspect of ethanol production should be given considerable attention since it affects its ultimate yield.

This book identified the main microorganisms used to ferment sugars to produce bioethanol: *Saccharomyces cerevisiae*, *Kluyveromyces marxianus*, *Zymomonas mobilis*, and *Escherichia coli*. The advantages and disadvantages of these microorganisms were discussed. Modeling and optimization of the fermentation process through conventional techniques were reviewed expensively. A brief review of the application of machine learning tools to modeling and optimization of the process is presented in the book. The book recognizes the limited application of the tools to the bioethanol production process and recommends more research that uses machine learning or data-driven tool because of their ability to capture the nonlinear nature of the bioethanol production system.

The recovery of ethanol after fermentation is a crucial step to remove water through distillation. This book reviews conventional and non-conventional tools used for ethanol dehydration. To circumvent problems associated with conventional distillation (extractive and azeotropic distillation), membrane and adsorption methods are being adopted nowadays due to low operational energy requirements and environmental friendliness. The suggested method for industrial ethanol production is the pervaporation-membrane method. Future investigations should focus on distillation-membrane pervaporation hybridization. Also, research on material choice and fabrication methods is needed to improve the recovery process and make it cost-effective.

Commercial ethanol on an industrial scale is limited to a few countries due to material availability, processing technologies, and expensive production process. All these challenges are addressed in the book. Consolidated bioprocessing (CBP) is an integrated process combining both hydrolysis and fermentation simultaneously. Only one microbial population is used to produce enzymes, perform hydrolysis, and ferment carbohydrates. Future research should focus on developing an efficient microbial population that can handle the various aspects of CBP. A breakthrough in scaling up the CBP to the industrial level will make bioethanol production cheaper and more accessible.

The successful development and implementation of bioethanol production can only be completed with a full assessment of the impact and/or footprint of a typical process/system. Global warming acidification, eutrophication, potential, toxicity, ecotoxicity, and resource depletion are the impact categories typically used for life cycle impact assessment (LCA) since they represent the cause-effect relationship between different system stages and their environmental impacts. Presently, social impact assessments are virtually absent in bioethanol LCA studies, which should be among the focus of future studies on this subject.

Printed in the United States
by Baker & Taylor Publisher Services